# Dynamics Reported

## Volume 1

*Dynamics Reported*

**Board of Editors:** H. Amann (Zürich), P. Brunovský (Bratislava), S. N. Chow (East Lansing), R. L. Devaney (Boston), U. Kirchgraber (Zürich), J. Mawhin (Louvain-la-Neuve), R. D. Nussbaum (New Brunswick), H. O. Walther (München), G. Wanner (Genève), J. C. Willems (Groningen), E. Zehnder (Zürich).

**Advisory Board:** L. Arnold, H. Broer, N. Borderies, A. Coppel, B. Fiedler, J. Grasman, E. Hairer, J. Henrard, R. A. Johnson, H. Kielhöfer, A. Lasota, M. Misiurewicz, R. Moeckel, R. E. O'Malley, K. J. Palmer, H. O. Peitgen, F. Przytycki, K. Rybakowski, D. G. Saari, J. A. Sanders, A. N. Sarkovskii, J. Scheurle, K. Schmitt, A. Vanderbauwhede, J. Waldvogel, J. A. Yorke.

The rapid development in the area of Dynamical Systems continually calls for comprehensive presentations of the current topics, in particular because Dynamical Systems are closely related to many other areas of mathematics and are of utmost interest to engineers, scientists, etc.

*Dynamics Reported* is a book series. Each volume contains about four or five longer articles of up to 60 pages. Each article treats an important subject of current research on a high scientific *and* didactic level. The main results are accompanied by carefully written proofs to ensure an adequate degree of completeness of each article. *Dynamics Reported* is published bi-annually or annually.

Authors of *Dynamics Reported* will receive a page payment of £5 (five pounds sterling) per printed page for their contribution. Authors will receive 25 reprints of their article free of charge.

Manuscripts (typewritten, double spaced) and correspondence should be addressed to the Managing Editors:

| | |
|---|---|
| U. Kirchgraber | H. O. Walther |
| *Applied Mathematics* | *Mathematics* |
| *Swiss Federal Institute of Technology (ETH)* | *Ludwig-Maximilians University* |
| *CH-8092 Zurich* | *D-8000 Munich* |
| *Switzerland* | *Federal Republic of Germany* |

# Dynamics Reported

## Volume 1

**U. Kirchgraber and H. O. Walther**
**Managing Editors**

Mather Sets for Twist Maps and Geodesics on Tori
*V. Bangert*

Connecting Orbits in Scalar Reaction Diffusion Equations
*P. Brunovský and B. Fiedler*

Qualitative Theory of Nonlinear Resonance by Averaging and
Dynamical Systems Methods
*James Murdock*

An Algorithmic Approach for Solving Singularly Perturbed
Initial Value Problems
*K. Nipp*

Exponential Dichotomies, the Shadowing Lemma and Transversal
Homoclinic Points
*Kenneth J. Palmer*

B. G. Teubner
Stuttgart

JOHN WILEY & SONS
Chichester · New York · Brisbane · Toronto · Singapore

*British Library Cataloguing in Publication Data:*

Dynamics reported. — Vol. 1–
  1. System analysis
   003              QA402

ISBN 978-3-519-02150-6          ISBN 978-3-322-96656-8 (eBook)
DOI 10.1007/978-3-322-96656-8

Typeset by MCS Ltd, Salisbury

# Contents

# Preface

*Dynamics Reported* reports on recent developments in dynamical systems theory.

Dynamical systems theory of course originated from ordinary differential equations. Today, dynamical systems theory covers a much larger area, including dynamical processes described by functional and integral equations, by partial and stochastic differential equations, etc. Dynamical systems theory has evolved remarkably rapidly in the recent years. A wealth of new phenomena, new ideas and new techniques proved to be of considerable interest to scientists in rather different fields. It is not surprising that thousands of publications on the theory itself and on its various applications have appeared and still will appear.

*Dynamics Reported* presents carefully written articles on major subjects in dynamical systems and their applications, addressed not only to specialists but also to a broader range of readers. Topics are advanced while detailed exposition of ideas, restriction to *typical* results, rather than to the *most general* ones, and last but not least lucid proofs help to gain an utmost degree of clarity.

It is hoped that *Dynamics Reported* will stimulate exchange of ideas among those working in dynamical systems and moreover will be useful for those entering the field.

<table>
<tr><td>Zürich and München,</td><td>Urs Kirchgraber</td><td>Hans-Otto Walther</td></tr>
<tr><td>October 1987</td><td colspan="2">Managing Editors</td></tr>
</table>

# Contributing Authors


**V. Bangert**, *Mathematisches Institut, Universitat Bern, Sidlerstr. 5, CH-3012 Bern, Switzerland*

**P. Brunovský**, *Universita Komenského, Ústav aplikovanoj matematiky, Mlynská dolina, 842 15 Bratislava 2, Czechoslovakia*

**B. Fiedler**, *Sonderforschungsbereich 123, Institut für Angewandte Mathematik, Im Neuenheimer Feld 294, 6900 Heidelberg, Federal Republic of Germany*

**J. Murdock**, *Department of Mathematics, Iowa State University, Ames, Iowa 50011, USA*

**K. Nipp**, *Applied Mathematics, ETH-Zentrum, CH-8092 Zurich, Switzerland*

**K. J. Palmer**, *Department of Mathematics and Computer Science, University of Miami, Coral Gables, Florida 33124 and Department of Mathematics, University of Melbourne, Parkville, Victoria, 3052, Australia*

*Dynamics Reported, Volume 1*
Edited by U. Kirchgraber and H. O. Walther
© 1988 John Wiley & Sons and B. G. Teubner

# 1

# Mather Sets for Twist Maps and Geodesics on Tori

## V. Bangert

*Mathematisches Institut, Universität Bern, Switzerland*

## CONTENTS



## INTRODUCTION

The title refers to a theory which is based on independent research in three different fields—differential geometry, dynamical systems and solid state physics—and which has attracted growing interest and research activity in the last few years. The objects of this theory are respectively:

(1) Geodesics on a 2-dimensional torus with Riemannian (or symmetric Finsler) metric.

(2) The dynamics of monotone twist maps of an annulus.
(3) The discrete Frenkel–Kontorova model.

While the results in case (1) go back to Hedlund [26] and Morse [43] and date from 1932 and 1924 the results in cases (2) and (3) were obtained quite recently, independently by Mather [35] in case (2) and by Aubry–LeDaeron [4] in case (3).

The principal aim of this paper is to give a survey of these results including complete proofs and to present the relations and differences between the problems in (1)–(3). In order to reduce the proofs to one common root we introduce a variational problem which can be interpreted in each of the three cases. While most of the results presented here already appeared in published form there is one notable exception: part of the results on geodesics, in particular Theorems (6.9) and (6.10), are new[1]. They complete the work of Morse and Hedlund to a certain degree. Given the relation between (1) and (3) these results follow from Aubry–LeDaeron [4].

From the historical point of view it is interesting to note that in the geometric setting Hedlund and Morse anticipated many of the results which were proved about fifty years later by Aubry and Mather in their respective areas. However, Hedlund's work never became popular. Maybe this is due to the fact that the order of trajectories (resp. the absence self-intersections for the geodesic problem) which is nowadays believed to be at the heart of the matter shows up only implicitly in his work.

In case (1) we study geodesics which—when considered on the universal cover of the torus—minimize arclength between any two of their points. Geometrically these minimal geodesics are the most natural and interesting ones. We present and extend results by Hedlund [26] and Morse [43] and thus give a precise description of the set of minimal geodesics. In the flat cases the universal cover is the euclidean plane and the minimal geodesics are the affine lines. It is of geometric interest to see that many of the properties of lines generalize to minimal geodesics.

In case (2) we study area-preserving monotone twist maps $\varphi: S^1 \times [0,1] \to S^1 \times [0,1]$ preserving the boundary components. Such maps occur frequently as sections maps of Hamiltonian systems with two degrees of freedom. The 'Mather Sets' from the title are particular $\varphi$-invariant subsets of $S^1 \times [0,1]$. In order to illustrate the interest in these sets we briefly review the fundamental dynamical problem for such maps $\varphi$: one would like to find closed $\varphi$-invariant curves which separate $S^1 \times \{0\}$ from $S^1 \times \{1\}$. The existence of such a curve $C$ is related to the stability of the system $\{\varphi^n\}_{n \in \mathbb{Z}}$: every orbit of $\varphi$ remains on one side of $C$ so that there are no orbits wandering from one boundary component to the other. Actually the converse is also true: if such an invariant curve does not exist there exists $p \in S^1 \times (0,1)$ such that $\varphi^n(p)$

---

[1] cf the remark at the end of Section 10

converges to $S^1 \times \{0\}$ for $n \to -\infty$ and to $S^1 \times \{1\}$ for $n \to +\infty$. This generalization of a theorem by Birkhoff, see e.g. [28], I, §3, was proved by Mather [41] using a refinement of the variational ideas which are also presented in this paper. The simplest twist maps $\varphi$ are the integrable ones; for these the annulus is foliated by closed invariant curves on which $\varphi$ acts as rotation. KAM-Theory shows that for maps $\varphi$ as above which are sufficiently $C^k$-close to an integrable $\varphi_0$ many of the invariant curves persist. More precisely, there exists an invariant curve on which $\varphi$ is conjugate to a rotation by $\alpha$ if $\alpha$ cannot be approximated by rationals too rapidly; for an exact formulation cf. [44], p. 52 (for $k \geqslant 5$) or [28], IV, §5 (for $k \geqslant 3 + \varepsilon$). On the other hand there are simple examples without any invariant curve separating the boundary components, see e.g. (7.10). So the invariant curves are destroyed when we get too far away from the integrable situation and the Mather sets $M_\alpha$ are the most important remnants of the invariant curves of irrational rotation numbers $\alpha$. They are 'Lipschitz Cantor curves', more precisely: $M_\alpha = \{ (\xi, \Psi_\alpha(\xi)) \mid \xi \in A_\alpha \}$ where $A_\alpha$ is a Cantor set and $\Psi_\alpha : A_\alpha \to (0, 1)$ is Lipschitz. We present alternative proofs for the results in [35], see also [31]. In Section 9 we construct examples which are closely related to the ones obtained in [36].

In case (3) we describe the minimum energy configurations of a 1-dimensional model in solid state physics: the discrete Frenkel–Kontorova model. For a short introduction see Section 8. The results presented in Sections 3–5 coincide with those proved in [4]. Our proofs are different, but similar in spirit. We use slightly weaker hypotheses.

The equivalence of the dynamics of (2) and (3) up to some technical problems is well known, cf. [4] and [40]. It relies on the description of the orbits of (2) by means of a generating function. The relation between (1) and (2) or (3) is less obvious and not widely known. In general one cannot find a section which allows one to describe the geodesic flow of a torus by an area-preserving map of an annulus. However, one can characterize the minimal geodesics by a variational principle which is similar to the one in (3). For details and more comments we refer to Sections 1, 6 and the end of Section 7.

The proofs of all these results are elementary but not obvious and sometimes tricky. The key ingredients are the $\mathbb{Z}^2$-periodicity of the variational principle and the fact that the orbits have codimension one in the configuration space which allows us to use order arguments.

Finally we describe how the paper is organized: In Section 1 we define the basic variational principle and outline how it can be specialized to the applications (1)–(3). Section 2 gives some elementary facts from the Denjoy theory of circle homeomorphisms. In Sections 3–5 we prove the results on the minimal trajectories of the variational principle. These results are specialized to minimal geodesics in Section 6 and to Mather sets for twist maps in Section 7. Section 8 briefly describes the Frenkel–Kontorova model and provides a dictionary which allows one to translate the results of Sections 3–5 into physical terms. In Section 9 we present geometric examples and related results. The last section provides a brief review of some of the recent developments. Sections 1–5 are elementary and self-contained—except for some simple facts on circle maps in Section 2. Sections 6, 7 and 9 presuppose a small amount of knowledge on geodesics and on area-preserving maps.


# 1  THE VARIATIONAL PROBLEM

In this section we define the variational problem which will be investigated in Sections 3–5. Subsequently, in order to motivate the abstract setting, we briefly show how the more concrete problems mentioned in the introduction fit into this framework.

We begin by fixing some notation. We consider the space $\mathbb{R}^{\mathbb{Z}} = \{x \mid x: \mathbb{Z} \to \mathbb{R}\}$ of bi-infinite sequences of real numbers with the product topology. An element $x \in \mathbb{R}^{\mathbb{Z}}$ will also be denoted by $(x_i)_{i \in \mathbb{Z}}$ and will sometimes be called a trajectory. Convergence of a sequence $x^n \in \mathbb{R}^{\mathbb{Z}}$ to $x \in \mathbb{R}^{\mathbb{Z}}$ means that $\lim_{n \to \infty} x_i^n = x_i$ for all $i \in \mathbb{Z}$. We will frequently use the following simple version of Tychonow's Theorem which can be proved by a diagonal sequence argument:

(1.1)    For every $a \in \mathbb{R}^{\mathbb{Z}}$ the set $\{x \in \mathbb{R}^{\mathbb{Z}} \mid |x_i| \leqslant a_i \text{ for all } i \in \mathbb{Z}\}$ is compact.

Given a function $H: \mathbb{R}^2 \to \mathbb{R}$ we extend $H$ to arbitrary finite segments $(x_j, \ldots, x_k), j < k$, of trajectories $x \in \mathbb{R}^{\mathbb{Z}}$ by

$$H(x_j, \ldots, x_k) := \sum_{i=j}^{k-1} H(x_i, x_{i+1})$$

We say that the segment $(x_j, \ldots, x_k)$ is *minimal* with respect to $H$ if

$$H(x_j, \ldots, x_k) \leqslant H(x_j^*, \ldots, x_k^*)$$

for all $(x_j^*, \ldots, x_k^*)$ with $x_j = x_j^*$ and $x_k = x_k^*$. The objects we are interested in satisfy the following 'global' minimality condition:

(1.2)    DEFINITION    $x \in \mathbb{R}^{\mathbb{Z}}$ is *minimal* (with respect to $H$) if every finite segment of $x$ is minimal (with respect to $H$).

The set of minimal trajectories $x \in \mathbb{R}^{\mathbb{Z}}$ will be denoted by $\mathcal{M} = \mathcal{M}(H)$. The aim of Sections 3–5 is to give a reasonably complete description of $\mathcal{M}$.

Obviously we have to impose some restrictions on $H$ in order to get results. Throughout we will assume that $H$ is *continuous* and satisfies the following conditions $(H_1)$–$(H_4)$ which will soon be motivated:

$(H_1)$ 'periodicity condition':    For all $(\xi, \eta) \in \mathbb{R}^2 : H(\xi + 1, \eta + 1) = H(\xi, \eta)$

($H_2$) 'condition at infinity':  $\displaystyle\lim_{|\eta|\to\infty} H(\xi, \xi+\eta) = \infty$ uniformly in $\xi$.

(H) 'ordering condition:  If $\underline{\xi} < \bar{\xi}, \underline{\eta} < \bar{\eta}$ then
$$H(\underline{\xi}, \underline{\eta}) + H(\bar{\xi}, \bar{\eta}) < H(\underline{\xi}, \bar{\eta}) + H(\bar{\xi}, \underline{\eta})$$

($H_4$) 'transversality condition':  If $(x_{-1}, x_0, x_1) \neq (x^*_{-1}, x^*_0, x^*_1)$ are minimal and $x_0 = x^*_0$ then $(x_{-1} - x^*_{-1})(x_1 - x^*_1) < 0$.

(1.3)  REMARK  If $H$ is $C^2$ and satisfies ($H_1$) and $D_2 D_1 H \leqslant \div \delta < 0$ then ($H_2$)–($H_4$) follow from $D_2 D_1 H \leqslant -\delta$:

To obtain ($H_2$) one integrates over the triangle with vertices $(\xi, \xi)$, $(\xi, \xi+\eta)$, $(\xi+\eta, \xi+\eta)$, to obtain ($H_3$) over the quadrangle $(\underline{\xi}, \underline{\eta})$, $(\bar{\xi}, \underline{\eta})$, $(\underline{\xi}, \bar{\eta})$, $(\bar{\xi}, \bar{\eta})$. $H_4$ follows from the monotonicity of $\eta \to D_1 H(\xi, \eta)$ and $\xi \to D_2 H(\xi, \eta)$.

DEFINITION  If $H \in C^2$ we say that $x \in \mathbb{R}^{\mathbb{Z}}$ is *stationary* if
$$D_2 H(x_{i-1}, x_i) + D_1 H(x_i, x_{i+1}) = 0 \quad \text{for all } i \in \mathbb{Z}.$$
Obviously each $x \in \mathscr{M}(H)$ is stationary with respect to $H$.

The following example corresponds to the flat resp. integrable situation in the applications:

(1.4)  EXAMPLE  Suppose $H(\xi, \eta) = h(\xi - \eta)$ where $h : \mathbb{R} \to \mathbb{R}$ is strictly convex, $h'' > 0$. Then $x = (x_i)_{i \in \mathbb{Z}}$ is stationary if
$$h'(x_{i-1} - x_i) = h'(x_i - x_{i+1}) \quad \text{for all} \quad i \in \mathbb{Z}.$$

Hence for every $x_0 \in \mathbb{R}$ and every $\alpha \in \mathbb{R}$ there is the stationary trajectory $x_i = x_0 + i\alpha$ and every stationary trajectory is of this type. This implies easily that all stationary trajectories are actually minimal, cf. Theorem (7.7).

All our results will rely on order properties of the minimal trajectories and on a $\mathbb{Z}^2$-action on $\mathbb{R}^{\mathbb{Z}}$ which preserves minimality. We define the relevant notions: $\mathbb{R}^{\mathbb{Z}}$ is partially ordered by
$$x < x^* \quad \text{if and only if} \quad x_i < x^*_i \quad \text{for all} \quad i \in \mathbb{Z}.$$

(1.5)  DEFINITION  $x \in \mathbb{R}^{\mathbb{Z}}$ and $x^* \in \mathbb{R}^{\mathbb{Z}}$ *cross*
  (a) *at $i \in \mathbb{Z}$* if $x_i = x^*_i$ and $(x_{i-1} - x^*_{i-1})(x_{i+1} - x^*_{i+1}) < 0$.
  (b) *between $i$ and $i+1$* if $(x_i - x^*_i)(x_{i+1} - x^*_{i+1}) < 0$.

According to the transversality condition ($H_4$) trajectories $x, x^* \in \mathscr{M}$ either cross or are comparable (i.e. $x < x^*$ or $x = x^*$ or $x > x^*$). It is useful to keep the following picture in mind: one can identify each $x \in \mathbb{R}^{\mathbb{Z}}$ with its graph $\{(i, x_i) \mid i \in \mathbb{Z}\} \subset \mathbb{Z} \times \mathbb{R} \subset \mathbb{R}^2$.

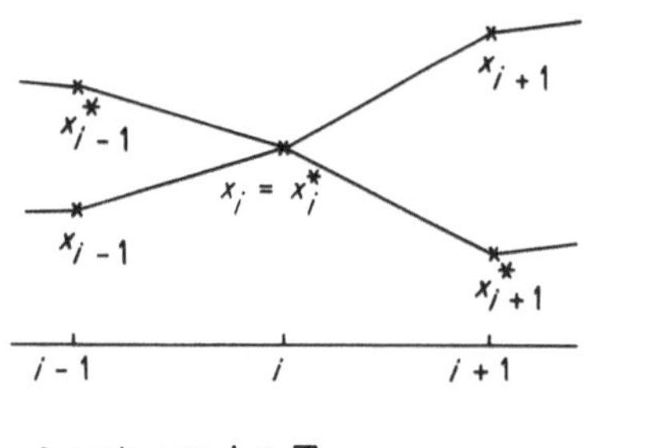

Crossing at $i \in \mathbb{Z}$

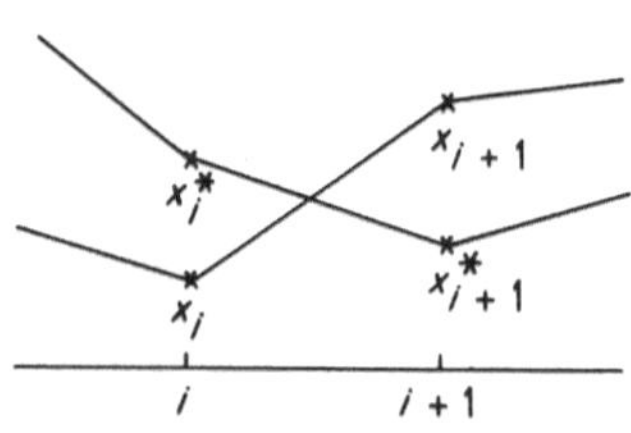


Crossing between $i$ and $i+1$

**Fig. 1**

If one joins successive points of the graph of $x \in \mathbb{R}^{\mathbb{Z}}$ by straight line segments in $\mathbb{R}^2$ one obtains curves in $\mathbb{R}^2$; $x$ and $x^*$ cross if and only if the corresponding curves intersect (Fig. 1).

In particular it is clear how one counts the number of crossings. A related concept is:

(1.6)   DEFINITION   $x \in \mathbb{R}^{\mathbb{Z}}$ and $x^* \in \mathbb{R}^{\mathbb{Z}}$ are

$\alpha$-*asymptotic* if $\lim\limits_{i \to -\infty} |x_i - x_i^*| = 0$

$\omega$-*asymptotic* if $\lim\limits_{i \to \infty} |x_i - x_i^*| = 0$

*asymptotic* if they are both $\alpha$- and $\omega$-asymptotic.

There is an action $T$ of the group $\mathbb{Z}^2$ on $\mathbb{R}^{\mathbb{Z}}$ by order-preserving homeomorphisms: if $(a, b) \in \mathbb{Z}^2$ and $x \in \mathbb{R}^{\mathbb{Z}}$ then

$$T_{(a,b)}x = x^* \quad \text{where} \quad x_i^* = x_{i-a} + b$$

The action of $T_{(a,b)}$ on $x$ corresponds to translation of $\text{graph}(x) \subseteq \mathbb{R}^2$ by $(a, b) \in \mathbb{Z}^2$. Similarly one can talk about crossings of segments and $T_{(a,b)}$ maps segments from $j$ to $k$ to segments from $j + a$ to $k + a$.

In the applications the elements of a $T$-orbit $\{T_{(a,b)}x \mid (a, b) \in \mathbb{Z}^2\}$ are identified. This motivates

(1.7)   DEFINITION   $x \in \mathbb{R}^{\mathbb{Z}}$ is *periodic with period* $(q, p) \in (\mathbb{Z}\setminus\{0\}) \times \mathbb{Z}$
if $T_{(q,p)}x = x$.

The periods of a periodic $x \in \mathbb{R}^{\mathbb{Z}}$ are the non-trivial elements of a cyclic subgroup of $\mathbb{Z}^2$ whose generators are called the prime periods of $x$.

We now state a few elementary facts which follow directly from the definitions and which we will use without reference: as a consequence of the periodicity condition $(H_1)$ we have $H(x) = H(T_{(a,b)}(x))$ for every segment $x = (x_j, ..., x_k), k > j$, and every $(a,b) \in \mathbb{Z}^2$. In particular $T_{(a,b)}$ maps minimal segments to minimal ones and $\mathcal{M}$ onto itself. The continuity of $H$ implies that $\mathcal{M}$ is closed in $\mathbb{R}^\mathbb{Z}$. Using the condition at infinity $(H_2)$ one can easily prove that for all $(\xi, \eta) \in \mathbb{R}^2$ and all $j < k$ there exists a minimal segment $(x_j, ..., x_k)$ with $x_j = \xi$, $x_k = \eta$. If $(x_j, ..., x_k)$ is minimal then so is every subsegment $(x_l, ..., x_m)$ where $l \geqslant j, m \leqslant k$.

Now we describe how the problems mentioned in the introduction give rise to functions $H$ with $(H_1)$–$(H_4)$. For details we refer to Sections 6–8. We start with the most intuitive but technically most involved case:

**Geodesics on tori**

Suppose $g$ is a Riemannian metric on the torus $T^2 = \mathbb{R}^2/\mathbb{Z}^2$ or, equivalently, $g$ is a $\mathbb{Z}^2$-periodic Riemannian metric on $\mathbb{R}^2$. A non-constant geodesic $c: \mathbb{R} \to \mathbb{R}^2$ is called *minimal* if for every interval $[a,b] \subseteq \mathbb{R}$ the segment $c \,|\, [a,b]$ of $c$ is the shortest curve joining $c(a)$ and $c(b)$. So, by definition, $\text{length}_g(c \,|\, [a,b]) = \tilde{d}(c(a), c(b))$ where $\tilde{d}$ denotes the distance on $\mathbb{R}^2$ induced by $g$. Minimal geodesics were first investigated by Morse [43] (for metrics covering compact surfaces of genus $> 1$) and by Hedlund [26] in the present case. If $g$ is euclidean (i.e. $g$ is constant) then all geodesics ($=$ straight lines) are minimal and we will show that many of the qualitative properties of straight lines carry over to minimal geodesics with respect to an arbitrary $\mathbb{Z}^2$-periodic $g$. We are looking for a function $H$ satisfying $(H_1)$–$(H_4)$ so that $\mathcal{M}(H)$ corresponds to the set of minimal geodesics. In a first step we will show that one can introduce new coordinates on $\mathbb{R}^2/\mathbb{Z}^2$ so that the coordinate lines $s \to (i, s)$, $i \in \mathbb{Z}$, are minimal geodesics. This is not obvious, cf. the discussion in Section 6. Then we define

$$(1.7) \qquad H(\xi, \eta) = \tilde{d}((0, \xi), (1, \eta))$$

where for $p, q \in \mathbb{R}^2$

$$\tilde{d}(p, q) = \inf\{\text{length}_g(\gamma) \,|\, \gamma \text{ is a curve from } p \text{ to } q\}.$$

This $H$ is continuous but, in general, it is not everywhere differentiable: the existence of conjugate points on a geodesic joining $\{0\} \times \mathbb{R}$ to $\{1\} \times \mathbb{R}$ induces a discontinuity in the <u>first derivatives</u> of $H$. In the euclidean case we have $H(\xi, \eta) = h(\xi - \eta) = \sqrt{[1 + (\eta - \xi)^2]}$. In general $H$ satisfies $(H_1)$ since the translation by $(0, 1) \in \mathbb{Z}^2$ is an isometry of $g$. Properties $(H_2)$–$(H_4)$ follow from our choice of coordinates, cf. Section 6. Note that

$$(1.8) \qquad H(x_j, ..., x_k) = \sum_{i=j}^{k-1} \tilde{d}((i, x_i), (i+1, x_{i+1}))$$

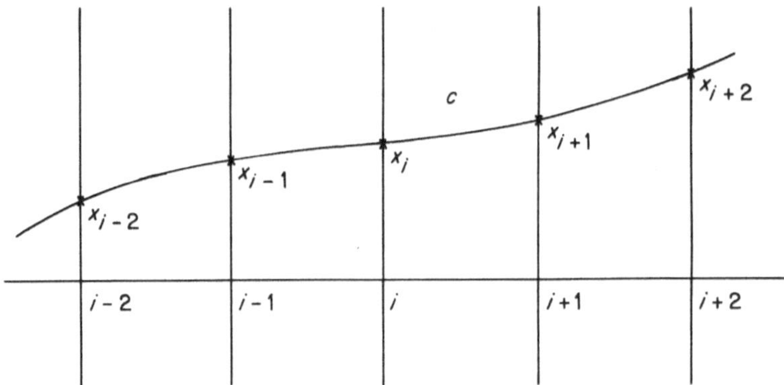


**Fig. 2.** Geodesic $c$ and the corresponding $x_i$

since $\tilde{d}((i, x_i), (i + 1, x_{i+1})) = \tilde{d}((0, x_i), (1, x_{i+1}))$ by $\mathbb{Z}^2$-invariance. Hence every minimal trajectory $x \in \mathcal{M}(H)$ determines a minimal geodesic passing through the points $(i, x_i)$ for all $i \in \mathbb{Z}$: because of (1.8) the minimal geodesic segments $c_i$ from $(i, x_i)$ to $(i + 1, x_{i+1})$, $i \in \mathbb{Z}$, join to form a minimal geodesic $c : \mathbb{R} \to \mathbb{R}^2$.

Conversely, the intersections of a minimal geodesic $c(s) = (\xi(s), \eta(s))$ with the lines $\{i\} \times \mathbb{R}$, $i \in \mathbb{Z}$, determine some $x \in \mathcal{M}(H)$ provided $\xi(s)$ is surjective. This last condition can always be fulfilled by a change of coordinates $(\xi, \eta)$ (Fig. 2).

So the study of minimal geodesics is equivalent to the study of $\mathcal{M}(H)$. In this geometric interpretation most of the proofs in Sections 3–5 rely on the simple fact that a curve which consists of geodesic segments pieced together with angle $\neq \pi$ cannot be a shortest connection between its endpoints.


## Monotone twist maps

More or less by definition an area-preserving monotone twist map $\varphi : S^1 \times [0, 1] \to S^1 \times [0, 1]$ can be described by a generating function $H : D \subseteq \mathbb{R}^2 \to \mathbb{R}$ in the following way:

$$(1.9) \qquad \varphi(x_0, y_0) = (x_1, y_1) \Leftrightarrow \begin{cases} -D_1H(x_0, x_1) = y_0 \\ D_2H(x_0, x_1) = y_1 \end{cases}$$

Here $D$ is a closed strip in $\mathbb{R}^2$ which is invariant under translation by $(1, 1) \in \mathbb{Z}^2$, and $H \in C^2$ satisfies (H$_1$) and $D_2D_1H \leqslant -\delta < 0$. $H$ can be $C^2$-extended to all of $\mathbb{R}^2$ so that (H$_1$)–(H$_4$) are satisfied, cf. (1.3) and Section 7. (1.9) implies that the orbits $(\varphi^i(x_0, y_0))_{i \in \mathbb{Z}}$ are in 1–1 correspondence with those stationary trajectories $x \in \mathbb{R}^{\mathbb{Z}}$ with respect to $H$ which satisfy $(x_0, x_1) \in D$. We will discuss those orbits of $\varphi$ which correspond to minimal trajectories. These constitute either $\varphi$-invariant circles separating $S^1 \times \{0\}$ from $S^1 \times \{1\}$ or the Cantor sets mentioned in the introduction. If $H(\xi, \eta) = h(\xi - \eta)$ then (1.9) implies that the second coordinate is an integral for $\varphi$, i.e. $\varphi(S^1 \times \{y\}) = S^1 \times \{y\}$ for all $y \in [0, 1]$. Hence $S^1 \times [0, 1]$ is foliated by invariant circles, cf. (7.5)(a). It should be noted that in this case the variational principle is only a tool which sorts out interesting orbits whereas in the other applications minimality is a value in itself.

### The discrete Frenkel–Kontorova model

This is a simple model for a 1-dimensional crystal: a 1-dimensional bi-infinite string of particles is described by the position $x_i \in \mathbb{R}$ of the $i$th particle. So $\mathbb{R}^{\mathbb{Z}}$ is the set of all 'states of the model'. For all $i \in \mathbb{Z}$ the particles numbered $i$ and $i + 1$ are coupled by a spring potential $\frac{1}{2}C(x_{i+1} - x_i)^2$ where $C > 0$ is independent of $i \in \mathbb{Z}$. Moreover a periodic potential $V(\xi) = V(\xi + 1)$ exerts forces $- V'(x_i)$ on the particles. If we define

$$(1.10) \qquad H(\xi, \eta) = \tfrac{1}{2}(C(\xi - \eta)^2 + V(\xi) + V(\eta))$$

then the stationary trajectories with respect to $H$ are precisely the equilibrium states of the model. Obviously $H$ satisfies $(H_1)$ and $D_2 D_1 H = - C < 0$, hence $(H_1)$–$(H_4)$ by (1.3). In physical terms the minimal trajectories $x \in \mathcal{M}(H)$ which we intend to study are the 'minimal energy configurations'. If $V \equiv 0$ then every equilibrium is a minimal energy configuration. Hence in this case a state is a minimal energy configuration if and only if the distance of neighboring particles does not depend on $i \in \mathbb{Z}$.

## 2 BASIC FACTS ON CIRCLE HOMEOMORPHISMS

In Section 3 we will prove that every minimal trajectory is an orbit of a circle homeomorphism. In this section we present some elementary results on such homeomorphisms which will be used to describe the minimal trajectories more closely.

We begin with a general outline of basic Denjoy theory; precise formulations will be given below. Let $G_+$ denote the group of orientation-preserving homeomorphisms of the circle $S^1 = \mathbb{R}/\mathbb{Z}$. For every $\varphi \in G_+$ one defines the *Poincaré rotation number* $\alpha(\varphi) \in S^1$ which can be interpreted geometrically as the 'average angle' by which $\varphi$ 'rotates' $S^1$. Homeomorphisms $\varphi \in G_+$ with $\alpha(\varphi)$ irrational differ radically from those with rational $\alpha(\varphi)$: we have $\alpha(\varphi) \in \mathbb{Q}/\mathbb{Z}$ if and only if $\varphi$ has a periodic point. If $\alpha(\varphi)$ is irrational the limit set of every orbit $\{\varphi^i(z) \mid i \in \mathbb{Z}\}$ is the unique smallest closed non-empty $\varphi$-invariant subset of $S^1$, i.e. the unique minimal set of $\varphi$. We denote it by

$\mathrm{Rec}(\varphi)$ and the points $z \in \mathrm{Rec}(\varphi)$ are called recurrent. There are two possibilities:

(i)  Every orbit is dense in $S^1$, i.e. $\mathrm{Rec}(\varphi) = S^1$. This is the case if and only if there exists an $h \in G_+$ such that $h \circ \varphi \circ h^{-1}$ is a rotation by $\alpha(\varphi)$.
(ii)  $\mathrm{Rec}(\varphi)$ is a Cantor set.

By Denjoy's theorem, cf. [19], case (i) occurs if $\varphi$ is a $C^1$-diffeomorphism with $\varphi'$ of bounded variation. In our application, however, the second case will be more frequent.

Note that here as well as in Sections 4, 6 and 7 Birkhoff's [9] notion of recurrence (which—in the compact case—means element of a minimal set) coincides with the (generally weaker) notion of recurrence which is mostly used nowadays: a point $p$ is recurrent if $p$ is in the closure of $\mathrm{orbit}(p)\backslash\{p\}$.

Now we give a precise formulation of some of these results in a form appropriate for our applications. For more details and proofs we refer to [1], 3. §11, or to [21], 27.2. Actually the proof of Lemma (2.3) below will also prove some of these statements. We let

$$\tilde{G}_+ = \{f \mid f \colon \mathbb{R} \to \mathbb{R} \text{ continuous, strictly increasing, } f(x+1) = f(x) + 1\}$$

denote the group of lifts to $\mathbb{R}$ of homeomorphisms $\varphi \in G_+$. The Poincaré rotation number $\alpha \colon G_+ \to S^1$ has a lift $\tilde{\alpha} \colon \tilde{G}_+ \to \mathbb{R}$ defined by

$$(2.1) \qquad \tilde{\alpha}(f) := \lim_{|i| \to \infty} \frac{f^i(x)}{i}$$

This limit exists and is independent of $x \in \mathbb{R}$. More precisely: For all $i \in \mathbb{Z}$ the periodic function $r_i(x) := f^i(x) - x - i\tilde{\alpha}(f)$ satisfies

$$(2.2) \qquad |r_i(x)| < 1 \quad \text{and there exists} \quad x_0 \in \mathbb{R} \text{ with } r_i(x_0) = 0$$

In particular (2.1) and (2.2) imply that $\tilde{\alpha}(f) = (p/q) \in \mathbb{Q}$ if and only if there exists $x_0 \in \mathbb{R}$ such that $f^q(x_0) = x_0 + p$. If $\tilde{\alpha}(f) \in \mathbb{R}\backslash\mathbb{Q}$ we define

$$\mathrm{Rec}(f) = \text{set of accumulation points of } \{f^i(x) + k \mid (i, k) \in \mathbb{Z}^2\}.$$

$\mathrm{Rec}(f)$ does not depend on the choice of $x \in \mathbb{R}$ and we obtain the same set if we restrict $(i, k)$ to lie in $\mathbb{N} \times \mathbb{Z}$. $\mathrm{Rec}(f)$ is either a periodic Cantor set or $\mathrm{Rec}(f) = \mathbb{R}$. If $f$ is a lift of $\varphi \in G_+$ then $\mathrm{Rec}(f)$ projects to $\mathrm{Rec}(\varphi)$.

The following lemma will be crucial in Section 4:

(2.3)  LEMMA  Suppose $f_0 \in \tilde{G}_+, f_1 \in \tilde{G}_+$ and $\tilde{\alpha}(f_0) = \tilde{\alpha}(f_1) = \alpha \in \mathbb{R}\backslash\mathbb{Q}$. Then either $\mathrm{Rec}(f_0) = \mathrm{Rec}(f_1)$ and $f_0 \mid \mathrm{Rec}(f_0) = f_1 \mid \mathrm{Rec}(f_1)$ or there exist $x_0 \in \mathrm{Rec}(f_0)$ and $x_1 \in \mathrm{Rec}(f_1)$ such that the orbits $(f_0^i(x_0))_{i \in \mathbb{Z}} \in \mathbb{R}^{\mathbb{Z}}$ and $(f_1^i(x_1))_{i \in \mathbb{Z}} \in \mathbb{R}^{\mathbb{Z}}$ cross infinitely often.

*Proof*  We use the following fundamental property of homeomorphisms

$f \in \tilde{G}_+$ with irrational rotation number $\alpha$: for every $x_0 \in \mathbb{R}$ the map

$$j\alpha + k \to f^j(x_0) + k \qquad (j, k) \in \mathbb{Z}^2$$

is strictly increasing, cf. [21], Corollary 27.2.2. For the corresponding $\varphi \in G_+$ this means that each orbit of $\varphi$ is ordered in the same way as the orbits of the rotation of $S^1$ by the angle $\alpha$. We define functions $x^+ (f, x_0) = x^+ : \mathbb{R} \to \mathbb{R}$ and $x^- : \mathbb{R} \to \mathbb{R}$ by

$$x^+ (t) = \inf\{f^j(x_0) + k \mid j\alpha + k > t\}$$
$$x^- (t) = \sup\{f^j(x_0) + k \mid j\alpha + k < t\}$$

Using the fact that $\{j\alpha + k \mid (j, k) \in \mathbb{Z}^2\}$ is dense in $\mathbb{R}$ it is easy to prove the following properties of $x^+$ and $x^-$:

(a) $x^+$ and $x^-$ are strictly increasing.
(b) $x^+$ is continuous from the right, $x^-$ from the left.
(c) $x^+$ and $x^-$ are continuous at the same points and coincide at such points.
(d) $x^\pm(t + 1) = x^\pm (t) + 1$
(e) $f \circ x^\pm(t) = x^\pm(t + \alpha)$
(f) $\mathrm{Rec}(f) = x^+ (\mathbb{R}) \cup x^- (\mathbb{R})$

If $x^+$ and $x^-$ are not continuous, then—by (a) and (e)—they have jumps on a countable, dense subset of $\mathbb{R}$. If $x^+ = x^- =: x$ is continuous then (a), (d) and (e) show that $x \in \tilde{G}_+$ and $(x^{-1} \circ f \circ x)(t) = t + \alpha$, i.e. $x$ conjugates $f$ to (a lift of a) rotation by $\alpha$.

Now choose functions $x_0^\pm$ and $x_1^\pm$ as above for $f_0$ and $f_1$. There are two possibilities:

(1) There exists $c \in \mathbb{R}$ such that $x_0^- (t + c) - x_1^- (t)$ changes sign.
(2) For all $c \in \mathbb{R}$ the function $x_0^- (t + c) - x_1^- (t)$ does not change sign, i.e. if $t_0 \in \mathbb{R}$ and $x_0^- (t_0 + c) < x_1^- (t_0)$ then $x_0^- (t + c) \leqslant x_1^- (t)$ for all $t \in \mathbb{R}$.

In case (1) we will find orbits of $f_0$ and $f_1$ which cross infinitely often: by (b) and (d) there are open intervals $I_1 \neq \varnothing$, $I_2 \neq \varnothing$ such that

$$
(*) \qquad
\begin{aligned}
x_0^- (t + c) &< x_1^- (t) \quad \text{if} \quad t \in I_1 \bmod \mathbb{Z} \\
x_0^- (t + c) &> x_1^- (t) \quad \text{if} \quad t \in I_2 \bmod \mathbb{Z}
\end{aligned}
$$

We define $x_0 = x_0^- (c)$ and $x_1 = x_1^- (0)$. By (f) we have $x_0 \in \mathrm{Rec}(f_0)$ and $x_1 \in \mathrm{Rec}(f_1)$. From (e) we obtain

$$f_0^j(x_0) = x_0^- (c + j\alpha), \qquad f_1^j(x_1) = x_1^- (j\alpha)$$

Since $\{j\alpha + k \mid (j, k) \in \mathbb{N} \times \mathbb{Z}\}$ is dense in $\mathbb{R}$ there exist infinitely many $j \in \mathbb{N}$ such that $j\alpha \in I_1 \bmod \mathbb{Z}$ and infinitely many $j \in \mathbb{N}$ such that $j\alpha \in I_2 \bmod \mathbb{Z}$. According to (*) this implies that $(f_0^j(x_0))_{j \in \mathbb{Z}}$ and $(f_1^j(x_1))_{j \in \mathbb{Z}}$ cross infinitely often.

Finally we prove that in case (2) the functions $x_0^\pm$ and $x_1^\pm$ coincide up to a phase. Let

$$c_0 = \sup\{c \mid x_0^-(t+c) \leqslant x_1^-(t) \text{ for all } t \in \mathbb{R}\}.$$

According to (b) we have $x_0^-(t+c_0) \leqslant x_1^-(t)$ for all $t \in \mathbb{R}$. If $x_0^-$ is continuous at $t_0 + c_0$ and $x_0^-(t_0 + c_0) < x_1^-(t_0)$ then there exists $c_1 > c_0$ such that $x_0^-(t_0 + c_1) < x_1^-(t_0)$. By (2) this contradicts the definition of $c_0$. Hence $x_0^-(t+c_0) = x_1^-(t)$ whenever $x_0^-$ is continuous in $t + c_0$. Now (a), (b) and (c) imply $x_0^\pm(t+c_0) = x_1^\pm(t)$ for all $t \in \mathbb{R}$. Finally (e) and (f) show that in this case the first alternative of our claim is true.

## 3  THE ROTATION NUMBER OF A MINIMAL TRAJECTORY

In this section we first prove the existence of periodic $x \in \mathcal{M}$ for every period $(q, p) \in (\mathbb{Z} \setminus \{0\}) \times \mathbb{Z}$. Then we shall see that every $x \in \mathcal{M}$ is an orbit of some circle map $f \in \tilde{G}_+$. This allows us to define a rotation number for every $x \in \mathcal{M}$. Finally we will prove that every real number is the rotation number of some $x \in \mathcal{M}$.

The proofs depend in a crucial way on the following simple, but fundamental consequence of properties $(H_3)$ and $(H_4)$:

(3.1)   LEMMA   Minimal trajectories cross at most once. If $x \in \mathcal{M}$ and $x^* \in \mathcal{M}$ coincide at $i \in \mathbb{Z}$ then $x$ and $x^*$ cross at $i \in \mathbb{Z}$.

*Remarks*   (1) Similar statements are true for minimal segments.
(2) For geodesics the first claim in (3.1) corresponds to the following well-known fact: two geodesics which issue from the same point and intersect cannot both be shortest beyond the first point of intersection.

*Proof*   The second claim follows from the transversality condition $(H_4)$. To prove the first statement we assume that $x$ and $x^*$ cross between $j$ and $j+1$ and between $k$ and $k+1$, $j < k$ (Fig. 3).

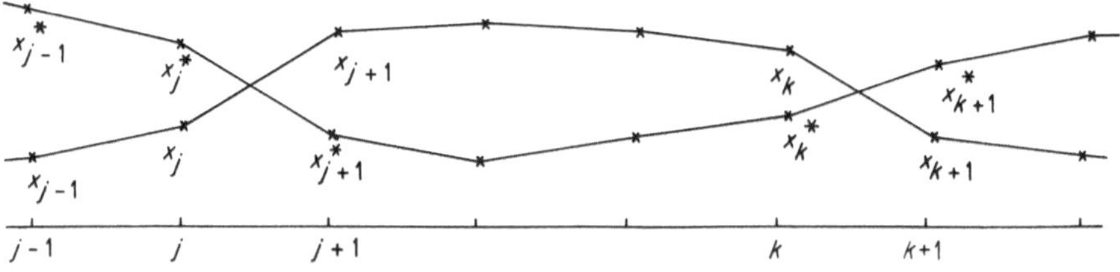

**Fig. 3.** $x$ and $x^*$ cross between $j$ and $j+1$ and between $k$ and $k+1$

The cases that one or both of the crossings take place at an integer can be treated similarly, cf. the proof of Lemma (3.9). We consider the segments $(x_j, x^*_{j+1}, ..., x^*_k, x_{k+1})$ and $(x^*_j, x_{j+1}, ..., x_k, x^*_{k+1})$. Using the ordering condition $(H_3)$ we see:

$$H(x_j, x^*_{j+1}, ..., x^*_k, x_{k+1}) + H(x^*_j, x_{j+1}, ..., x_k, x^*_{k+1})$$
$$= H(x_j, x^*_{j+1}) + H(x^*_{j+1}, ..., x^*_k) + H(x^*_k, x_{k+1}) + H(x^*_j, x_{j+1})$$
$$\quad + H(x_{j+1}, ..., x_k) + H(x_k, x^*_{k+1})$$
$$< H(x^*_j, x^*_{j+1}, ..., x^*_{k+1}) + H(x_j, x_{j+1}, ..., x_{k+1})$$

This contradicts the minimality of at least one of the segments $(x_j, ..., x_{k+1})$, $(x^*_j, ..., x^*_{k+1})$.

An easy consequence of Lemma (3.1) is

<br>

(3.2)　　COROLLARY　If $x \in \mathcal{M}$ and $x^* \in \mathcal{M}$ are periodic with the same period then $x$ and $x^*$ do not cross. If $x \in \mathcal{M}$ is periodic with minimal period $(q, p)$ then $q$ and $p$ are relatively prime.

<br>

*Proof*　If $x$ and $x^*$ have the same period and cross once then $x$ and $x^*$ cross infinitely often, in contradiction to (3.1). If $x \in \mathcal{M}$ is periodic with minimal period $(q, p)$ and $(q, p) = n(a, b)$ with $(a, b) \in \mathbb{Z}^2$ and $n > 1$ then $T_{(a, b)}x \neq x$. Since $x$ and $T_{(a, b)}x$ do not cross we have either $T_{(a, b)}x < x$ or $T_{(a, b)}x > x$. By induction we get $T_{(q, p)}\, x < x$ resp. $T_{(q, p)}x > x$, in contradiction to our hypothesis.

Next we prove that the object of our interest is not completely trivial:

<br>

(3.3)　　THEOREM　For all $(q, p) \in (\mathbb{Z} \setminus \{0\}) \times \mathbb{Z}$ there exists $x \in \mathcal{M}$ periodic with $(q, p)$.

<br>

*Proof*　We may assume $q > 0$. The idea of the proof is to consider the set $P_{q, p} = \{x \in \mathbb{R}^{\mathbb{Z}} \mid T_{(q, p)}x = x\}$ of trajectories periodic with $(q, p)$ and to minimize the function

$$H_{q, p} : P_{q, p} \to \mathbb{R}, \; H_{q, p}(x) = H(x_0, ..., x_q)$$

We will prove that arbitrary segments of a minimum $x$ of $H_{q, p}$ are minimal, i.e. $x \in \mathcal{M}$. This will follow easily once we have proved that no two minima of $H_{q, p}$ cross.

It is obvious from properties $(H_1)$ and $(H_2)$ that $H_{q, p}$ attains its infimum $H^{\min}_{q, p}$ on $P_{q, p}$. Assume two minima $x$ and $x^*$ of $H_{q, p}$ cross. This can only happen if $q \geqslant 2$. We define $x^+ \in \mathbb{R}^{\mathbb{Z}}$ and $x^- \in \mathbb{R}^{\mathbb{Z}}$ by $x^+_i = \max\{x_i, x^*_i\}$ and $x^-_i = \min\{x_i, x^*_i\}$. Then $x^+$ and $x^-$ are also periodic with $(q, p)$. Using $(H_3)$

we see that

$$(3.4) \qquad H_{q,p}(x^-) + H_{q,p}(x^+) \leqslant H_{q,p}(x) + H_{q,p}(x^*) = 2H_{q,p}^{\min}$$

with strict inequality if $x$ and $x^*$ cross between $i$ and $i+1$ for some $0 \leqslant i < q$. Since $x^-, x^+ \in P_{q,p}$ we have equality in (3.4) so that $x$ and $x^*$ cannot cross between $i$ and $i+1$ for any $0 \leqslant i < q$. By periodicity this implies that $x$ and $x^*$ cannot cross between $i$ and $i+1$ for any $i \in \mathbb{Z}$. Hence, if $x$ and $x^*$ cross they cross at some $i \in \mathbb{Z}$. We may assume that $i = 1$ since $T_{(i-1,0)}x$ and $T_{(i-1,0)}x^*$ are also minima of $H_{q,p}$. In this case (H$_4$) implies that not both $(x_0^-, x_1^-, x_2^-)$ and $(x_0^+, x_1^+, x_2^+)$ are minimal. So, by changing $x_1^-$ resp. $x_1^+$ we may reduce $H(x_0^-, x_1^-, x_2^-)$ resp. $H(x_0^+, x_1^+, x_2^+)$. Since $q \geqslant 2$ we can find $\tilde{x}^-$ resp. $\tilde{x}^+$ in $P_{q,p}$ which coincides with $x^-$ resp. $x^+$ except for $i = nq + 1$, $n \in \mathbb{Z}$, and such that

$$(3.5) \qquad H_{q,p}(\tilde{x}^-) < H_{q,p}(x^-) \text{ resp. } H_{q,p}(\tilde{x}^+) < H_{q,p}(x^+).$$

But (3.4) implies that $x^-$ and $x^+$ are minima of $H_{q,p}$ and this contradicts (3.5). Hence $x$ and $x^*$ cannot cross at all. In particular, we obtain:

(3.6)  If $x$ is a minimum of $H_{q,p}$ then $x$ does not cross any of its translates $T_{(j,k)}x$, $(j,k) \in \mathbb{Z}^2$.

Finally we prove that (3.6) implies our claim: suppose $n \geqslant 1$, $(q^*, p^*) = n(q, p)$ and $x \in P_{q^*,p^*}$ is a minimum of $H_{q^*,p^*}$. So (3.6) applies to $x$ with $(q, p)$ replaced by $(q^*, p^*)$. As in the proof of (3.2) we see that this implies $x \in P_{q,p}$. From this we conclude

$$(3.7) \qquad H_{q^*,p^*}^{\min} = nH_{q,p}^{\min}$$

since $H_{q^*,p^*}(\tilde{x}) = nH_{q,p}(\tilde{x})$ for all $\tilde{x} \in P_{q,p}$. Equation (3.7) implies that every minimum $x$ of $H_{q,p}$ is also a minimum of $H_{q^*,p^*}$ for every $(q^*, p^*) = n(q, p)$. In particular, if $x \in P_{q,p}$ is a minimum of $H_{q,p}$ we have:

(3.8)  $\qquad (x_0, \ldots, x_{nq})$ is a minimal segment for all $n \geqslant 1$.

Using the periodicity of $x$ we see that (3.8) implies that arbitrary segments of $x$ are minimal, i.e. $x \in \mathcal{M}$.

The following lemma shows that being $\alpha$- or $\omega$-asymptotic counts like a crossing:

(3.9)  LEMMA  Suppose $x \in \mathcal{M}$ and $x^* \in \mathcal{M}$ are $\alpha$- resp. $\omega$-asymptotic and $|x_{i+1} - x_i|$ is bounded for $i \to -\infty$ resp. $i \to \infty$. Then $x$ and $x^*$ do not cross.

*Remark*  In Corollary (3.16) we shall see that actually $|x_{i+1} - x_i|$ is always bounded if $x \in \mathcal{M}$.

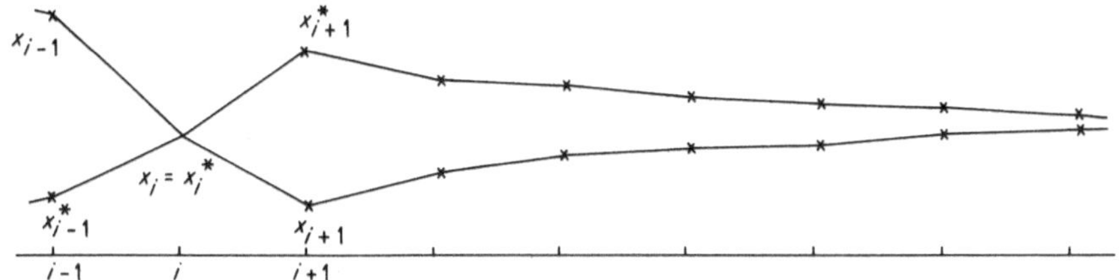


**Fig. 4.** $x$ and $x^*$ are $\omega$-asymptotic and cross at $i \in \mathbb{Z}$

*Proof* We treat the case that $x$ and $x^*$ are $\omega$-asymptotic under the assumption that $x$ and $x^*$ cross at $i \in \mathbb{Z}$ (Fig. 4). The remaining cases can be treated similarly.

Because of $(H_4)$ and $(x_{i-1} - x^*_{i-1})(x_{i+1} - x^*_{i+1}) < 0$ not both $(x_{i-1}, x_i, x^*_{i+1})$ and $(x^*_{i-1}, x_i, x_{i+1})$ can be minimal. Hence there exist $\bar{x}_i$ and $\tilde{x}_i$ such that

$$H(x_{i-1}, \bar{x}_i, x^*_{i+1}) + H(x^*_{i-1}, \tilde{x}_i, x_{i+1}) < H(x_{i-1}, x_i, x^*_{i+1}) + H(x^*_{i-1}, x_i, x_{i+1})$$

Using $x_i = x^*_i$ we obtain

$$(3.10) \quad H(x_{i-1}, \bar{x}_i, x^*_{i+1}) + H(x^*_{i-1}, \tilde{x}_i, x_{i+1}) < H(x_{i-1}, x_i, x^*_{i+1}) \\ + H(x^*_{i-1}, x^*_i, x^*_{i+1})$$

On the other hand the minimality of $(x_{i-1}, \ldots, x_{j+1})$ and $(x^*_{i-1}, \ldots, x^*_{j+1})$ imply

$$(3.11) \quad H(x_{i-1}, \bar{x}_i, x^*_{i+1}) + H(x^*_{i+1}, \ldots, x^*_j) + H(x^*_j, x_{j+1}) \\ \geqslant H(x_{i-1}, x_i, x_{i+1}) + H(x_{i+1}, \ldots, x_j) + H(x_j, x_{j+1})$$

and

$$(3.12) \quad H(x^*_{i-1}, \tilde{x}_i, x_{i+1}) + H(x_{i+1}, \ldots, x_j) + H(x_j, x^*_{j+1}) \\ \geqslant H(x^*_{i-1}, x^*_i, x^*_{i+1}) + H(x^*_{i+1}, \ldots, x^*_j) + H(x^*_j, x^*_{j+1})$$

Adding (3.11) and (3.12) we obtain a contradiction to (3.10) provided

$$\lim_{j \to \infty} |H(x^*_j, x_{j+1}) - H(x^*_j, x^*_{j+1})| = 0 = \lim_{j \to \infty} |H(x_j, x^*_{j+1}) - H(x_j, x_{j+1})|$$

We give the proof for the first equality. Using $(H_1)$ we can write

$$|H(x^*_j, x_{j+1}) - H(x^*_j, x^*_{j+1})| = |H(x^*_j - k_j, x_{j+1} - k_j) - H(x^*_j - k_j, x^*_{j+1} - k_j)|$$

with $k_j \in \mathbb{Z}$ such that $0 \leqslant x^*_j - k_j < 1$. By our hypothesis $|x_{j+1} - k_j|$ and $|x^*_{j+1} - k_j|$ are bounded for $j \to \infty$. Now

$$\lim_{j \to \infty} |H(x^*_j, x_{j+1}) - H(x^*_j, x^*_{j+1})| = 0 \text{ follows from } \lim_{j \to \infty} |x_j - x^*_j| = 0$$

and the uniform continuity of $H$ on compact sets.

The following theorem gives one of the fundamental properties of minimal trajectories:

(3.13)  THEOREM  Suppose $x \in \mathcal{M}$. Then $x$ and $T_{(a,b)}x$ do not cross for any $(a, b) \in \mathbb{Z}^2$.

*Remark*  A  different  way  to  express  (3.13)  is:  if  $x \in \mathcal{M}$  then $B_x = \{ T_{(a,b)}x \,|\, (a, b) \in \mathbb{Z}^2 \}$ is totally ordered.

*Proof*  The statement is trivial for $a = 0$. We assume that $x$ and $x^* = T_{(a,b)}x$ cross and, without loss of generality, that the crossing takes place at 0 or between 0 and 1. According to Lemma (3.1) $x$ and $x^*$ do not cross again. Interchanging $x$ and $x^*$ if necessary we may assume that

$$x_j^* < x_j \quad \text{for} \quad j < 0 \quad \text{and} \quad x_j^* > x_j \quad \text{for} \quad j > 0.$$

We are going to give the proof for the case $a > 0$; the case $a < 0$ can be treated similarly.

The preceding equations imply: for every $j \leqslant 0$ the sequence

$$v \in \mathbb{N} \to x_{j - va} + vb$$

is decreasing and for every $j > 0$ the sequence

$$v \in \mathbb{N} \to x_{j + va} - vb$$

is decreasing.

We compare $x$ to an $\bar{x} \in \mathcal{M}$ which is periodic with period $(a, b)$ and satisfies $\bar{x}_0 < x_0$. To obtain such $\bar{x}$ we use (3.3) and a translation $T_{(0, j)}$, if necessary.

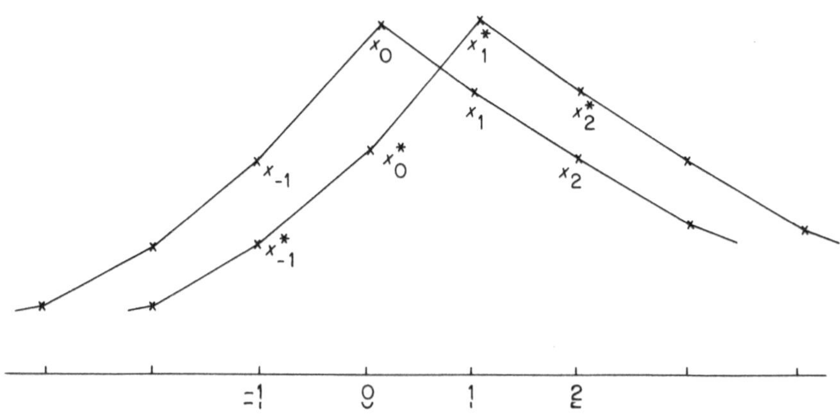



**Fig. 5.**  $x$ and $x^* = T_{(a, b)}x$ for $(a, b) = (1, 0)$

(In the case $a < 0$ we obtain increasing sequences and compare to an $\bar{x} \in \mathcal{M}$ which is periodic with $(a, b)$ and satisfies $\bar{x}_0 > x_0$.) By Lemma (3.1) we have $\bar{x}_j < x_j$ for $j \leqslant 0$ or $\bar{x}_j < x_j$ for $j \geqslant 0$.

Let us treat the case $\bar{x}_j < x_j$ for $j \leqslant 0$; the case $\bar{x}_j < x_j$ for $j \geqslant 0$ is completely analogous. For $j \leqslant 0$ the sequence $v \rightarrow x_{j-va} + vb$ is decreasing and bounded below by $\bar{x}_{j-va} + vb = \bar{x}_j$. Hence

$$\tilde{x}_j := \lim_{v \to \infty} (x_{j-va} + vb) = \lim_{v \to \infty} (T_{(va, vb)}x)_j$$

exists for $j \leqslant 0$ and $\tilde{x}_{j-a} + b = \tilde{x}_j$. From the periodicity of $(\tilde{x}_j)_{j \leqslant 0}$ one easily concludes that both $x$ and $x^*$ are $\alpha$-asymptotic to $(\tilde{x}_j)_{j \leqslant 0}$ and that $|x_{i+1} - x_i|$ is bounded for $j \rightarrow -\infty$. Now (3.9) contradicts our assumption that $x$ and $x^*$ cross.

We denote by $\bar{B}_x$ the closure of $B_x = \{T_{(a,b)}x \mid (a, b) \in \mathbb{Z}^2\} \subset \mathbb{R}^{\mathbb{Z}}$ and by $p_i : \mathbb{R}^{\mathbb{Z}} \rightarrow \mathbb{R}$ the projection $x \rightarrow x_i$. Obviously $p_i$ is continuous, open and order preserving.

(3.14)   LEMMA   Suppose $x \in \mathcal{M}$. Then $\bar{B}_x$ is totally ordered. The projection $p_0$ maps $\bar{B}_x$ homeomorphically onto a closed subset of $\mathbb{R}$.

*Proof*   The first claim follows easily from (3.13). Hence $p_0 \mid \bar{B}_x$ is injective, so that $p_0 \mid \bar{B}_x : \bar{B}_x \rightarrow p_0(\bar{B}_x)$ is a homeomorphism. It remains to prove that $p_0(\bar{B}_x)$ is closed. Suppose $x^n \in \bar{B}_x$ is a sequence such that $x_0^n = p_0(x^n)$ converges. Choose $a \in \mathbb{Z}$ and $b \in \mathbb{Z}$ such that $x_0 + a < x_0^n < x_0 + b$ for all $n \in \mathbb{N}$. Since $\bar{B}_x$ is totally ordered this implies $T_{(0, a)}x < x^n < T_{(0, b)}x$ for all $n \in \mathbb{N}$. Now the easy version (1.1) of Tychonow's theorem provides a convergent subsequence of $x^n$. Its limit $x^*$ is in $\bar{B}_x$ and $p_0(x^*) = \lim_{n \to \infty} p_0(x^n)$.

Now we are in the position to prove the main result of this section.

(3.15)   THEOREM   For every $x \in \mathcal{M}$ there exists a circle map $f \in \tilde{G}_+$ such that $x_{i+1} = f(x_i)$ for all $i \in \mathbb{Z}$.

For the definition of $\tilde{G}_+$ see Section 2.

*Proof*   First we define $f$ on the closed set $A := p_0(\bar{B}_x)$ by $f := p_1 \circ (p_0 \mid \bar{B}_x)^{-1}$ or, equivalently, by $f(x_0^*) := x_1^*$ if $x^* \in \bar{B}_x$. By (3.14) $f$ is a strictly increasing homeomorphism of $A$ onto itself. Obviously $f$ satisfies $f(t + 1) = f(t) + 1$ for all $t \in A$. We extend $f$ from $A$ to $\mathbb{R}$ in an affine way; i.e. if $\mathbb{R} \backslash A = \cup (a_n, b_n)$ then $f((1 - t)a_n + tb_n) := (1 - t)f(a_n) + tf(b_n)$ for $t \in [0, 1]$. One easily sees that $f \in \tilde{G}_+$ and $f(x_i) = f((T_{(-i, 0)}x)_0) = x_{i+1}$.

Combining (3.15) with the results (2.1) and (2.2) on circle maps we obtain

(3.16)   COROLLARY   There exists a continuous map $\tilde{\alpha} : \mathcal{M} \to \mathbb{R}$ with the following properties:

  (a) For all $x \in \mathcal{M}$, $i \in \mathbb{Z}$ we have $|x_i - x_0 - i\tilde{\alpha}(x)| < 1$, in particular $\tilde{\alpha}(x) = \lim_{|i| \to \infty} x_i / i$.
  (b) If $x \in \mathcal{M}$ is periodic with $(q, p)$ then $\tilde{\alpha}(x) = p/q$.
  (c) $\tilde{\alpha}$ is invariant under $T$; i.e. $\tilde{\alpha}(T_{(a,b)}x) = \tilde{\alpha}(x)$ for all $(a, b) \in \mathbb{Z}^2$.

*Remark*   We call $\tilde{\alpha}(x)$ the rotation number of $x \in \mathcal{M}$.

*Proof*   For $x \in \mathcal{M}$ we choose $f \in \tilde{G}_+$ according to (3.15) and define $\tilde{\alpha}(x) := \tilde{\alpha}(f)$. Then (a) follows from (2.1) and (2.2). In particular $\tilde{\alpha}(x)$ is well-defined. The continuity of $\tilde{\alpha}$ as well as (b) and (c) are immediate consequences of (a).

In conjunction with (3.16) the existence result (3.3) easily implies:

(3.17)   THEOREM   For all $\alpha \in \mathbb{R}$ the set $\mathcal{M}_\alpha := \{x \in \mathcal{M} \mid \alpha(x) = \alpha\}$ is not empty.

*Proof*   By (3.3) and (3.16) (b) we know that $\mathcal{M}_\alpha \neq \emptyset$ if $\alpha \in \mathbb{Q}$. Choose a sequence $\alpha_n \in \mathbb{Q}$ such that $\lim \alpha_n = \alpha$ and $x^n \in \mathcal{M}_{a_n}$ with $x_0^n \in [0, 1]$. Then $|\alpha_n| \leqslant C$ for some $C > 0$. From (3.16) (a) we obtain the estimate

$$|x_i^n| \leqslant 2 + |i| C \qquad \text{for all } n \in \mathbb{N}, i \in \mathbb{Z}.$$

Now (1.1) provides a point of accumulation $x^* \in \mathcal{M}$ of the sequence $x^n$. By the continuity of $\tilde{\alpha}$ we have $\tilde{\alpha}(x^*) = \alpha$.

We note a few further consequences of (3.16). According to (a) every $x \in \mathcal{M}_\alpha$ grows almost linearly with slope $\alpha$. In particular if $x \in \mathcal{M}$, $x^* \in \mathcal{M}$ and $\tilde{\alpha}(x) \neq \tilde{\alpha}(x^*)$ then $x$ and $x^*$ cross exactly once. The proof of (3.17) shows:

(3.18)   For every $C > 0$ the set $\{x \in \mathcal{M} \mid {}' |x_0| \leqslant C$ and $|\tilde{\alpha}(x)| \leqslant C\}$ is compact.

Finally we note that the circle maps $f \in \tilde{G}_+$ occurring in (3.15) are bi-Lipschitz if $H$ is $C^2$, satisfies $(\mathrm{H}_1)$ and $D_2 D_1 H < 0$. Actually this is true not only for minimal trajectories: Let $\mathcal{N}$ be a $T$-invariant set of trajectories $x = (x_i)_{i \in \mathbb{Z}} \in \mathbb{R}^{\mathbb{Z}}$ which are stationary with respect to $H$, i.e. $D_2 H(x_{i-1}, x_i) + D_1 H(x_i, x_{i+1}) = 0$ for all $x \in \mathcal{N}$ and all $i \in \mathbb{Z}$. Suppose no two elements of $\mathcal{N}$ cross. Then $p_0 | \mathcal{N} : \mathcal{N} \to p_0(\mathcal{N}) =: B \subseteq \mathbb{R}$ is one-to-one and

$p_1 \circ (p_0 \mid \mathcal{N})^{-1} : B \to B$, $x_0 \to x_1$, is defined. We prove

(3.19)   The map $f := p_1 \circ (p_0 \mid \mathcal{N})^{-1} : B \to B$ is bi-Lipschitz.

*Proof*   First note that $|x_1 - x_0|$ and $|x_0 - x_{-1}|$ are uniformly bounded for all $x \in \mathcal{N}$: If we fix some $\tilde{x} \in \mathcal{N}$ then $|x_1 - x_0| \leqslant |\tilde{x}_1 - \tilde{x}_0| + 1$ and $|x_0 - x_{-1}| < |\tilde{x}_0 - \tilde{x}_{-1}| + 1$. Now let $x$ and $\bar{x}$ be arbitrary elements in $\mathcal{N}$. By $(H_1)$ and the $T$-invariance of $\mathcal{N}$ it suffices to consider the case $0 \leqslant x_0 \leqslant \bar{x}_0 < 2$. Then $x_{-1} < \bar{x}_{-1}$, $x_1 < \bar{x}_1$ and, by the uniform bound mentioned above, all the points $x_{-1}, x_0, x_1, \bar{x}_{-1}, \bar{x}_0, \bar{x}_1$ are contained in some compact interval $I$. By assumption there exist $\delta > 0$ and $L > 0$ such that $D_2 D_1 H \leqslant -\delta < 0$ on $I \times I$ and $D_1 H, D_2 H$ are Lipschitz on $I \times I$ with constant $L$. Then we can estimate:

$$\delta((\bar{x}_{-1} - x_{-1}) + (\bar{x}_1 - x_1)) \leqslant D_2 H(x_{-1}, x_0) - D_2 H(\bar{x}_{-1}, x_0) + D_1 H(x_0, x_1)$$
$$- D_1 H(x_0, x_1) = D_2 H(\bar{x}_{-1}, \bar{x}_0) - D_2 H(\bar{x}_1, x_0) + D_1 H(\bar{x}_0, \bar{x}_1)$$
$$- D_1 H(x_0, \bar{x}_1) \leqslant 2L(\bar{x}_0 - x_0).$$

This proves our claim.


# 4   STRUCTURE OF THE SET OF MINIMAL TRAJECTORIES WITH IRRATIONAL ROTATION NUMBER

In this section we study $\mathcal{M}_\alpha$ for $\alpha \in \mathbb{R} \backslash \mathbb{Q}$. Theorem (3.17) shows that $\mathcal{M}_\alpha$ is not empty. At first sight it looks quite possible that the structure of $\mathcal{M}_\alpha$ is complicated in the following sense: there might be many minimal trajectories $x$ in $\mathcal{M}_\alpha$ so that the closures of their $T$-orbits $\bar{B}_x$ are pairwise disjoint. Though the structure of each $\bar{B}_x$ is simple since $\bar{B}_x$ is a closed set of orbits of a circle map the structure of all of $\mathcal{M}_\alpha$ would be very complicated. However, all this cannot happen: $\mathcal{M}_\alpha$ can be described by a single circle map and, in particular, every $\bar{B}_x \subseteq \mathcal{M}_\alpha$ contains the set $\mathcal{M}_\alpha^{\mathrm{rec}}$ of recurrent trajectories in $\mathcal{M}_\alpha$,

$$\mathcal{M}_\alpha^{\mathrm{rec}} := \left\{ x \in \mathcal{M}_\alpha \mid \text{There exist } k_i \in \mathbb{Z}^2 \backslash \{0\} \text{ such that } x = \lim_{i \to \infty} T_{k_i} x \right\}$$

The crucial result is:


(4.1)   THEOREM   Suppose $\alpha$ is irrational. Then $\mathcal{M}_\alpha$ is totally ordered.

*Proof*   Choose $x \in \mathcal{M}_\alpha$ and $f \in \tilde{G}_+$ according to (3.15), i.e. $x_i = f^i(x_0)$ for all $i \in \mathbb{Z}$. First we want to show that every recurrent orbit of $f$ is in $\mathcal{M}_\alpha$, i.e. if $x_0^* \in \mathrm{Rec}(f)$ then $x_i^* = f^i(x_0^*)$ defines an element of $\mathcal{M}_\alpha$. According to Section

2 there exists a sequence $(i_n, k_n) \in \mathbb{Z}^2$ such that

$$x_0^* = \lim_{n \to \infty} (x_{i_n} + k_n)$$

Then

$$x_i^* = f^i(x_0^*) = \lim_{n \to \infty} f^i(x_{i_n} + k_n) = \lim_{n \to \infty} (x_{i_n + i} + k_n)$$

so that $x^* = \lim_{n \to \infty} T_{(-i_n, k_n)} x \in \mathcal{M}_\alpha$. Now suppose $x^0$ and $x^1$ are in $\mathcal{M}_\alpha$ and $f_0, f_1$ are corresponding maps in $\tilde{G}_+$. By Lemma (3.1) minimal trajectories cross only once. So the preceding argument shows that no two recurrent orbits of $f_0$ and $f_1$ can intersect more than once. Hence Lemma (2.3) implies that $f_0$ and $f_1$ coincide on $\mathrm{Rec}(f_0) = \mathrm{Rec}(f_1)$. So, if $x_0^0 \in \mathrm{Rec}(f_0)$ and $x_0^1 \in \mathrm{Rec}(f_1)$ the trajectories $x^0$ and $x^1$ will not cross. In the general case let $x_0^\pm, x_1^\pm : \mathbb{R} \to \mathbb{R}$ denote the strictly increasing maps constructed in (2.3) from the orbits $x_j^0 = f_0^j(x_0^0)$ and $x_j^1 = f_1^j(x_0^1)$. According to the proof of (2.3) there exists $c \in \mathbb{R}$ such that $x_0^\pm(t + c) = x_1^\pm(t)$ for all $t \in \mathbb{R}$. By the definition of $x_0^\pm$ and $x_1^\pm$

$$x_0^-(j\alpha) \leqslant x_j^0 \leqslant x_0^+(j\alpha)$$

and

$$x_1^-(j\alpha) \leqslant x_j^1 \leqslant x_1^+(j\alpha)$$

Hence $x^0$ and $x^1$ could only cross if $c = 0$, i.e. if $x_0^\pm = x_1^\pm$. But in this case $x^0$ and $\mathrm{x}^1$ are asymptotic since

$$\sum_{j \in \mathbb{Z}} |x_j^1 - x_j^0| \leqslant \sum_{j \in \mathbb{Z}} (x_0^+(j\alpha) - x_0^-(j\alpha)) \leqslant x_0^+(1) - x_0^+(0) = 1.$$

So $x^0$ and $x^1$ do not cross according to Lemma (3.9).

We list some consequences of Theorem (4.1) which provide a detailed picture of $\mathcal{M}_\alpha$ for $\alpha \in \mathbb{R} \backslash \mathbb{Q}$.

(4.2)   There exists a circle map $f \in \tilde{G}_+$ with $\tilde{\alpha}(f) = \alpha$ and a closed $f$-invariant set $A_\alpha \subseteq \mathbb{R}$ such that $\mathcal{M}_\alpha$ consists of the orbits of $f$ contained in $A_\alpha$; i.e. $x \in \mathcal{M}_\alpha$ if and only if $x_0 \in A_\alpha$ and $x_i = f^i(x_0)$ for all $i \in \mathbb{Z}$. The projection $p_0$ maps $\mathcal{M}_\alpha$ homeomorphically onto $A_\alpha$.

For the construction of $f$ cf. (3.15) and (3.14).

(4.3)   A minimal trajectory $x \in \mathcal{M}_\alpha$ is recurrent if and only if $x_0$ is a recurrent point for $f$, i.e. $p_0(\mathcal{M}_\alpha^{\mathrm{rec}}) = \mathrm{Rec}(f)$. There are the following alternatives:

(a) $\mathrm{Rec}(f) = \mathbb{R}$. Then for every $\xi \in \mathbb{R}$ there exists $x \in \mathcal{M}_\alpha^{\mathrm{rec}}$ such that $x_0 = \xi$. Moreover if a sequence $x^n$ in $\mathcal{M}_\alpha^{\mathrm{rec}}$ converges to $x \in \mathcal{M}_\alpha^{\mathrm{rec}}$ then this convergence is uniform, i.e. for every $\varepsilon > 0$ there exists $n_0 \in \mathbb{N}$ such that $|x_i^n - x_i| < \varepsilon$ for all $n \geqslant n_0$ and all $i \in \mathbb{Z}$.

(b) $\mathrm{Rec}(f)$ is a Cantor set. In this case there exist three different kinds of trajectories in $\mathcal{M}_\alpha^{\mathrm{rec}}$. There is the set of $x \in \mathcal{M}_\alpha^{\mathrm{rec}}$ which can be approximated by elements of $\mathcal{M}_\alpha^{\mathrm{rec}}$ from above and from below. This set has the power of the continuum and corresponds to those points in $\mathrm{Rec}(f)$ which are not endpoints of components of $\mathbb{R}\backslash\mathrm{Rec}(f)$. Finally there are the sets of $x \in \mathcal{M}_\alpha^{\mathrm{rec}}$ which can be approximated by elements of $\mathcal{M}_\alpha^{\mathrm{rec}}$ only from above or only from below. These sets are countable and correspond to the right resp. left endpoints of components of $\mathbb{R}\backslash\mathrm{Rec}(f)$. $x \in \mathcal{M}_\alpha^{\mathrm{rec}}$ and $x^* \in \mathcal{M}_\alpha^{\mathrm{rec}}$ are asymptotic if and only if $x_0$ and $x_0^*$ are the endpoints of some component of $\mathbb{R}\backslash\mathrm{Rec}(f)$. In this case we have $\sum_{i \in \mathbb{Z}} |x_i - x_i^*| \leqslant 1$. Convergence in $\mathcal{M}_\alpha^{\mathrm{rec}}$ is never uniform.

These facts follow from (4.1), (4.2) in conjunction with the results on circle maps described in the first part of Section 2. As we shall see in Section 9 it is by no means exceptional that $\mathcal{M}_\alpha^{\mathrm{rec}}$ is a Cantor set for all $\alpha \in \mathbb{R}\backslash\mathbb{Q}$. We give a more precise version of the last statement in (4.3):

(4.4)   Suppose $\mathcal{M}_\alpha^{\mathrm{rec}}$ is homeomorphic to a Cantor set. Define

$$a_\alpha := \max\{\, |x_i - x_i^*| \mid x \in \mathcal{M}_\alpha^{\mathrm{rec}} \ \text{ and } \ x^* \in \mathcal{M}_\alpha^{\mathrm{rec}} \text{ asymptotic}, i \in \mathbb{Z}\}$$

If $\underline{x} \in \mathcal{M}_\alpha^{\mathrm{rec}}$, $\bar{x} \in \mathcal{M}_\alpha^{\mathrm{rec}}$ are not asymptotic and $\underline{x} < \bar{x}$ then

$$\inf(\bar{x}_i - \underline{x}_i) > 0, \ \limsup_{i \to \infty}(\bar{x}_i - \underline{x}_i) > a_\alpha \quad \text{and} \quad \limsup_{i \to -\infty}(\bar{x}_i - \underline{x}_i) > a_\alpha.$$

*Proof*   Choose functions $x^\pm$ for $f$ as in (2.3). Then $a_\alpha = \max(x^+(t) - x^-(t))$. There exist $\underline{c} \in \mathbb{R}$, $\bar{c} \in \mathbb{R}$ such that

$$\underline{x}_i = x^+(i\alpha + \underline{c}) \quad \text{or} \quad \underline{x}_i = x^-(i\alpha + \underline{c})$$

and

$$\bar{x}_i = x^+(i\alpha + \bar{c}) \quad \text{or} \quad \bar{x}_i = x^-(i\alpha + \bar{c}).$$

Our hypothesis implies $\underline{c} < \bar{c}$. Using properties (2.3) (a), (c) and (d) of $x^\pm$ we conclude that $x^-(t + \bar{c}) - x^+(t + \underline{c})$ is bounded below by some $\delta > 0$. Hence $\inf(\bar{x}_i - \underline{x}_i) \geqslant \delta > 0$. Choose $t_0 \in \mathbb{R}$ such that $x^+(t_0) - x^-(t_0) = a_\alpha$. Since $\{i\alpha + k \mid (i, k) \in \mathbb{N} \times \mathbb{Z}\}$ is dense in $\mathbb{R}$ there exists a sequence $(i_n, k_n) \in \mathbb{N} \times \mathbb{Z}$ such that $i_n \to \infty$, $i_n\alpha - k_n + \underline{c} < t_0 < i_n\alpha - k_n + \bar{c}$ and $\lim(i_n\alpha - k_n + \underline{c}) = t_0$. Then

$$\lim_{n \to \infty}(\bar{x}_{i_n} - \underline{x}_{i_n}) \geqslant a_\alpha + \delta.$$

The proof for $i \to -\infty$ is analogous.

(4.5)   Every $x \in \mathcal{M}_\alpha^{\mathrm{rec}}$ can be approximated by periodic minimal trajectories.

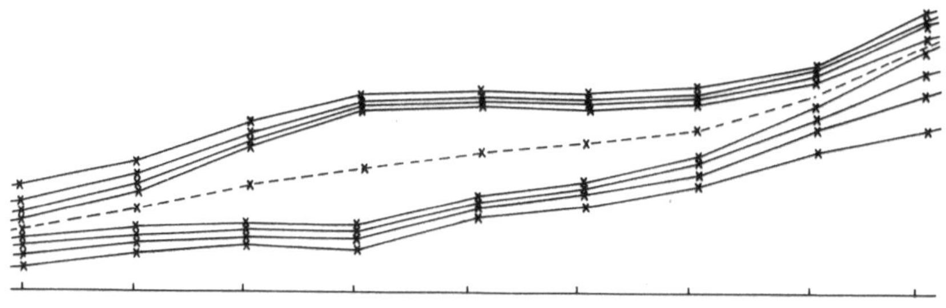

**Fig. 6.** $\mathcal{M}_\alpha$ for $\alpha \in \mathbb{R}\backslash\mathbb{Q}$

*Proof*  Let $\tilde{\mathcal{M}}_\alpha \subseteq \mathcal{M}_\alpha$ denote the set of $x \in \mathcal{M}_\alpha$ which can be so approximated. Then $\tilde{\mathcal{M}}_\alpha$ is closed and $T$-invariant. In (3.17) we actually proved $\tilde{\mathcal{M}}_\alpha \neq \emptyset$. Hence $\mathcal{M}_\alpha^{\mathrm{rec}} \subseteq \tilde{\mathcal{M}}_\alpha$.

(4.6)  For every $x \in \mathcal{M}_\alpha\backslash\mathcal{M}_\alpha^{\mathrm{rec}}$ there exist $\underline{x} \in \mathcal{M}_\alpha^{\mathrm{rec}}$, $\bar{x} \in \mathcal{M}_\alpha^{\mathrm{rec}}$ such that $\underline{x} < x < \bar{x}$ and $x$ and $\bar{x}$ are asymptotic.

Inasmuch as this is possible Fig. 6 gives a picture of $\mathcal{M}_\alpha$ in case $\alpha \in \mathbb{R}\backslash\mathbb{Q}$ and $p_0(\mathcal{M}_\alpha) \neq \mathbb{R}$. The dotted line represents an $x \in \mathcal{M}_\alpha\backslash\mathcal{M}_\alpha^{\mathrm{rec}}$.

One deficiency in this qualitative description of $\mathcal{M}_\alpha$ is that we do not say much about the size of $\mathcal{M}_\alpha\backslash\mathcal{M}_\alpha^{\mathrm{rec}}$. One expects that for generic $H$ one will have $\mathcal{M}_\alpha^{\mathrm{rec}} = \mathcal{M}_\alpha$ for most $\alpha \in \mathbb{R}\backslash\mathbb{Q}$, cf. the end of Section 9.

At this point we make some brief remarks on the invariant curve problem mentioned in the introduction. In our setting the existence of an invariant curve corresponds to the case that $p_0(\mathcal{M}_\alpha) = \mathbb{R}$. The integrable cases correspond to functions $\bar{H}$ of the form $\bar{H}(\xi, \eta) = h(\xi - \eta)$, cf. (1.4). The results of KAM-Theory mentioned in the introduction say that $p_0(\mathcal{M}_\alpha) = \mathbb{R}$ and $\mathcal{M}_\alpha^{\mathrm{rec}} = \mathcal{M}_\alpha$ if $H$ is sufficiently $C^{k+1}$-close to an integrable $\bar{H}$ and if $\alpha$ cannot be approximated by rationals too rapidly.

In the general case there is a criterion by J. Mather [38] which is closely related to the physically motivated 'Peierl's energy barrier', cf. [40], §14. It can be easily adapted to the present setting: suppose $r = (p/q) \in \mathbb{Q}$ with $p$ and $q$ relatively prime. Using the notation introduced in (3.3) we let $\Delta H_r$ denote the infimum of all $\varepsilon > 0$ such that every two periodic $x^0$, $x^1 \in \mathcal{M}_r$ can be joined by a continuous family $x^t$ of $(q, p)$-periodic elements $x^t \in P_{q, p}$ satisfying $H_{q, p}(x^t) \leqslant H_{q, p}(x^0) + \varepsilon$. Note that $H_{q, p}(x^0) = H_{q, p}(x^1)$ by (5.1). Then for every $\alpha \in \mathbb{R}\backslash\mathbb{Q}$ the limit $\lim_{r \to \alpha} \Delta H_r := \Delta H_\alpha$ exists and $p_0(\mathcal{M}_\alpha) = \mathbb{R}$ if and only if $\Delta H_\alpha = 0$.

## 5    STRUCTURE OF THE SET OF MINIMAL TRAJECTORIES WITH RATIONAL ROTATION NUMBER

In this section we study $\mathcal{M}_\alpha$ for rational $\alpha$, say $\alpha = p/q$ with $p$ and $q$ relatively prime. We start with $\mathcal{M}_\alpha^{\mathrm{per}}$, the set of periodic trajectories in $\mathcal{M}_\alpha$. The essential properties of $\mathcal{M}_\alpha^{\mathrm{per}}$ follow easily from the results in Section 3.

**(5.1)** THEOREM   $\mathcal{M}_\alpha^{\mathrm{per}}$ is non-empty, closed and totally ordered. Every $x \in \mathcal{M}_\alpha^{\mathrm{per}}$ has minimal period $(q, p)$. If $x \in \mathcal{M}_\alpha^{\mathrm{per}}$ then $x$ is a minimum of $H_{q,p} : P_{q,p} \to \mathbb{R}$, in particular $H_{q,p}(x) = H_{q,p}^{\min}$ for all $x \in \mathcal{M}_\alpha^{\mathrm{per}}$. Recall that $P_{q,p} = \{ x \in \mathbb{R}^{\mathbb{Z}} \mid T_{(q,p)}x = x \}$, $H_{q,p}(x) = H(x_0, ..., x_q)$.

*Proof*   According to (3.3) we have $\mathcal{M}_\alpha^{\mathrm{per}} \neq \varnothing$. The second assertion in (3.2) shows that every $x \in \mathcal{M}_\alpha^{\mathrm{per}}$ has minimal period $(q, p)$. The first assertion in (3.2) then implies that $\mathcal{M}_\alpha^{\mathrm{per}}$ is totally ordered. The last statement in (5.1) is proved by contradiction. Assume there exist $x \in P_{q,p}$ and $x^* \in \mathcal{M}_\alpha^{\mathrm{per}}$ such that

$$\varepsilon := H(x_0^*, ..., x_q^*) - H(x_0, ..., x_q) > 0$$

Choose $n \in \mathbb{N}$ so large that $n\varepsilon > H(x_0^*, x_1) - H(x_0, x_1) + H(x_{q-1}, x_q^*) - H(x_{q-1}, x_q)$. Using $H(x_{q-1}, x_q^*) = H(x_{nq-1}, x_{nq}^*)$ and $H(x_{q-1}, x_q) = H(x_{nq-1}, x_{nq})$ we obtain

$$H(x_0^*, x_1, ..., x_{nq-1}, x_{nq}^*) < H(x_0, x_1, ..., x_{nq-1}, x_{nq}) + n\varepsilon = H(x_0^*, ..., x_{nq}^*)$$

This contradicts the minimality of $x^*$.

Two elements of $\mathcal{M}_\alpha^{\mathrm{per}}$ are called neighboring if there does not exist an element of $\mathcal{M}_\alpha^{\mathrm{per}}$ between them.

**(5.2)** DEFINITION   Suppose $x^- < x^+$ are neighboring elements of $\mathcal{M}_\alpha^{\mathrm{per}}$. Then

$$\mathcal{M}_\alpha^+(x^-, x^+) := \{ x \in \mathcal{M}_\alpha \mid x \text{ is } \alpha\text{-asymptotic to } x^-$$
$$\text{and } \omega\text{-asymptotic to } x^+ \}.$$

$$\mathcal{M}_\alpha^-(x^-, x^+) := \{ x \in \mathcal{M}_\alpha \mid x \text{ is } \omega\text{-asymptotic to } x^-$$
$$\text{and } \alpha\text{-asymptotic to } x^+ \}.$$

We denote by $\mathcal{M}_\alpha^+$ resp. $\mathcal{M}_\alpha^-$ the union of the sets $\mathcal{M}_\alpha^+(x^-, x^+)$ resp. $\mathcal{M}_\alpha^-(x^-, x^+)$ extended over all pairs of neighboring elements $(x^-, x^+)$.

Note: By (3.9) every $x \in \mathcal{M}_\alpha^-(x^-, x^+) \cup \mathcal{M}_\alpha^+(x^-, x^+)$ satisfies $x^- < x < x^+$.

Next we prove that every $x \in \mathcal{M}_\alpha \backslash \mathcal{M}_\alpha^{\mathrm{per}}$ is $\alpha$- resp. $\omega$-asymptotic to neighboring periodic minimals.

**(5.3)** THEOREM   If $\alpha$ is rational $\mathcal{M}_\alpha$ is the disjoint union of $\mathcal{M}_\alpha^{\mathrm{per}}$, $\mathcal{M}_\alpha^+$ and $\mathcal{M}_\alpha^-$.

*Proof*   Suppose $x \in \mathcal{M}_\alpha \backslash \mathcal{M}_\alpha^{\mathrm{per}}$. To obtain the neighboring $x^-$ and $x^+$ we use the property of circle maps with rational rotation number which corresponds to (5.3). According to (3.15) there exists $f \in \tilde{G}_+$ such that $f(x_i) = x_{i+1}$. Then $\tilde{\alpha}(f) = \alpha = p/q$. As in (2.2) we consider the periodic function $r_q(t) := f^q(t) - t - p$. Since $x \notin \mathcal{M}_\alpha^{\text{per}}$ we have $r_q(x_0) \neq 0$, say $r_q(x_0) > 0$. According to (2.2) there exist $x_0^-$ and $x_0^+$ such that $r_q(x_0^-) = r_q(x_0^+) = 0$, $x_0 \in (x_0^-, x_0^+)$ and $r_q \,|\, (x_0^-, x_0^+) > 0$. This implies that the sequence $\{f^{nq}(x_0) - np\}_{n \in \mathbb{Z}}$ is strictly increasing and $\lim_{n \to \infty} (f^{nq}(x_0) - np) = x_0^+$, $\lim_{n \to -\infty} (f^{nq}(x_0) - np) = x_0^-$. If we define $x^+$ and $x^-$ by $x_i^+ := f^i(x_0^+)$, $x_i^- := f^i(x_0^-)$ then $T_{(q,p)} x^+ = x^+$, $T_{(q,p)} x^- = x^-$ and $\lim_{n \to \infty} T_{(-nq,\,-np)} x = x^+$, $\lim_{n \to -\infty} T_{(-nq,\,-np)} x = x^-$. Hence $x^- \in \mathcal{M}_\alpha^{\text{per}}$, $x^+ \in \mathcal{M}_\alpha^{\text{per}}$ and $x^- < x < x^+$. Moreover for all $0 \leqslant j \leqslant q$:

$$\lim_{n \to \infty} |\, x_{j+nq} - x_{j+nq}^+ \,| = \lim_{n \to \infty} |\, x_{j+nq} - np - x_j^+ \,|$$

$$= \lim_{n \to \infty} |\, (T_{(-nq,\,-np)} x)_j - x_j^+ \,| = 0$$

Hence $x$ is $\omega$-asymptotic to $x^+$ and, similarly, $x$ is $\alpha$-asymptotic to $x^-$. If $r_q(x_0) < 0$ an analogous proof gives $x^- < x^+$ in $\mathcal{M}_\alpha^{\text{per}}$ such that $x$ is $\omega$-asymptotic to $x^-$ and $\alpha$-asymptotic to $x^+$. It remains to prove that $x^-$ and $x^+$ are neighbors: Assume there exists $x^* \in \mathcal{M}_\alpha^{\text{per}}$ such that $x^- < x^* < x^+$. Then $x$ and $x^*$ cross, say between $i-1$ and $i$ (see Fig. 7). The case that $x$ and $x^*$ cross at some $i \in \mathbb{Z}$ can be treated similarly.

For $j > i$ we define segments $(\tilde{x}_{i-1}, \ldots, \tilde{x}_{j+q})$ by $\tilde{x}_{i-1} := x_{i-1}$, $\tilde{x}_m := x_m^*$ for $i \leqslant m \leqslant i + q - 1$, $\tilde{x}_m := x_{m-q} + p$ for $i + q \leqslant m \leqslant j + q - 1$ and $\tilde{x}_{j+q} := x_{j+q}$. We want to show that for large $j$ we have

(5.4)     $$H(\tilde{x}_{i-1}, \ldots, \tilde{x}_{j+q}) < H(x_{i-1}, \ldots, x_{j+q})$$

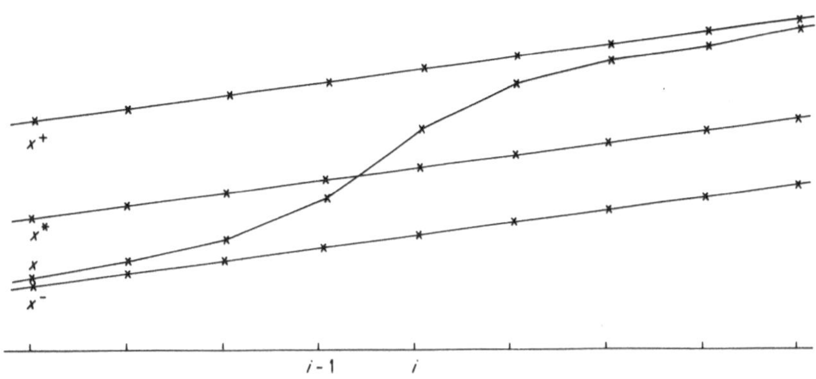


**Fig. 7.** $x^- < x^* < x^+$ in $\mathcal{M}_\alpha^{\text{per}}$

in contradiction to the minimality of $x$. According to the definition above

$$H(\tilde{x}_{i-1}, ..., \tilde{x}_{j+q}) = H(x_{i-1}, x_i^*) + H(x_i^*, ..., x_{i+q}^*) - H(x_{i+q-1}^*, x_{i+q}^*)$$
$$+ H(x_{i+q-1}^*, x_i + p) + H(x_i, ..., x_{j-1}) + H(x_{j-1} + p, x_{j+q})$$

Since $x$ and $x^*$ cross between $i - 1$ and $i$ and since $x^*$ is periodic with $(q, p)$ we have

$$H(x_{i-1}, x_i) > H(x_{i-1}, x_i^*) + H(x_{i-1}^*, x_i) - H(x_{i+q-1}^*, x_{i+q}^*)$$

Inserting this into the preceding equation we obtain

$$(5.5) \quad H(\tilde{x}_{i-1}, ..., \tilde{x}_{j+q}) = H(x_{i-1}, ..., x_{j-1}) + H(x_i^*, ..., x_{i+q}^*)$$
$$+ H(x_{j-1} + p, x_{j+q}) - \varepsilon$$

for some $\varepsilon > 0$. Since $x$ is $\omega$-asymptotic to $x_+ \in \mathcal{M}_\alpha^{\mathrm{per}}$ one easily sees

$$(5.6) \qquad \lim_{j \to \infty} | H(x_{j-1}, x_j) - H(x_{j-1} + p, x_{j+q}) = 0$$

and

$$\lim_{j \to \infty} | H(x_j, ..., x_{j+q}) - H(x_0^+, ..., x_q^+) | = 0$$

By the last assertion in (5.1) we have $H(x_0^+, ..., x_q^+) = H(x_i^*, ..., x_{i+q}^*)$, hence

$$(5.7) \qquad \lim_{j \to \infty} | H(x_j, ..., x_{j+q}) - H(x_i^*, ..., x_{i+q}^*) | = 0$$

Now (5.5), (5.6) and (5.7) imply that (5.4) holds for large $j$. Hence there cannot exist $x^* \in \mathcal{M}^{\mathrm{per}}$ such that $x^- < x^* < x^+$, i.e. $x^-$ and $x^+$ are neighbors.

Finally we approach the question of existence:

(5.8)    THEOREM    Suppose $x^- < x^+$ are neighboring elements of $\mathcal{M}_\alpha^{\mathrm{per}}$. Then both $\mathcal{M}_\alpha^+(x^-, x^+)$ and $\mathcal{M}_\alpha^-(x^-, x^+)$ are not empty.

*Proof*   Choose $\alpha_n \in \mathbb{R}$ such that $\alpha_n > \alpha$, $\lim \alpha_n = \alpha$ and $x^n \in \mathcal{M}_{\alpha_n}$. We intend to find appropriate translates of the $x^n$ which converge to an $x \in \mathcal{M}_\alpha^+(x^-, x^+)$. Let $\varepsilon := \min_{i \in \mathbb{Z}}(x_i^+ - x_i^-)$. Choose $i(n) \in \mathbb{Z}$ such that

$$x_i^n \leqslant x_i^- + \frac{\varepsilon}{2} \quad \text{for} \quad i < i(n) \quad \text{and} \quad x_{i(n)}^n > x_{i(n)}^- + \frac{\varepsilon}{2}$$

This is possible since $\alpha_n > \alpha$, cf. (3.16)(a). Define $k(n) \in \mathbb{Z}$ such that $j(n) := i(n) + k(n)q \in \{0, ..., q - 1\}$ and set $\tilde{x}^n := T_{k(n)(q, p)}x^n$. Then we have

$$\tilde{x}_i^n = x_{i-k(n)q}^n + k(n)p \leqslant x_i^- + \frac{\varepsilon}{2} \quad \text{for} \quad i < j(n)$$

and

$$\tilde{x}_i^n > x_i^- + \frac{\varepsilon}{2} \quad \text{for} \quad i = j(n).$$

From (3.18) we conclude that there exists a convergent subsequence of $\tilde{x}^n$ with limit $x \in \mathcal{M}_\alpha$. If $j \in \{0, \ldots, q-1\}$ appears infinitely often as $j(n)$ in this subsequence then

$$x_i \leqslant x_i^- + \frac{\varepsilon}{2} < x_i^+ \quad \text{for} \quad i < j \quad \text{and} \quad x_j \geqslant x_j^- + \frac{\varepsilon}{2} > x_j^-.$$

According to (5.3) this implies $x \in \mathcal{M}_\alpha^+(x^-, x^+)$. Starting with a sequence $\alpha_n < \alpha$, $\lim \alpha_n = \alpha$, we obtain an $x \in \mathcal{M}_\alpha^-(x^-, x^+)$.

From theorems (5.1), (5.3) and (5.8) we obtain the following picture of $\mathcal{M}_\alpha$ for $\alpha = (p/q) \in \mathbb{Q}$: By the projection $p_0$ we can identify $\mathcal{M}_\alpha^{\mathrm{per}}$ to a closed subset $A_\alpha^{\mathrm{per}} \subseteq \mathbb{R}$. There is the exceptional case that $A_\alpha^{\mathrm{per}} = \mathbb{R}$, i.e. for every $\xi \in \mathbb{R}$ there exists $x \in \mathcal{M}_\alpha^{\mathrm{per}}$ such that $x_0 = \xi$. Then $\mathcal{M}_\alpha = \mathcal{M}_\alpha^{\mathrm{per}}$ is totally ordered. Otherwise there will exist neighboring elements in $\mathcal{M}_\alpha^{\mathrm{per}}$ which are joined by heteroclinic orbits in $\mathcal{M}_\alpha^+$ and in $\mathcal{M}_\alpha^-$. If $x \in \mathcal{M}_\alpha^+$ and $x^* \in \mathcal{M}_\alpha^-$ lie between the same pair of neighboring elements of $\mathcal{M}_\alpha^{\mathrm{per}}$ then $x$ and $x^*$ cross; this is the only case that two trajectories in $\mathcal{M}_\alpha$ cross. In particular $\mathcal{M}_\alpha^{\mathrm{per}} \cup \mathcal{M}_\alpha^+$ and $\mathcal{M}_\alpha^{\mathrm{per}} \cup \mathcal{M}_\alpha^-$ are totally ordered and closed. Again $\mathcal{M}_\alpha^{\mathrm{per}} \cup \mathcal{M}_\alpha^+$ and $\mathcal{M}_\alpha^{\mathrm{per}} \cup \mathcal{M}_\alpha^-$ can be described as closed sets of orbits or circle maps $f^+$ and $f^-$ in $\tilde{G}_+$ with $\tilde{\alpha}(f^+) = \tilde{\alpha}(f^-) = \alpha$. Examples in Section 9 will show that these are the only restrictions on the sets $\mathcal{M}_\alpha^{\mathrm{per}}$, $\mathcal{M}_\alpha^+$ and $\mathcal{M}_\alpha^-$. Generically one expects the following situation: $\mathcal{M}_\alpha^{\mathrm{per}}$, $\mathcal{M}_\alpha^+$ and $\mathcal{M}_\alpha^-$ each consist of exactly one $T$-orbit, cf. the end of Section 9.

## 6  APPLICATIONS TO GEODESICS

In this section we interpret the results from Sections 3–5 for geodesics on a 2-dimensional torus or, equivalently, on its universal cover $\mathbb{R}^2$. In view of the approach we took in Sections 3–5 it is natural to formulate the properties of geodesics that we need in an abstract way. We can then specialize to geodesics of Riemannian or symmetric Finsler metrics on $T^2$. The results obtained will include all those stated in [26], but we will also make considerable progress over Morse's and Hedlund's work.

   The following notions are taken from [13] and [24] and are only repeated for convenience. We start with a complete, locally compact metric space $(X, d)$ with symmetric distance function, $d(x, y) = d(y, x)$. We can define the *length* $L_d(\gamma) \in [0, \infty]$ of a continuous curve $\gamma: [a, b] \to X$ by

$$L_d(\gamma) = L(\gamma) = \sup\left\{ \sum_{i=1}^{n} d(\gamma(t_{i-1}), \gamma(t_i)) \,\middle|\, n \in \mathbb{N}, \right.$$
$$\left. a = t_0 \leqslant t_1 \leqslant \cdots \leqslant t_n = b \right\}$$

This length has the usual properties, cf. [24], Chapter 1. A *minimal geodesic segment* is a curve $c: [a, b] \to X$ such that

$$d(c(s), c(s')) = |s - s'|$$

for all $s, s' \in [a, b]$. A *geodesic* is a curve $c: \mathbb{R} \to X$ such that for all $s \in \mathbb{R}$ there exists $\varepsilon > 0$ such that $c \,|\, [s - \varepsilon, s + \varepsilon]$ is a minimal geodesic segment. The objects of our interest are geodesics $c$ such that $c \,|\, [a, b]$ is a minimal geodesic segment for all compact intervals $[a, b] \subseteq \mathbb{R}$. Obviously such geodesics can only exist if $X$ is noncompact. They have various names in the literature; we call them *minimal geodesics*. They were first studied by Morse [43] in the case that $X$ is the universal cover of a compact Riemannian surface of genus greater than one.

Our first postulate on $(X, d)$ is that it be a length space in the sense of [24], i.e.:

For all $x, y \in X$ we have:

$(\mathrm{G}_1)$   $d(x, y) = \inf\{ L_d(\gamma) \,|\, \gamma: [a, b] \to X \text{ continuous}, \gamma(a) = x, \gamma(b) = y \}$

The Arzela–Ascoli theorem together with $(\mathrm{G}_1)$, the completeness and local compactness of $(X, d)$ imply the following version of part of the Hopf–Rinow theorem, cf. [24]:

(6.1)   THEOREM   Two arbitrary points $x, y \in X$ can be joined by a minimal geodesic segment.

Our second postulate concerns the uniqueness of prolongations of geodesic segments:

$(\mathrm{G}_2)$   Suppose $c$ and $\bar{c}$ are minimal geodesic segments defined on $[a, b] \subseteq \mathbb{R}$. If $c \,|\, [a, a + \varepsilon] = \bar{c} \,|\, [a, a + \varepsilon]$ for some $\varepsilon > 0$ then $c = \bar{c}$.

Obviously $(\mathrm{G}_2)$ is a local condition, i.e. it suffices to require $(\mathrm{G}_2)$ for arbitrarily small $b - a$. In the following we will often identify minimal geodesic segments with their images in $X$. A useful consequence of $(\mathrm{G}_2)$ is

(6.2)   LEMMA.   Suppose two minimal geodesic segments have two points $p_1 \neq p_2$ in common. Then either their intersection is a minimal geodesic segment or $p_1$ and $p_2$ are the endpoints of both segments.

*Proof* Suppose $c: [a, b] \to X$, $\bar{c}: [a, \bar{b}] \to X$ are minimal geodesic segments satisfying $c(a) = \bar{c}(a)$ and $c(s) = \bar{c}(s)$ for some $s$, $a < s \leqslant \min\{b, \bar{b}\}$. If we do not have $s = b = \bar{b}$, say $s < \bar{b}$, then the minimal geodesic segments $c \,|\, [a, s]$ and $\bar{c} \,|\, [s, \bar{b}]$ join to form another minimal geodesic segment from $\bar{c}(a)$ to $\bar{c}(\bar{b})$. If we traverse both segments in the opposite sense ($G_2$) implies that $c$ and $\bar{c}$ coincide on $[a, \min\{b, \bar{b}\}]$.

Finally we treat coverings: if $(X, d)$ is a complete, locally compact metric space satisfying ($G_1$) and if $\pi: \tilde{X} \to X$ is a (topological) covering with connected $\tilde{X}$ then there is a unique metric $\tilde{d}$ on $\tilde{X}$ such that $\pi$ is a local isometry and ($G_1$) is satisfied: one defines the length of a curve on $\tilde{X}$ as the length of its projection to $X$ and takes ($G_1$) as definition for $\tilde{d}$. If a d-ball $B(x, 2r) \subseteq X$ is evenly covered by $\pi$ and $\tilde{x} \in \pi^{-1}(x)$ then $\pi: (\tilde{B}(\tilde{x}, r), \tilde{d}) \to (B(x, r), d)$ is an isometry. In particular, the length with respect to $\tilde{d}$ coincides with the length of the projection so that ($G_1$) is satisfied for $\tilde{d}$. Moreover $\tilde{d}$ generates the topology of $\tilde{X}$ and $(\tilde{X}, \tilde{d})$ is complete. The covering transformations of $\pi: \tilde{X} \to X$ are isometries of $(\tilde{X}, \tilde{d})$. The local condition ($G_2$) is true for $(\tilde{X}, \tilde{d})$ if true for $(X, d)$.

Our general hypothesis is that $(X, d)$ satisfies ($G_1$) and ($G_2$) and that $(X, d)$ is homeomorphic to $T^2$ or, equivalently, that we have a distance function $d$ on $T^2$ which generates the usual topology and satisfies ($G_1$) and ($G_2$). Completeness and local compactness are obvious in this case. Note that we do not require prolongability of geodesics. In particular we allow Riemannian metrics with cone-like singularities and they yield an interesting example in Section 9. We want to describe the minimal geodesics on $(\mathbb{R}^2, \tilde{d})$ where $\tilde{d}$ is induced from $d$ by some covering $\pi: \mathbb{R}^2 \to T^2$. We will always assume that the covering transformations are the translations $T_k: \mathbb{R}^2 \to \mathbb{R}^2$, $q \to q + k$ with $k \in \mathbb{Z}^2$. This is a slight abuse of notation since in Sections 1 and 3–5 the letter $T$ denotes the action of $\mathbb{Z}^2$ on $\mathbb{R}^\mathbb{Z}$. However, this abuse is natural since the translation by $k \in \mathbb{Z}^2$ of a minimal geodesic corresponds to the action of $T_k$ on the corresponding $x \in \mathcal{M}$, cf. Fig. 2.

The most important examples for such spaces $(T^2, d)$ are given by Riemannian or symmetric Finsler metrics. Finsler metrics generalize Riemannian metrics in much the same way that Hamiltonian systems with $H_{yy} > 0$ generalize classical systems with Hamiltonian 'kinetic + potential energy', cf. also the remarks around (7.11). For sake of completeness we define the relevant notions, describe the relation to the previously defined objects and give references for the proofs. To be as elementary as possible we start on $\mathbb{R}^2$. A Finsler metric $F$ on $\mathbb{R}^2$ is a non-negative continuous function on the tangent bundle $T\mathbb{R}^2 = \mathbb{R}^2 \times \mathbb{R}^2$ of $\mathbb{R}^2$ which is smooth off the zero-section $\mathbb{R}^2 \times \{0\}$ and satisfies the following conditions:

($F_1$)  Homogeneity: For all $(q, \dot{q}) \in \mathbb{R}^2 \times \mathbb{R}^2$ and all $\lambda > 0$ we have
$$F(q, \lambda \dot{q}) = \lambda F(q, \dot{q}).$$

(F$_2$)   Legendre condition: For all $(q, \dot{q}) \in \mathbb{R}^2 \times (\mathbb{R}^2 \backslash \{0\})$ the matrix

$$\left( \frac{\partial^2 (F^2)}{\partial \dot{q}_i \partial \dot{q}_j} (q, \dot{q}) \right)_{1 \leqslant i, j \leqslant 2} \quad \text{is positive definite.}$$

A Finsler metric $F$ is symmetric if $F(q, \dot{q}) = F(q, -\dot{q})$ for all $(q, \dot{q}) \in \mathbb{R}^2 \times \mathbb{R}^2$. A well-known class of symmetric Finsler metrics are the Riemannian metrics which—by definition—are quadratic in $\dot{q}$, i.e. $F^2(q, \dot{q}) = \Sigma_{i, j=1}^2 \, g_{ij}(q)\dot{q}_i\dot{q}_j$ where $(g_{ij}(q))$ is symmetric and positive definite. To obtain a Finsler metric on $T^2 = \mathbb{R}^2 / \mathbb{Z}^2$ we require that all translations $T_k, k \in \mathbb{Z}^2$, be isometries of $F$, i.e.

(F$_3$)   For all $(q, \dot{q}) \in \mathbb{R}^2 \times \mathbb{R}^2$ and all $k \in \mathbb{Z}^2$ we have $F(q + k, \dot{q}) = F(q, \dot{q})$.

The $F$-length of a piecewise $C^1$-curve $\gamma : [a, b] \to \mathbb{R}^2$ is defined by

$$L_F(\gamma) := \int_b^a F(\gamma(t), \dot{\gamma}(t)) \, \mathrm{d}t$$

The $F$-length of a piecewise $C^1$-curve in $\mathbb{R}^2 / \mathbb{Z}^2$ is the $F$-length of any of its lifts to $\mathbb{R}^2$. Now we can define a distance function $d$ on $\mathbb{R}^2 / \mathbb{Z}^2$ given a symmetric Finsler metric $F$ on $\mathbb{R}^2 / \mathbb{Z}^2$,

$$d(p, q) = \inf \{ L_F(\gamma) \,|\, \gamma : [a, b] \to \mathbb{R}^2 / \mathbb{Z}^2 \text{ piecewise } C^1, \, \gamma(a) = p, \gamma(b) = q \},$$

and similarly a distance function $\tilde{d}$ on $\mathbb{R}^2$. Then $d$ and $\tilde{d}$ are related by the lifting process described above and generate the usual topologies. It is neither obvious nor completely trivial that this $d$ satisfies (G$_1$) and (G$_2$); to prove this one needs the local existence and regularity theory for curves which minimize the integral $\int F^2(\gamma(t), \dot{\gamma}(t)) \, \mathrm{d}t$. We simply state that $d$ satisfies (G$_1$) and (G$_2$), cf. [22], Kap. V, or [16] §§383–93. The $\tilde{d}$-geodesics are precisely those maximal solutions of the Euler–Lagrange equations

$$(F^2)_q(c, \dot{c}) - \frac{\mathrm{d}}{\mathrm{d}t} ((F^2)_{\dot{q}}(c, \dot{c})) = 0$$

which in addition satisfy $F(c, \dot{c}) = 1$. Knowing this we see that the unique prolongation property (G$_2$) is a simple consequence of the fact that solutions of ODEs are determined by their initial data.

Now we explain how we get a function $H : \mathbb{R}^2 \to \mathbb{R}$ with properties (H$_1$)–(H$_4$) from Section 1 if we are given a torus $(T^2, d)$ satisfying (G$_1$) and (G$_2$). Let $\pi : \mathbb{R}^2 \to \mathbb{R}^2 / \mathbb{Z}^2 \simeq T^2$ be a covering and $\tilde{d}$ the induced distance function on $\mathbb{R}^2$. Assume for the moment that the coordinates on $\mathbb{R}^2$ have been so chosen that the coordinate line $s \to (0, s)$ is a minimal $\tilde{d}$-geodesic. Since the $T_k, k \in \mathbb{Z}^2$, are isometries all lines $s \to (i, s)$, $i \in \mathbb{Z}$, are minimal geodesics and, by the assumed symmetry, the same is true for $s \to (i, -s)$. Under the assumption made above we define $H : \mathbb{R}^2 \to \mathbb{R}$ by

$$H(\xi, \eta) := \tilde{d}((0, \xi), (1, \eta))$$

Before we start to prove that $H$ has properties $(H_1)$–$(H_4)$ we mention:

(6.3)  LEMMA   Let $(x_j, ..., x_k)$ be a minimal segment with respect to $H$ and $k - j \geqslant 2$. Then there exists a unique minimal geodesic segment $c: [0, L] \to \mathbb{R}^2$ which joins $(j, x_j)$ to $(k, x_k)$ passing through the points $(i, x_i), j < i < k$. Conversely, every minimal geodesic segment which is not part of a line $\{i\} \times \mathbb{R}, i \in \mathbb{Z}$, intersects each line $\{i\} \times \mathbb{R}$ at most once and the points of intersection form the graph of a minimal segment with respect to $H$.

*Proof*  This follows immediately from (6.1) and (6.2) and the fact that $\tilde{d}((0, \xi), (1, \eta)) = \tilde{d}((i, \xi), (i + 1, \eta))$ for all $i \in \mathbb{Z}$.

(6.4)  LEMMA   $H$ satisfies $(H_1)$–$(H_4)$.

*Proof*  Since $T_{(0, 1)}$ is a $\tilde{d}$-isometry we have $(H_1)$:

$$H(\xi + 1, \eta + 1) = H(\xi, \eta)$$

The condition at infinity, $(H_2)$, can be seen as follows:

$$H(\xi, \xi + \eta) \geqslant \tilde{d}((1, \xi), (1, \xi + \eta)) - \tilde{d}((1, \xi), (0, \xi))$$

Since $s \to (1, s)$ is a minimal geodesic we have

$$\tilde{d}((1, \xi), (1, \xi + \eta)) = |\eta|$$

This implies $(H_2)$ since $\xi \to \tilde{d}((1, \xi), (0, \xi))$ is a continuous periodic function. For $(H_3)$ we have to prove:
If $\underline{\xi} < \bar{\xi}$ and $\underline{\eta} < \bar{\eta}$ then

$$H(\underline{\xi}, \underline{\eta}) + H(\bar{\xi}, \bar{\eta}) < H(\underline{\xi}, \bar{\eta}) + H(\bar{\xi}, \underline{\eta})$$

According to (6.1) we can choose minimal geodesic segments $c_0$ from $(0, \underline{\xi})$ to $(1, \bar{\eta})$ and $c_1$ from $(0, \bar{\xi})$ to $(1, \underline{\eta})$. By (6.3) both segments are contained in $(0, 1) \times \mathbb{R}$ except for the endpoints. Since $\underline{\xi} < \bar{\xi}$ and $\underline{\eta} < \bar{\eta}$ the segments intersect inside $(0, 1) \times \mathbb{R}$. Hence we can join parts of $c_0$ and $c_1$ to obtain curves $\gamma$ from $(0, \underline{\xi})$ to $(1, \underline{\eta})$ and $\bar{\gamma}$ from $(0, \bar{\xi})$ to $(1, \bar{\eta})$ (Fig. 8).

Then $L(\gamma) + L(\bar{\gamma}) = L(c_0) + L(c_1) = H(\underline{\xi}, \bar{\eta}) + H(\bar{\xi}, \underline{\eta})$. Now $L(\gamma) + L(\bar{\gamma}) \leqslant H(\underline{\xi}, \underline{\eta}) + H(\bar{\xi}, \bar{\eta})$ only if $\gamma$ and $\bar{\gamma}$ are minimal geodesics segments. But this would contradict $(G_2)$. Hence $(H_3)$ is true.

To prove $(H_4)$ we consider two minimal segments $(x_{-1}, x_0, x_1) \neq (x_{-1}^*, x_0^*, x_1^*)$ with respect to $H$ which satisfy $x_0 = x_0^*$. We want to show $(x_{-1} - x_{-1}^*)(x_1 - x_1^*) < 0$.

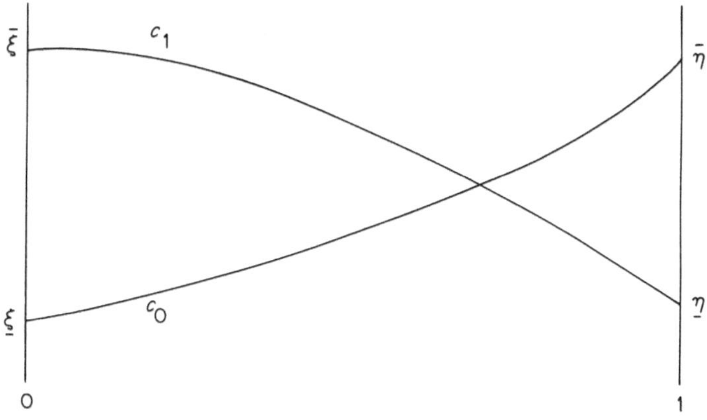



**Fig. 8**

According to (6.3) there are minimal geodesic segments $c, c^*$ in $[-1, 1] \times \mathbb{R}$ with initial points $(-1, x_{-1})$ resp. $(-1, x_{-1}^*)$ and endpoints $(1, x_1)$ resp. $(1, x_1^*)$ such that $c$ and $c^*$ intersect in $(0, x_0)$. (G$_2$) and (6.2) imply $(x_{-1} - x_{-1}^*)$ $(x_1 - x_1^*) \neq 0$. Let us assume $(x_{-1} - x_{-1}^*)(x_1 - x_1^*) > 0$ and let $\bar{c}$ be a minimal geodesic segment from $(-1, x_{-1})$ to $(1, x_1^*)$. By (6.3) the segment $\bar{c}$ is contained in $[-1, 1] \times \mathbb{R}$ and by (6.2) it does not intersect $c$ or $c^*$ inside $(-1, 1) \times \mathbb{R}$. A simple topological argument shows that this is not possible. This proves (H$_4$).

Finally we have to show that our assumption '$s \to (0, s)$ is a minimal geodesic' is no restriction of generality.

Here our approach gets a little circuitous since we essentially have to reprove Theorem (3.3). We show that there exists a minimal $\bar{d}$-geodesic $c$ which is invariant under the translation $T_{(0, 1)}$. Then $c$ does not intersect any of its translates $T_{(k, 0)}c, k \neq 0$. It is a well-known topological fact that there exists a homeomorphism $\Phi : \mathbb{R}^2 \to \mathbb{R}^2$ such that $\Phi(0, s) = c(s)$ and $\Phi(p + k) = \Phi(p) + k$ for all $p \in \mathbb{R}^2$, $k \in \mathbb{Z}^2$, cf. [50], Thm 6.2.5. Hence $s \to (0, s)$ is a minimal geodesic with respect to the metric $\bar{d}(p, q) := \tilde{d}(\Phi(p), \Phi(q))$ which is the lift of $d$ by the covering $\bar{\pi} = \pi \circ \Phi : \mathbb{R}^2 \to T^2$. To obtain such a minimal $\bar{d}$-geodesic $c$ choose a point $(\xi, \eta) \in \mathbb{R}^2$ which is an absolute minimum of the function $(x, y) \to \tilde{d}((x, y), (x, y + 1))$. Let $c$ be the curve formed by all the translates $T_{(0, j)}c_0, j \in \mathbb{Z}$, where $c_0$ is a minimal geodesic segment from $(\xi, \eta)$ to $(\xi, \eta + 1)$. The proof that this curve $c$ is a minimal geodesic can be found in [14] (for the case of G-spaces) and it carries over to the present situation. The idea is the same as the one used to prove (3.3).

It is clear from (6.3) that minimal trajectories $x \in \mathcal{M} = \mathcal{M}(H)$ correspond to minimal $\bar{d}$-geodesics $c : \mathbb{R} \to \mathbb{R}^2$. More precisely, for every $x \in \mathcal{M}$ there exists a minimal geodesic $c$ passing through the points $(i, x_i)_{i \in \mathbb{Z}}$; $c$ is unique up to parametrization. On the other hand if $c(s) = (\xi(s), \eta(s))$ is a minimal geodesic and $\xi(s)$ is surjective then $c$ intersects $\mathbb{Z} \times \mathbb{R} \subseteq \mathbb{R}^2$ in the graph of some $x \in \mathcal{M}$.

Here we have to pay tribute to the fact that our approach is not invariant: minimal geodesics $c(s) = (\xi(s), \eta(s))$ such that $\xi(s)$ is not surjective need special attention. Since $c$ intersects each of the lines $\{i\} \times \mathbb{R}, i \in \mathbb{Z}$, at most once we see that in this case $\bar{\xi}(s) := \xi(s) + \eta(s)$ is surjective.

Hence in the new coordinates $(\bar{\xi}, \bar{\eta}) := (\xi + \eta, \eta)$ on $\mathbb{R}^2$, which correspond to a different covering $\bar{\pi}: \mathbb{R}^2 \to T^2, c$ is no longer special. Actually (5.3) tells us that there exists $i \in \mathbb{Z}$ such that $c(s) = (\xi(s), \eta(s))$ is contained in $[i, i+1) \times \mathbb{R}$.

We are now in the position to interpret the results of Sections 3–5 as statements on minimal $\tilde{d}$-geodesics where $\tilde{d}$ is a lift of a metric $d$ on $\mathbb{R}^2/\mathbb{Z}^2$ satisfying $(G_1)$ and $(G_2)$. In particular all of the following statements are true for symmetric Finsler metrics. Note that the statements are invariant under coordinate changes $\Phi: \mathbb{R}^2 \to \mathbb{R}^2$ satisfying $\Phi(p + k) = \Phi(p) + k$ for all $p \in \mathbb{R}^2, k \in \mathbb{Z}^2$. From (3.13), (3.16) and (3.17) we obtain:

(6.5)    THEOREM    Let $c(s) = (\xi(s), \eta(s))$ be a minimal geodesic in $(\mathbb{R}^2, \tilde{d})$. If $c$ and a translate $T_k c, k \in \mathbb{Z}^2$, have a point in common then $c$ and $T_k c$ coincide up to parametrization. If $\xi(s)$ is surjective then the average slope $\alpha(c) := \lim_{|s| \to \infty} \eta(s)/\xi(s)$ exists in $(-\infty, \infty)$. For every $\alpha \in \mathbb{R}$ there exists a minimal geodesic $c$ with $\alpha(c) = \alpha$.

*Remarks* (1)    If $\xi(s)$ is not surjective then $\xi(s)$ is bounded. In this case we define $\alpha(c) = \infty$.

(2) Actually (3.16)(a) implies more than simply the existence of the limit in (6.5). Using a compactness argument we obtain Theorem VII from [26]: let $\tilde{d}$ be a metric on $\mathbb{R}^2$ which is a lift of a metric $d$ on $\mathbb{R}^2/\mathbb{Z}^2$ satisfying $(G_1)$ and $(G_2)$. Then there exists a constant $D > 0$ such that every minimal $\tilde{d}$-geodesic is contained in a strip of width $D$ bounded by two parallel straight lines in $\mathbb{R}^2$.

At first sight there seems to be a problem to obtain $D$ valid for arbitrarily large average slope $\alpha$. However, this difficulty disappears if we use a second coordinate system whose vertical axis projects to a closed geodesic in a different homology class.

(3) The first statement in (6.5) says that the projection of a minimal geodesic to $T^2 = \mathbb{R}^2/\mathbb{Z}^2$ does not have transverse self-intersections.

(4) From the invariant standpoint the number $\alpha(c)$ depends on the choice of a basis for $H_1(T^2, \mathbb{Z})$ which determines a covering $\pi: \mathbb{R}^2 \to T^2$ (with $\mathbb{Z}^2$-translations as deck transformations) up to coordinate changes $\Phi$ as above. In particular, only the rationality resp. irrationality of $\alpha(c)$ is a completely invariant property of the corresponding geodesic on $T^2$.

To remain in the historical order of discovery we continue with the results from Section 5. A geodesic $c: \mathbb{R} \to \mathbb{R}^2$ is periodic with period $(q, p) \in \mathbb{Z}^2 \setminus \{0\}$ if $T_{(q, p)} c$ and $c$ coincide up to parametrization. So periodic geodesics correspond to homotopically non-trivial closed geodesics on $T^2$. Using (3.2), the proof of (3.3) and (5.1) we conclude:

(6.6)  THEOREM   Periodic minimal geodesics are exactly the lifts of closed geodesics on $T^2$ which have minimal length in their free homotopy class. Two different periodic minimal geodesics with the same period do not intersect. The minimal period $(q, p)$ of a periodic minimal geodesic is relatively prime.

From (5.3) we conclude

(6.7)  THEOREM   A minimal geodesic $c$ with $\alpha(c) \in \mathbb{Q}$ is either periodic or is contained in a strip between two periodic minimal geodesics $c^-$, $c^+$ with $\alpha(c^-) = \alpha(c^+) = \alpha(c)$. In each direction $c$ is asymptotic to exactly one of $c^-$ and $c^+$. There are no periodic minimal geodesics in the strip between $c^-$ and $c^+$.

Finally the existence result (5.8) says:

(6.8)  THEOREM   In every strip between two periodic minimal geodesics $c^-$, $c^+$ which does not contain other periodic minimal geodesics there exist minimal geodesics $c$ and $c^*$ such that $c$ is $\alpha$-asymptotic to $c^-$ and $\omega$-asymptotic to $c^+$ (and vice versa for $c^*$).

Hedlund's results [26] are contained in (6.5) and (6.6) while (6.7) (6.8) are essentially due to Morse [43]. The intersections of geodesics on surfaces with minimal length in their homotopy class have been studied in [23].

Now we come to the results in Section 4 which—when applied to geodesics—go beyond the previous knowledge. The set of minimal geodesics with average slope $\alpha$ will also be denoted by $\mathcal{M}_\alpha$. The set of those geodesics in $\mathcal{M}_\alpha$ which—when considered on $T^2$—are recurrent is denoted by $\mathcal{M}_\alpha^{\mathrm{rec}}$. On $\mathbb{R}^2$ these are the $c \in \mathcal{M}_\alpha$ which can be approximated by their translates $T_k c, k \in \mathbb{Z}^2$. The main problem which is not solved in [26] and [43] is the following: Can there exist intersecting $c$ and $c'$ in $\mathcal{M}_\alpha$ for $\alpha \in \mathbb{R}\backslash\mathbb{Q}$? If so the structure of $\mathcal{M}_\alpha$ for $\alpha \in \mathbb{R}\backslash\mathbb{Q}$ could be very complicated. However, Theorem (4.1) implies:

(6.9)  THEOREM   Two different geodesics in $\mathcal{M}_\alpha$, $\alpha \in \mathbb{R}\backslash\mathbb{Q}$, do not intersect. The remaining results of Section 4 combine to:

(6.10)  THEOREM   Suppose $\alpha$ is irrational. Either there is a $c \in \mathcal{M}_\alpha^{\mathrm{rec}}$ through every point of $\mathbb{R}^2$ ($c$ is unique up to parametrization) or $\mathcal{M}_\alpha^{\mathrm{rec}}$ intersects every periodic minimal geodesic in a Cantor set. Every $c \in \mathcal{M}_\alpha \backslash \mathcal{M}_\alpha^{\text{rec}}$ is contained in a strip bounded by two asymptotic geodesics in $\mathcal{M}_\alpha^{\text{rec}}$. Every $c \in \mathcal{M}_\alpha^{\text{rec}}$ can be approximated by periodic minimal geodesics.

Briefly one can say that $\mathcal{M}_\alpha^{\text{rec}}$ is either a foliation or a lamination of $\mathbb{R}^2$. For a (necessarily sketchy) picture of a lamination see Fig. 6. We note the surprising analogy to the euclidean situation: The sets $\mathcal{M}_\alpha^{\text{rec}}$, $\alpha \in \mathbb{R} \cup \{\infty\}$, correspond to the families of parallel lines, $\mathcal{M}_\alpha^{\text{rec}} = \mathcal{M}_\alpha^{\text{per}}$ for $\alpha \in \mathbb{Q} \cup \{\infty\}$. No two elements of the same family intersect while two elements of different families intersect exactly once.

In Section 9 we will give examples of smooth Riemannian metrics on $T^2$ such that $\mathcal{M}_\alpha^{\text{rec}}$ corresponds to a Cantor set for all $\alpha \in \mathbb{R} \backslash \mathbb{Q}$. The set of vectors tangent to the projection to $T^2$ of some $c \in \mathcal{M}_\alpha^{\text{rec}}$ form a minimal set for the geodesic flow on the unit tangent bundle of $T^2$; this is a consequence of (3.15) and the corresponding property for circle maps. So we find many examples of complicated minimal sets for a smooth geodesic flow of a torus. The construction of such minimal sets for geodesic flows of surfaces of higher genus is one of the main results in [42].

We note the following analogue to (3.19): Let $\mathcal{N}$ be a set of geodesics of a Finsler metric on a surface such that no two elements of $\mathcal{N}$ intersect. Let $N$ be the set of points through which elements of $\mathcal{N}$ pass. Then the set of tangent vectors to geodesics of $\mathcal{N}$ forms a Lipschitz section in the unit tangent bundle over $N$. Using the fact that sufficiently short geodesic segments depend smoothly on their endpoints one can prove this Lipschitz continuity in a completely elementary way.

Finally we remark that Zaustinsky [51] generalized many of Morse's [43] and Hedlund's [26] results to more general and in particular non-symmetric metrics. In the approach presented here such generalization is not possible: To describe non-self-intersecting minimal geodesics by circle maps we need a periodic geodesic which is minimal when traversed in either sense. A priori it is not clear if (6.9) and (6.10) are true in the non-symmetric case. However, there is a way to define the rotation number of a non-self-intersecting minimal geodesic which does not rely on circle maps, cf. [5], Sect. 4. Using this concept and Zaustinsky's results it should be possible to generalize (6.9) and (6.10) to non-symmetric metrics.


## 7  APPLICATIONS TO MONOTONE TWIST MAPS

In this section we deduce Mather's theory of invariant sets for monotone twist maps from the results in Sections 3–5. A generating function for the monotone twist map will play the role of the function $H$ from Section 1. In the definitions of the term 'monotone twist map' the regularity and the domain of definition vary with the applications. In order to minimize technical difficulties we will use the following fairly restrictive conditions. Later we will discuss how they can be relaxed.

(7.1)  DEFINITION  A monotone twist map is an orientation preserving $C^1$-diffeomorphism $\varphi : S^1 \times [0, 1] \to S^1 \times [0, 1]$ of an annulus which admits a lift $\tilde{\varphi} = (f, g) : \mathbb{R} \times [0, 1] \to \mathbb{R} \times [0, 1]$ with the following properties:
(a) $\tilde{\varphi}$ preserves (Lebesgue) area.
(b) Twist condition: $D_2 f > 0$.
(c) $g(\xi, 0) = 0,\ g(\xi, 1) = 1$.

*Remark*  Instead of (a) we could require $\det(\tilde{\varphi}') = 1$. Condition (c) means that $\varphi$ does not commute the boundary components. We assume that the translation $T_{(1,0)}$, $T_{(1,0)}(\xi, y) = (\xi + 1, y)$, generates the group of covering transformations. Then $f(\xi + 1, y) = f(\xi, y) + 1$, $g(\xi + 1, y) = g(\xi, y)$.

Monotone twist maps turn up as Poincaré maps of Hamiltonian systems with two degrees of freedom at various occasions, see [44], Chapter II.4. In (7.5) we present three concrete examples, one of which is the billiard in a strictly convex bounded domain.

A fundamental property of a monotone twist map is that it can be globally described by a 'generating function'

$$H : D \to \mathbb{R} \quad \text{where} \quad D = \{\, (\xi, \eta) \in \mathbb{R}^2 \mid f(\xi, 0) \leqslant \eta \leqslant f(\xi, 1) \,\}.$$

Up to an additive constant $H$ is uniquely determined by

$$(7.2) \qquad \tilde{\varphi}(x_0, y_0) = (x_1, y_1) \Leftrightarrow \begin{cases} -D_1 H(x_0, x_1) = y_0 \\ \phantom{-}D_2 H(x_0, x_1) = y_1 \end{cases}$$

To construct $H$ let $a, b : D \to \mathbb{R}$ be defined by

$$a(\xi, f(\xi, y)) := y,\ b(\xi, \eta) := g(\xi, a(\xi, \eta))$$

Then (7.2) is equivalent to

$$dH = - a\, d\xi + b\, d\eta$$

From $\det(\tilde{\varphi}') = 1$ one concludes that the 1-form $\omega = - a\, d\xi + b\, d\eta$ is closed. Since $D$ is simply connected there exists $H : D \to \mathbb{R}$ such that (7.2) is true. $H$ is $C^2$ in the interior of $D$ and $D_1 H, D_2 H$ and $D_2 D_1 H$ extend continuously to $D$. Moreover (7.1)(b) implies $D_2 D_1 H = - D_2 a \leqslant - \delta$ for some $\delta > 0$. Finally $H(\xi + 1, \eta + 1) = H(\xi, \eta)$, i.e. $\omega$ is even exact on the cylinder 'D modulo translations $T_{(j,j)}, j \in \mathbb{Z}'$ : $\omega$ vanishes along the curve $\xi \rightarrow (\xi, f(\xi, 0))$ from $(0, f(0, 0))$ to $(1, f(0,0) + 1)$. Now it is easy to extend $H$ to a $C^2$-function on $\mathbb{R}^2$ such that

$$(H_1) \qquad\qquad\qquad H(\xi + 1, \eta + 1) = H(\xi, \eta)$$

and

$$D_2 D_1 H \leqslant -\delta/2 < 0$$

Hence, according to (1.3), the extended $H : \mathbb{R}^2 \rightarrow \mathbb{R}$ satisfies $(H_1)$–$(H_4)$ from Section 1. Taking (7.2) as definition we can extend $\tilde{\varphi}$ resp. $\varphi$ to $\mathbb{R}^2$ resp. $S^1 \times \mathbb{R}$. According to (7.2) we have a variational principle for the orbits of the extended $\tilde{\varphi}$. We recall that a sequence $(x_i)_{i \in \mathbb{Z}}$ is stationary with respect to $H$ if $D_2 H(x_{i-1}, x_i) + D_1 H(x_i, x_{i+1}) = 0$ for all $i \in \mathbb{Z}$, cf. (1.3).

(7.3)  If $(x_i)_{i \in \mathbb{Z}}$ is stationary with respect to $H$ then $(x_i, y_i)_{i \in \mathbb{Z}}$, $y_i = -D_1 H(x_i, x_{i+1})$, is an orbit of $\tilde{\varphi}$, i.e. $\tilde{\varphi}(x_i, y_i) = (x_{i+1}, y_{i+1})$. Conversely, if $(x_i, y_i)_{i \in \mathbb{Z}}$ is an orbit of $\tilde{\varphi}$ then $(x_i)_{i \in \mathbb{Z}}$ is stationary with respect to $H$.

Since the $(x_i)_{i \in \mathbb{Z}} \in \mathcal{M}(H)$ are stationary with respect to $H$ every statement on minimal trajectories can be interpreted as a statement on certain types of orbits of the extended $\tilde{\varphi}$. We can return to the original $\tilde{\varphi} : \mathbb{R} \times [0, 1] \rightarrow \mathbb{R} \times [0, 1]$ in the following way: Let $f_0, f_1 \in \tilde{G}_+$ be defined by $f_0(\xi) = f(\xi, 0)$, $f_1(\xi) = f(\xi, 1)$ and let $\alpha_0, \alpha_1$ be the rotation number of $f_0, f_1$. The interval $[\alpha_0, \alpha_1]$ is called the *twist interval* of $\tilde{\varphi}$. In (7.9) we will see that $\alpha_0 < \alpha_1$.

(7.4)  LEMMA   Suppose $x \in \mathcal{M}_\alpha$ and $\alpha \in (\alpha_0, \alpha_1)$, i.e. $\alpha$ is in the interior of the twist interval of $\tilde{\varphi}$. Then the orbit $(x_i, y_i)_{i \in \mathbb{Z}}$, $y_i = -D_1 H(x_i, x_{i+1})$, is contained in $\mathbb{R} \times (0, 1)$.
The proof of (7.4) is given after (7.7).

(7.5)  EXAMPLES

(a) The integrable case: Set $\tilde{\varphi}(\xi, y) = (\xi + r(y), y)$ for some function $r : [0, 1] \rightarrow \mathbb{R}$ with $r' > 0$. Then the second coordinate is an integral for $\tilde{\varphi}$. Generating functions are of the type $H(\xi, \eta) = h(\xi - \eta)$ discussed in (1.4) where $h$ satisfies $h'(r(y)) = y$. The twist interval is $[r(0), r(1)]$.

(b) A much studied model problem is the 'standard diffeomorphism'

$$\tilde{\varphi}(\xi, y) = (\xi + y + s(\xi), y + s(\xi))$$

where $s$ satisfies $s(\xi + 1) = s(\xi)$ and $\int_0^1 s(\xi)\, d\xi = 0$, cf. [28], [34], [39].

For $s \neq 0$ this $\tilde{\varphi}$ does not leave $\mathbb{R} \times [0, 1]$ invariant, but it satisfies (7.1)(a) and (b). A generating function $H$ for $\tilde{\varphi}$ in the sense of (7.2) is

$$H(\xi, \eta) = \tfrac{1}{2}(\xi - \eta)^2 + S(\xi) \quad \text{where} \quad S'(\xi) = s(\xi).$$

So minimal trajectories in $\mathcal{M}(H)$ yield orbits of $\tilde{\varphi}$ in this case as well.

(c)  The billiard in a bounded convex domain $C \subseteq \mathbb{R}^2$ is a fascinating system which can be described by a monotone twist map. One is interested in the possible orbits of a particle which moves at constant speed on straight lines in $C$ and is reflected on $\partial C$ according to the law 'angle of incidence = angle of reflection'. We assume that $c : S^1 \to \partial C$ is a $C^3$-parametrization of $\partial C$ by arc-length and that $\partial C$ does not contain straight line segments. We define a 'monotone twist map' $\varphi : S^1 \times (-1, 1) \to S^1 \times (-1, 1)$ as follows: suppose the particle hits $\partial C$ consecutively in $c(s_0)$ and in $c(s_1)$. Let $\tau_0$ resp. $\tau_1$ be the angles between $\dot{c}(s_0)$ resp. $\dot{c}(s_1)$ and the outgoing direction of the particle in $c(s_0)$ resp. $c(s_1)$. Then

$$\varphi(s_0, -\cos \tau_0) := (s_1, -\cos \tau_1)$$

$\varphi$ extends continuously to $S^1 \times [-1, 1]$ by $\varphi \,|\, S^1 \times \{-1, 1\} = \text{id}$. Elementary geometry shows that

$$H(\xi, \eta) := -\,|\,c(\xi) - c(\eta)\,|$$

is a generating function for $\tilde{\varphi}$ on the set $\mathring{D} = \{\,(\xi, \eta) \in \mathbb{R}^2 \,|\, 0 < \eta - \xi < 1\}$ and that (7.1)(b) is satisfied. This implies that $\varphi$ is a monotone twist map in the sense of (7.5) except that the extension to $S^1 \times [-1, 1]$ is not $C^1$. However the remarks following (7.9) show that the results of Sections 3–5 apply also in this case. For more information see [18], [20] and [37].

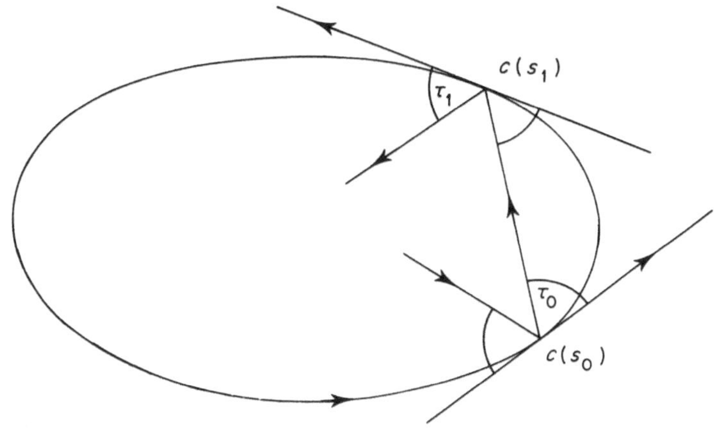


**Fig. 9**

The orbits of the billiard can be interpreted as the geodesics of a flattened Riemannian sphere which doubly covers $C$ with an edge along $\partial C$. It is surprising that we can describe certain types of geodesics on flattened spheres and on tori by the same method.

Now we proceed to interpret the results from Sections 3–5 as statements on orbits of monotone twist maps: for rational $\alpha \in (\alpha_0, \alpha_1)$ Theorem (3.3) yields the so-called Birkhoff periodic orbits of minimum type: if $x \in \mathcal{M}_\alpha^{\text{per}}$ and $\alpha = p/q$ with $(p, q) = 1$ then the corresponding orbit $(x_i, y_i)_{i \in \mathbb{Z}}$ of $\tilde{\varphi}$ satisfies $(x_{i+q}, y_{i+q}) = (x_i + p, y_i)$. Theorem (5.8) proves the existence of heteroclinic (resp. homoclinic) orbits connecting neighboring Birkhoff periodic orbits of minimum type. For irrational $\alpha \in (\alpha_0, \alpha_1)$ we obtain the existence of Mather sets, cf. [35]:

(7.6)    THEOREM    For every irrational $\alpha \in (\alpha_0, \alpha_1)$ there exists a $\varphi$-invariant set $M_\alpha \subseteq S^1 \times (0, 1)$ with the following properties:

(a)    $M_\alpha$ is the graph of a Lipschitz function $\psi_\alpha : A_\alpha \to (0, 1)$ defined on a closed set $A_\alpha \subseteq S^1$.

(b)    $\varphi$ has rotation number $\alpha(\text{mod } \mathbb{Z})$ on $M_\alpha$, i.e. there exists $h \in G_+$ with $\alpha(h) \equiv \alpha \bmod \mathbb{Z}$ such that $h(A_\alpha) = A_\alpha$ and

$$\varphi(\xi, \psi_\alpha(\xi)) = (h(\xi), \psi_\alpha(h(\xi)) \quad \text{for all} \quad \xi \in A_\alpha$$

(c)    The set $M_\alpha^{\text{rec}}$ of recurrent points in $M_\alpha$ projects either to a Cantor set in $S^1$ or to all of $S^1$. In the latter case $M_\alpha$ is a $\varphi$-invariant Lipschitz curve winding once around the annulus $S^1 \times [0, 1]$ and $\varphi \mid M_\alpha$ is topologically conjugate to a rotation.

*Proof*    We prove $(a)$–$(c)$ for $\tilde{\varphi}$ instead of $\varphi$; everything is invariant under $(\xi, \eta) \to (\xi + 1, \eta)$. Keeping (7.4) in mind we define

$$\widetilde{M}_\alpha = \{ (\xi, \eta) \in \mathbb{R} \times (0, 1) \mid \text{there exists}$$

$$x \in \mathcal{M}_\alpha \text{ such that } \xi = x_0, \eta = -D_1 H(x_0, x_1)\}.$$

According to (7.2) and (7.3) the set $\widetilde{M}_\alpha$ is $\tilde{\varphi}$-invariant. By (4.1) and (4.2) the set $\widetilde{M}_\alpha$ has one-to-one projection onto a closed set $\tilde{A}_\alpha \subseteq \mathbb{R}$. According to (3.19) there exists a Lipschitz continuous $\tilde{h} \in \tilde{G}_+$ of rotation number $\alpha$ such that $\tilde{h}(x_0) = x_1$ for all $x \in \mathcal{M}_\alpha$. Hence $\tilde{\psi}_\alpha(\xi) := -D_1 H(\xi, \tilde{h}(\xi))$ for $\xi \in \tilde{A}_\alpha$ is Lipschitz as well. This proves (a) and (b). Now (c) is a consequence of the properties of circle maps stated in Section 2.

While there may be a continuum of disjoint $\varphi$-invariant sets satisfying (7.6)(a) and (b) for some fixed $\alpha$ (cf. [36] and (9.15)) $M_\alpha$ is uniquely determined by (a) and (b) if $A_\alpha = S^1$:

(7.7) THEOREM If $C$ is a $\varphi$-invariant curve winding once around $S^1 \times [0, 1]$ and if $\varphi \,|\, C$ has rotation number $\alpha$ then $C \subseteq M_\alpha$ and $C = M_\alpha$ if $\alpha$ is irrational.

*Proof* According to Birkhoff's Theorem [8], §44, $C$ is the graph of a (Lipschitz) function $\psi : S^1 \to [0, 1]$. Hence

$$\varphi(\xi, \psi(\xi)) = (h(\xi), \psi(h(\xi))) \quad \text{for some } h \in G_+ \text{ of rotation number } \alpha.$$

Let

$$\mathscr{C} := \{ x \in \mathbb{R}^{\mathbb{Z}} \,|\, x_{i+1} = f(x_i, \tilde{\psi}(x_i)) = \tilde{h}(x_i) \text{ for all } i \in \mathbb{Z} \},$$

where $\tilde{\psi}, \tilde{h}$ are lifts of $\psi, h$ to $\mathbb{R}$. $\mathscr{C}$ is a closed totally ordered set of stationary trajectories w.r.t. $H$, cf. (7.3). For every $\xi \in \mathbb{R}$ there exists exactly one $x \in \mathscr{C}$ with $x_0 = \xi$. We are going to show that this implies $\mathscr{C} \subseteq \mathscr{M}(H)$ which proves our claim. Let $(x_i, ..., x_j)$ be a segment of some $x \in \mathscr{C}$. Let $(\bar{x}_i, ..., \bar{x}_j)$ be a minimal segment from $\bar{x}_i = x_i$ to $\bar{x}_j = x_j$. There exists a maximal $x^* \in \mathscr{C}$ such that $(x_i^*, ..., x_j^*) \leqslant (\bar{x}_i, ..., \bar{x}_j)$. Then $x_k^* = \bar{x}_k$ for some $k, i \leqslant k \leqslant j$. The same reasoning as in (1.3) shows that $(H_4)$ holds also for stationary segments if $H \in C^2$ and $D_2 D_1 H < 0$. Hence $x_k^* = \bar{x}_k$ and $(x_i^*, ..., x_j^*) \leqslant (\bar{x}_i, ..., \bar{x}_j)$ imply that $x_i^* = \bar{x}_i$ or $x_j^* = \bar{x}_j$. By our assumption on $\mathscr{C}$ we obtain $x = x^*$, hence

$$(x_i, ..., x_j) \leqslant (\bar{x}_i, ..., \bar{x}_j)$$

Similarly $(x_i, ..., x_j) \geqslant (\bar{x}_i, ..., \bar{x}_j)$ so that $(x_i, ..., x_j) = (\bar{x}_i, ..., \bar{x}_j)$ is minimal. If $\alpha \in \mathbb{R} \backslash \mathbb{Q}$ then $\mathscr{C} = \mathscr{M}_\alpha$ since $\mathscr{M}_\alpha$ is totally ordered, cf. (4.1).

Now the proof of (7.4) is almost obvious:

By (7.7) we have for $s = 0, 1$:

$$(7.8) \qquad \{ x \in \mathbb{R}^{\mathbb{Z}} \,|\, x_{i+1} = f_s(x_i) \text{ for all } i \in \mathbb{Z} \} \subseteq \mathscr{M}_{\alpha_s}$$

Suppose $x \in \mathscr{M}_\alpha$ and $\alpha_0 < \alpha < \alpha_1$. Define $\underline{x} \in \mathscr{M}_{\alpha_0}$, $\bar{x} \in \mathscr{M}_{\alpha_1}$ by $\underline{x}_i = f_0^i(x_0)$, $\bar{x}_i = f_1^i(x_0)$. Since $\underline{x}, x$ and $\bar{x}$ cross only in $0$ and $\tilde{\alpha}(\underline{x}) = \alpha_0 < \tilde{\alpha}(x) = \alpha < \tilde{\alpha}(\bar{x}) = \alpha_1$ we have $\underline{x}_1 < x_1 < \bar{x}_1$. Hence

$$0 = -D_1 H(\underline{x}_0, \underline{x}_1) < y_0 = -D_1 H(x_0, x_1) < -D_1 H(\bar{x}_0, \bar{x}_1) = 1$$

Note that the right-hand sides of (7.8) are disjoint for $s = 0$ and $s = 1$. Hence the following is an immediate consequence of (7.8), (7.7), (5.1) and (5.3).

(7.9) LEMMA For every monotone twist map the twist interval has non-empty interior.

This was first proved by Birkhoff, cf. [10], §4. Using Theorem (7.7) we can give a simple example for which invariant curves do not exist at all. Note that (7.7) and its proof are also valid for the maps considered in (7.5)(b). The following observation is a special case of more elaborate estimates by Aubry [3], Mather [39] and MacKay–Percival [34]:

(7.10)  COROLLARY  For a standard diffeomorphism $\varphi : S^1 \times \mathbb{R} \to S^1 \times \mathbb{R}$ from (7.5)(b) choose the function $s$ such that $s'(\xi_0) < -2$ for some $\xi_0 \in \mathbb{R}$. Then there is no $\varphi$-invariant curve winding around $S^1 \times \mathbb{R}$.

*Proof*  We know from (7.7) that the existence of a $\varphi$-invariant curve winding around $S^1 \times \mathbb{R}$ implies the existence of an $x \in \mathcal{M}(H)$ with $x_0 = \xi_0$ where $H(\xi, \eta) = \frac{1}{2}(\xi - \eta)^2 + S(\xi), S' = s$. Now

$$D_{22}H(x_{-1}, x_0, x_1) = D_{22}H(x_{-1}, x_0) + D_{11}H(x_0, x_1) = 2 + s'(\xi_0) < 0$$

But this contradicts the minimality of the segment $(x_{-1}, x_0, x_1)$.

Finally we remark that the results of this section (with the exception of the Lipschitz continuity in (7.6)) remain true if we relax the conditions (7.1) on a monotone twist map $\varphi$ in the following way:

(a) It is sufficient that $\varphi$ is a homeomorphism instead of being a $C^1$-diffeomorphism.
(b) Instead of $D_2 f > 0$ it suffices that $y \to f(\xi, y)$ is strictly increasing for all $\xi \in \mathbb{R}$.

Unless $f_0$ and $f_1$ are rotations the author does not know how to extend $H$ resp. $\varphi$ under these weaker assumptions. Maybe the easiest way to obtain the results of this section under these hypotheses is to modify the proofs given before:

(1) To obtain a generating function $H : D \to \mathbb{R}$ of class $C^1$ one uses (7.1)(a) directly, without differentiating. The 1-form $\omega$ vanishes along $(\xi, f(\xi, 0))$. Since $\xi \to (\xi, f(\xi, 0))$ is rectifiable this implies $H(\xi + 1, \eta + 1) = H(\xi, \eta)$.
(2) One extends $H$ to those segments and sequences $(x_i)$ which satisfy $(x_i, x_{i+1}) \in D$ for all $i$. Then one can prove that $(H_3)$ and $(H_4)$ hold. If a minimal trajectory $(x_i)_{i \in \mathbb{Z}}$ has one segment $(x_j, x_{j+1})$ on $\partial D$, say $x_{j+1} = f(x_j, 0)$, then $x_{i+1} = f(x_i, 0)$ for all $i \in \mathbb{Z}$.
(3) Finally one has to convince oneself that the proofs for (3.1) (3.3), (3.9) and (3.13) go through with minor modifications. Of course in (3.3) one can only prove the existence of $(q, p)$-periodic minimal trajectories if $p/q \in [\alpha_0, \alpha_1]$. One needs an easy estimate to ensure the existence of $(q, p)$-periodic sequences $x \in \mathbb{R}^{\mathbb{Z}}$ such that $(x_i, x_{i+1}) \in D$ for all $i \in \mathbb{Z}$.

We add a remark due to J. Moser which gives a less technical relation between monotone twist maps and the variational problems of Section 6. The essence is that the relation between a monotone twist map and its generating function is a discrete version of the Legendre transformation. Look at the nonparametric version $L$ of a Finsler metric $F$ on $\mathbb{R}^2$:

$$L(t, x, \dot{x}) := F((t, x), (1, \dot{x}))$$

Then curves $(t, x(t))$ which are extremals of $\int L(t, x(t), \dot{x}(t))\,dt$ become $F$-geodesics if they are reparametrized by arc-length. By Legendre transformation extremals of $\int L$ correspond to solutions of a time-dependent Hamiltonian system.

$$\begin{aligned} \dot{x}(t) &= \phantom{-}H_y(t, x(t), y(t)) \\ \dot{y}(t) &= -H_x(t, x(t), y(t)) \end{aligned} \tag{7.11}$$

The system (7.11) induces area-preserving maps $\varphi_{n,i}$ by

$$\varphi_{n,i}\left[x\left(\frac{i}{n}\right), y\left(\frac{i}{n}\right)\right] = \left[x\left(\frac{i+1}{n}\right), y\left(\frac{i+1}{n}\right)\right]$$

where $(x(t), y(t))$ is a solution of (7.11). In the limit $n \to \infty$ the Legendre condition $L_{\dot{x}\dot{x}} > 0$ implies that the $\varphi_{n,i}$ are monotone twist maps (i.e. satisfy (7.1)(ii)). So the extremals of $\int L$ correspond to orbits of a product of monotone twist maps (which is not monotone twist in general). Note, however, that this description does not work for $F$-geodesics $(t(s), x(s))$ which have points with $\dot{t} = 0$. For more precise information and for the converse problem—how to interpolate a monotone twist map by a Hamiltonian system—we refer to [45].

Finally we sketch the variational principle proposed by Percival [49] which leads to different proofs for the results presented in this section. It was used by Mather in [35] where the statements below are rigorously proved.

One looks for a minimum of the functional

$$I_\alpha(x) = \int_0^1 H(x(t), x(t + \alpha))\,dt \tag{7.12}$$

ranging over all increasing $x : \mathbb{R} \to \mathbb{R}$ which are continuous from the left and satisfy $x(t + 1) = x(t) + 1$. Formally the Euler equation of (7.12) is

$$D_1 H(x(t), x(t + \alpha)) + D_2 H(x(t - \alpha), x(t)) = 0 \tag{7.13}$$

If we define $y(t) = -D_1 H(x(t), x(t + \alpha))$ then (7.13) and (7.2) imply

$$\tilde{\varphi}(x(t), y(t)) = (x(t + \alpha), y(t + \alpha))$$

Assuming the existence of a minimum $x$ satisfying (7.13) we obtain for irrational $\alpha$: if $x$ is continuous the set $\{(x(t), -D_1 H(x(t), x(t + \alpha))) \mid t \in \mathbb{R}\}$ is a $\varphi$-invariant curve winding around $S^1 \times [0, 1]$. Otherwise the closure of this set is a $\varphi$-invariant Cantor set. So we obtain Theorem (7.6). For continuous $x$ we can define $h \in G_+$ by $h(x(t)) = x(t + \alpha) \bmod \mathbb{Z}$ and this is exactly the $h$ which appears in (7.6)(b). So $x$ is a circle map which conjugates $h$ to a rotation. If

$x$ is not continuous the analogous statement holds on $A_\alpha^{\text{rec}}$ = closure of range $(x)$, and—up to a phase—$x$ is the map $x^-$ appearing in (2.3). In particular, the jumps of $x$ correspond to the gaps of $A_\alpha^{\text{rec}}$.

An alternate approach to overcome the technical problems connected with these ideas has been proposed by Moser [46]:

One regularizes Percival's variational principle and minimizes

$$I_{\alpha,\varepsilon}(x) = I_\alpha(x) + \varepsilon \int_0^1 |x'(t)|^2 \, dt$$

for functions $x(t)$ such that $\tilde{x}(t) = x(t) - t \in H^{1,2}(S^1)$. Now $I_{\alpha,\varepsilon}$ can be treated by conventional methods and for $\varepsilon \to 0$ one should obtain Mather's solution.

## 8   THE DISCRETE FRENKEL–KONTOROVA MODEL

In this section we briefly describe the relation to Aubry and LeDaeron's work [4] which—under stronger hypotheses and in different form—contains the same results as Sections 3–5. The discrete Frenkel–Kontorova model is this: a 1-dimensional bi-infinite chain of particles described by the position $x_i \in \mathbb{R}$ of the $i$th particle, $i \in \mathbb{Z}$. So the configuration space of the model is $\mathbb{R}^\mathbb{Z}$. For all $i \in \mathbb{Z}$ the particles numbered $i$ and $i+1$ are coupled by a spring potential $\frac{1}{2}C(x_{i+1} - x_i)^2$ where $C > 0$ is constant. Moreover for all $i \in \mathbb{Z}$ a periodic potential $V(\xi) = V(\xi + 1)$ exerts a force $-V'(x_i)$ on the $i$th particle. A configuration $x \in \mathbb{R}^\mathbb{Z}$ is stationary if for all $i \in \mathbb{Z}$ the sum of the forces acting on the $i$th particle is zero, i.e. if

$$-C(x_i - x_{i-1}) + C(x_{i+1} - x_i) - V'(x_i) = 0 \text{ for all } i \in \mathbb{Z}.$$

Using the notation introduced in (1.3) we can say that a configuration is stationary if and only if it is stationary with respect to $H$ where

$$(8.1) \qquad H(\xi, \eta) = \tfrac{1}{2}(C(\eta - \xi)^2 + V(\xi) + V(\eta))$$

Note that $H$ satisfies $H(\xi + 1, \eta + 1) = H(\xi, \eta)$ and $D_2 D_1 H = -C < 0$ so that the results of Sections 3–5 apply. Actually [4] treats the general case of $C^2$-functions $H$ satisfying $(H_1)$ and $D_2 D_1 H < 0$, not only those of type (8.1).

Now we give a dictionary from the notions used in Sections 3–5 to the physical terminology used in [4] which allows us to translate the statements of Sections 3–5 into the statements of [4] and vice versa. For the physical interpretation we refer to [4].

A configuration $x \in \mathbb{R}^\mathbb{Z}$ is in $\mathcal{M} = \mathcal{M}(H)$ if every segment $(x_j, ..., x_k)$ of $x$ minimizes the energy

$$\sum_{i=j}^{k-1} \tfrac{1}{2}C(x_{i+1} - x_i)^2 + \sum_{i=j}^{k} V(x_i)$$

compared with all segments $(\bar{x}_j, ..., \bar{x}_k)$ satisfying the same boundary conditions $\bar{x}_j = x_j, \bar{x}_k = x_k$. Configurations $x \in \mathcal{M}(H)$ are obviously stationary and are called minimal energy (abbreviated m.e.) configurations. Sections 3–5 give a pretty complete picture of the structure of the set of m.e. configurations. Since $H(\xi, \eta) = H(\eta, \xi)$ we can map $\mathcal{M}_\alpha$ to $\mathcal{M}_{-\alpha}$ by $x \in \mathcal{M}_\alpha \to \tilde{x} \in \mathcal{M}_{-\alpha}$ with $\tilde{x}_i = x_{-i}$. Hence it suffices to consider $\mathcal{M}_\alpha$ for $\alpha \geqslant 0$. The rotation number $\tilde{\alpha}(x) \geqslant 0$ of $x \in \mathcal{M}$ is interpreted as the atomic mean distance of the configuration $x$ since (3.16)(a) implies

$$\tilde{\alpha}(x) = \lim_{i \to \infty} \frac{1}{2i}(x_i - x_{-i})$$

In physical terminology (3.17) says that for every $\alpha \in \mathbb{R}$ there exists a m.e. configuration with atomic mean distance $\alpha$. A recurrent m.e. configuration $x \in \mathcal{M}^{\text{rec}}$ is called a ground-state of the model while a non-recurrent $x \in \mathcal{M} \backslash \mathcal{M}^{\text{rec}}$ is an elementary defect. A m.e. configuration $x \in \mathcal{M}$ is commensurate if $\tilde{\alpha}(x) \in \mathbb{Q}$ and incommensurate if $\tilde{\alpha}(x) \in \mathbb{R} \backslash \mathbb{Q}$. The sets $\mathcal{M}_\alpha^+$ and $\mathcal{M}_\alpha^-$ of commensurate elementary defects described in Theorems (5.3) and (5.8) are called advanced resp. delayed elementary discommensurations.


## 9  EXAMPLES AND MISCELLANEOUS RESULTS

Finally we present examples of Riemannian metrics on tori which display the two extremal possibilities for the set of minimal geodesics. We start with the well-known integrable geodesic flows for which this set is very large. Then we give several classes of metrics for which this set is small in various respects. In these examples the minimal geodesics can be regarded as part of a surface of higher genus and—in some cases—of negative curvature so that we are in a thoroughly studied situation, see e.g. Morse [42], [43], Hedlund [27] and E. Hopf [29]. One should note that examples of Riemannian metrics do not automatically yield examples of monotone twist maps with similar properties and that this does not work in the opposite direction either. However, the general impression is that in both cases the same phenomena occur. For the case of monotone twist maps we refer to the much more sophisticated work by Herman [28].

We consider Riemannian metrics $\tilde{g}$ on $\mathbb{R}^2$ for which the $\mathbb{Z}^2$-action by translations is isometric or—equivalently—which cover a Riemannian torus $(\mathbb{R}^2/\mathbb{Z}^2, g)$. Again we denote the set of minimal $\tilde{g}$-geodesics with average slope $\alpha$ by $\mathcal{M}_\alpha = \mathcal{M}_\alpha(\tilde{g})$ and we set $\mathcal{M} = \bigcup_{\alpha \in \mathbb{R} \cup \{\infty\}} \mathcal{M}_\alpha$.


### 9.1  Flat tori

The simplest Riemannian metrics for which the $\mathbb{Z}^2$-action is isometric are the euclidean metrics (i.e. $\tilde{g}$ does not depend on $p \in \mathbb{R}^2$). Their $\mathbb{Z}^2$-quotients are the flat tori. The geodesics are the straight lines and every geodesic is minimal. Every geodesic is recurrent and its image on $\mathbb{R}^2/\mathbb{Z}^2$ is either a simple closed curve or dense in $\mathbb{R}^2/\mathbb{Z}^2$. The average slope $\alpha(c)$ of $c \in \mathcal{M}$ is simply the slope of the corresponding straight line. The geodesics in $\mathcal{M}_\alpha$ cover $\mathbb{R}^2$ simply. According to E. Hopf's beautiful theorem [30] the flat metrics are characterized among the Riemannian metrics covering a torus by the property that every geodesic is minimal.

## 9.2   Tori with a 1-parameter group of isometries

If we revolve a plane closed curve around an axis which does not intersect the curve we obtain a torus in $\mathbb{R}^3$ which inherits a Riemannian metric from the euclidean structure of $\mathbb{R}^3$. The rotations around the axis restrict to isometries of this torus. The geodesics of such tori of revolution have been studied by Kimball [32]. This has been generalized to tori with an arbitrary 1-parameter group of isometries (within the framework of $G$-spaces) by Busemann and Pedersen [14]. Figures 10 and 11 show two typical geodesics on a torus of revolution. We state the results for the minimal geodesics of the universal Riemannian cover $(\mathbb{R}^2, \tilde{g})$ with 1-parameter group $\tilde{\varphi}_s$:

For all $\alpha \in \mathbb{R} \cup \{\infty\}$ except possibly for one $\alpha_0 \in \mathbb{Q} \cup \{\infty\}$ the sets $\mathcal{M}_\alpha$ consist of the images under $\tilde{\varphi}_s, s \in \mathbb{R}$, of a fixed geodesic in $\mathcal{M}_\alpha$. In particular the geodesics in $\mathcal{M}_\alpha$ cover $\mathbb{R}^2$ simply and $\mathcal{M}_\alpha^{\mathrm{per}} = \mathcal{M}_\alpha$ for $\alpha \in \mathbb{Q} \cup \{\infty\}$, $\alpha \neq \alpha_0$. $\mathcal{M}_{\alpha_0}^{\mathrm{per}}$ consists of those orbits $s \to \tilde{\varphi}_s(p)$ for which the Killing field $\tilde{X}(p) = (\partial/\partial s)\tilde{\varphi}_s(p)$ has minimal length. Every geodesic intersecting one of these orbits is minimal. Unless the torus is flat the geodesics in $\mathcal{M}_{\alpha_0}^{\mathrm{per}}$ do not cover $\mathbb{R}^2$. The strip between two neighboring geodesics $c^-$ and $c^+$ in $\mathcal{M}_{\alpha_0}^{\mathrm{per}}$ is simply covered by the family $\{\tilde{\varphi}_s(c) \mid s \in \mathbb{R}\} = \mathcal{M}^-(c^-, c^+)$ where $c \in \mathcal{M}^-(c^-, c^+)$ and similarly for $\mathcal{M}^+(c^-, c^+)$.

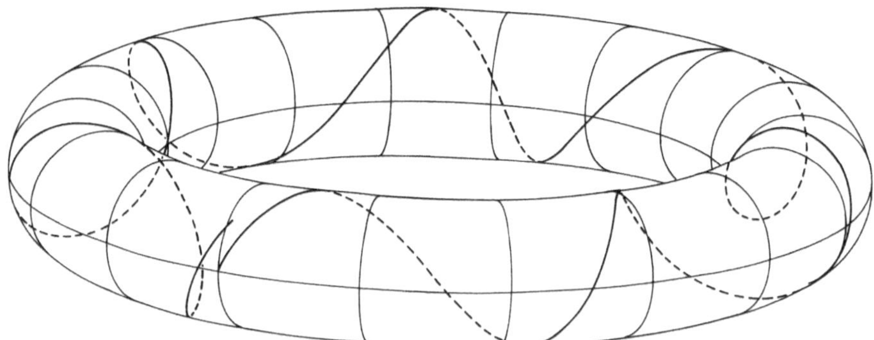

**Fig. 10.** Geodesic on a torus of revolution whose lift to the universal cover is minimal

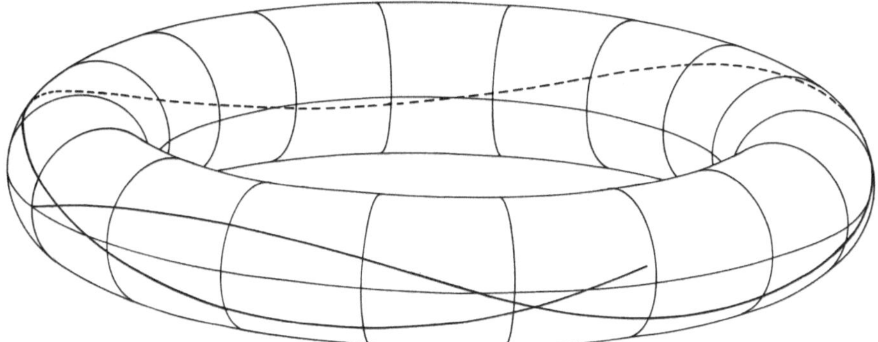

**Fig. 11.** Geodesic on a torus of revolution whose lift is not minimal

We briefly describe the situation from the dynamical standpoint: By Noether's theorem the 1-parameter group of isometries induces a second integral for the geodesic flow, the Clairaut integral, cf. [18], p. 60. Using a $c \in \mathcal{M}_\alpha^{\mathrm{per}}$, $\alpha \in \mathbb{Q}$, as base curve for a cross-section we can describe the geodesic flow of the torus by an integrable area-preserving map $\varphi$ of an annulus. For the torus of revolution of a strictly convex curve we obtain the well-known picture where $\varphi$ has one elliptic and one hyperbolic fixed point corresponding to the longest and shortest orbit of revolution respectively. The separatrices correspond to the geodesics in $\mathcal{M}_{\alpha_0}^- \cup \mathcal{M}_{\alpha_0}^+$. They divide the annulus into three components and the vectors tangent to geodesics whose lifts are minimal make up the components which do not contain the elliptic fixed point.

An alternative method to obtain tori with a 1-parameter group of isometries is as follows: Let $r(s) = r(s+1)$ be a positive periodic function and consider the immersion

$$F: \mathbb{R}^2 \to \mathbb{R}^3, \; F(s, t) = (s, r(s)\cos 2\pi t, r(s)\sin 2\pi t)$$

which parametrizes the surface of revolution with profile curve $(s, r(s))$. Then, in addition to the translations $T_k, k \in \mathbb{Z}^2$, the translations $T_{(0, t)}, t \in \mathbb{R}^2$, are isometries of the metric $\tilde{g}$ induced by $F$ on $\mathbb{R}^2$. In this case $\alpha_0 = \infty$ and $\mathcal{M}_\infty^{\mathrm{per}}$ consists of those lines $\{s_0\} \times \mathbb{R}$ for which $r(s_0)$ is an absolute minimum of $r$. This can be used to make the following observations:

(9.3)   There exist non-flat tori without any hyperbolic closed geodesics.

One simply has to choose $r$ such that $r''(s_0) = 0$ for all absolute minimum points $s_0$ of $r$. Since the set of absolute minimum points of a real function is an arbitrary closed set we obtain:

(9.4)   For every $\mathbb{Z}$-periodic closed set $A_\infty^{\mathrm{per}} \subseteq \mathbb{R}$ we can find $\tilde{g}$ as above such that the geodesics in $\mathcal{M}_\infty^{\mathrm{per}}(\tilde{g})$ intersect $\mathbb{R} \times \{0\}$ precisely in $A_\infty^{\mathrm{per}} \times \{0\}$.

Finally let $s_0 < s_1$ be consecutive absolute minimum points of $r$ so that $\{s_0\} \times \mathbb{R}$ and $\{s_1\} \times \mathbb{R}$ represent neighboring geodesics $c^-$ and $c^+$ in $\mathcal{M}_\infty^{\text{per}}$. Increasing $\tilde{g}$ along parts of a fixed line $\{s\} \times \mathbb{R}$, $s_0 < s < s_1$, one can achieve the following for arbitrary $\mathbb{Z}$-periodic closed sets $A^-, A^+$ in $\mathbb{R}$:

(9.5)   The geodesics in $\mathcal{M}^-(c^-, c^+)$ intersect $\{s\} \times \mathbb{R}$ in the set $\{s\} \times A^-$, those in $\mathcal{M}^+(c^-, c^+)$ intersect $\{s\} \times \mathbb{R}$ in the set $\{s\} \times A^+$.

Now we present classes of Riemannian metrics for which all the sets $\mathcal{M}_\alpha$ have gaps. A simple way to introduce gaps in the $\mathcal{M}_\alpha$ is to let a 'big bump' grow on the corresponding torus. Since we will frequently need this intuitive notion we formalize it:

(9.6)   A big bump on a torus $(T^2, g)$ is a closed disk $B \subseteq T^2$ with the following property: there exists $p \in B$ so that the (Riemannian) distance from $p$ to $\partial B$ is larger than $\frac{1}{4}$ length $(\partial B)$. The open set of all such $p$ will be called the center of $B$.

It is easy to construct metrics with big bumps: if $g$ is a metric on $T^2$ and $\phi \neq U \subseteq T^2$ is open one can deform $g$ inside $U$ so as to create an arbitrary finite number of big bumps.

The notion of a 'big bump' was introduced to have the following obvious consequences: let $c: [a, b] \to T^2$ be a geodesic on a Riemannian torus $(T^2, g)$ with big bump $B$. Suppose $c(a) \notin B, c(b) \notin B$ and $c$ has minimal length in its homotopy class. Then $c$ does not intersect the center of $B$. In particular, no

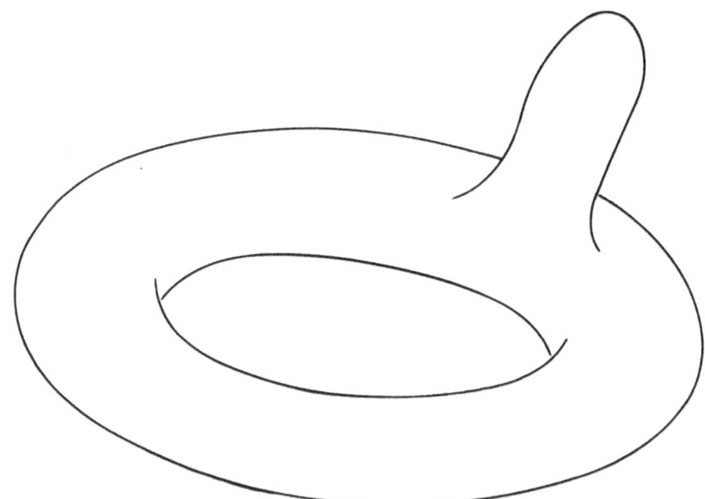

**Fig. 12.** Torus with a 'big bump'

geodesic on $T^2$ which lifts to a minimal geodesic intersects the center of $B$. These remarks and (6.10) show:

(9.7)  PROPOSITION  Suppose $(\mathbb{R}^2, \tilde{g})$ covers a torus $(\mathbb{R}^2/\mathbb{Z}^2, g)$ with a big bump. Then for all $\alpha \in \mathbb{R} \cup \{\infty\}$ the geodesics in $\mathcal{M}_\alpha$ do not cover $\mathbb{R}^2$. For $\alpha \in \mathbb{R} \backslash \mathbb{Q}$ the set $\mathcal{M}_\alpha^{\mathrm{rec}}$ is homeomorphic to a Cantor set.

More precise information on the size of the sets $\mathcal{M}_\alpha$ can be obtained for tori $(T^2, g)$ satisfying

(9.8)  All geodesics which lift to minimal geodesics are contained in the region of $T^2$ in which the Gaussian curvature $K$ is negative.

Since $\int_{T^2} K = 0$ the geodesics which lift to minimal geodesics cannot cover $T^2$ if (9.8) is satisfied. Before we start to investigate consequences of (9.8) we describe two classes of examples:

(9.9)  Riemannian tori with big bump $B$ so that the curvature is negative outside the center of $B$.

The remark preceding (9.7) shows that such tori satisfy (9.8). To obtain such tori we start with a metric $g_0$ of negative curvature on a surface $X$ with boundary $\partial X$ which is homeomorphic to $T^2$ with an open disk removed. It is well known that such metrics $g_0$ exist on $X$ and that they can have the following additional property: there exists a closed annulus $A \subseteq X$ with boundary components $A_0 = \partial X$ and $A_1$ such that the distance from $A_0 = \partial X$ to $A_1$ is larger than $\frac{1}{4}$ length $(A_1)$. Adding a cap $C$ to $\partial X$ and extending $g_0$ to a Riemannian metric on $X \cup C = T^2$ we obtain a Riemannian torus $(T^2, g)$ with big bump $B = A \cup C$ the center of which contains $C$. There are various possibilities to obtain metrics $g_0$ as above; a neat description how one can construct examples by hyperbolic geometry is contained in [15], Kap. 3. A more intuitive approach is as follows: we start with the non-smooth surface $S$ in $\mathbb{R}^3$ which is shown in Fig. 13.

Now one has to convince oneself that one can approximate $S$ by smooth surfaces $S_\varepsilon$ with the following properties:

(a) $S_\varepsilon$ has non-positive curvature except in a small neighborhood of the top of the cylinder and $K = 0$ only on a set of arbitrarily small measure.

(b) Copies of $S_\varepsilon$ can be glued together to form a smooth $\mathbb{Z}^2$-periodic surface.

If the approximation is not too bad the induced metrics will have a big bump (corresponding to the cylinder) and will have non-positive and mostly negative curvature outside the center of the big bump.

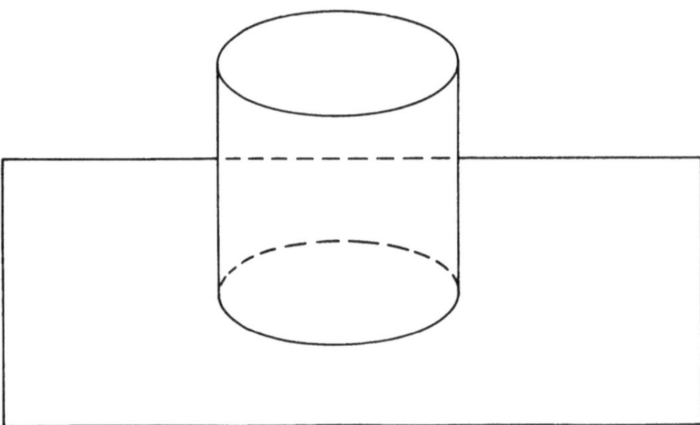

**Fig. 13.** The surface $S$

*Remark*  The examples (9.9) include a non-empty $C^2$-open set of Riemannian metrics. Hence property (9.8) is not exceptional.

The following examples are a little more sophisticated:

(9.10)  PROPOSITION  There exists a sequence $(T^2, g_n)$ satisfying (9.8) such that $(g_n)$ converges to a flat metric in the $C^1$-topology.

*Proof*  We start with an isosceles triangle $\Delta$ in a hyperbolic plane of curvature $-n^{-2}$. We assume that $\Delta$ has a side $c$ of length 1 and that the angles at $c$ are $\pi/4$. We glue four copies of $\Delta$ together to obtain a torus with Riemannian metric of curvature $-n^{-2}$ which is smooth except for a conical singularity in the point $p_0$ which corresponds to the vertex of $\Delta$ opposite to $c$. Alternatively we could erect a circular cone on a flat torus, smooth the edge at the base of the cone and let the height of the cone converge to zero. In both ways we obtain metrics on $T^2$ which satisfy properties $(G_1)$ and $(G_2)$ from Section 6. Obviously a geodesic which hits the singularity $p_0$ cannot be continued beyond $p_0$. Hence there exists an open ball $B(p_0, \delta_n)$ such that no minimal geodesic hits $B(p_0, \delta_n)$. Now we smooth the metric inside $B(p_0, \delta_n)$ without decreasing distances. Obviously this can be done so that the resulting Riemannian metrics $g_n$ converge to a flat metric in the $C^1$-topology. Moreover the minimal geodesics of $g_n$ coincide with the original ones so that (9.8) is satisfied.

Now we mention some consequences of condition (9.8). Using recent results by Birman and Series [11] we can conclude that under the hypothesis (9.8) the set $T\mathcal{M}$ of vectors tangent to geodesics which lift to minimal geodesics is very small.

(9.11)  THEOREM  Suppose $(T^2, g)$ satisfies (9.8). Then $T\mathcal{M}$ has Hausdorff dimension one.


*Remarks*  (1) Since $T\mathcal{M}$ is invariant under the geodesic flow its Hausdorff dimension is at least one.
(2) The differentiable structure of the unit tangent bundle distinguishes a family of equivalent metrics on the unit tangent bundle: metrics which are locally equivalent to the euclidean metric in some chart. We use the Hausdorff dimension with respect to this natural class of metrics.
(3) As a consequence of (9.11) the set $T\mathcal{M}$ has (Lebesgue) measure zero in the unit tangent bundle; see also the proof below.


*Proof*  Let $C \subseteq T^2$ be a compact set which contains all projections of minimal geodesics and on which the curvature is negative. It is not difficult to see that a neighborhood of $C$ can be isometrically embedded into a compact surface $N$ of negative curvature. This maps $T\mathcal{M}$ to a closed set in the unit tangent bundle of $N$ which is invariant under the geodesic flow of $N$. We note that the ergodicity of this flow implies that $T\mathcal{M}$ has measure zero. However, we do not use this fact. Instead we remember that by (6.5) a geodesic which is tangent to some $v \in T\mathcal{M}$ does not have (transverse) self-intersections.

Since the same is true for the images in $N$ we can apply [11], Theorem III, and obtain that $T\mathcal{M}$ has Hausdorff dimension 1. Actually this is proved in [11] only in case $N$ has constant negative curvature but for the case of variable negative curvature one simply replaces some hyperbolic geometry by comparison arguments.


*Remark*  From (9.11) or from the arguments in [11] one can conclude the following: if $s \to (0, s)$ is a minimal geodesic for the lifted metric $\tilde{g}$ on $\mathbb{R}^2$ let $A = \{s \in \mathbb{R} \mid$ There exists $c \in \mathcal{M} \backslash \mathcal{M}_\infty$ through $(0, s)\}$. Then $A$ has Hausdorff dimension 0. In particular $A$ has Lebesgue measure zero.

If for some $\alpha \in \mathbb{R} \cup \{\infty\}$ the geodesics in $\mathcal{M}_\alpha$ cover $\mathbb{R}^2$ then the set of vectors tangent to these geodesics separates the unit tangent bundle. Using ideas of Morse and Birkhoff we show that in the examples (9.9) the geodesic flow can wander freely through the gaps in the $\mathcal{M}_\alpha$. For monotone twist maps a related example was constructed by D. L. Goroff (Hyperbolic sets for twist maps, *Ergod. Th. Dynam. Sys.*, **5** (1985), 337–9).


(9.12)  THEOREM  Let $(\mathbb{R}^2, \tilde{g})$ cover a torus $(\mathbb{R}^2/\mathbb{Z}^2, g)$ of type (9.9). Let $\tilde{T}\mathcal{M}$ denote the set of vectors tangent to geodesics in $\mathcal{M}(g)$. Then $\tilde{T}\mathcal{M}$ is contained in a hyperbolic invariant subset on which the geodesic flow is topologically transitive.

*Remark*  In particular for every pair $c_1, c_2 \in \mathcal{M}(\tilde{g})$ there exists a geodesic $c$ which is $\alpha$-asymptotic to $c_1$ and $\omega$-asymptotic to $c_2$.

*Proof*  In slightly different form this statement is contained in [42]. We sketch a proof using a symbolism introduced in [9]. Let $B \subseteq \mathbb{R}^2/\mathbb{Z}^2$ be a big bump such that the curvature is negative outside the center of $B$. To avoid some technicalities we give the proof only for the following situation: there exists a disk $C \subseteq B$ contained in the center of $B$ and bounded by a closed geodesic such that the curvature is negative on $X = \mathbb{R}^2/\mathbb{Z}^2 \backslash \text{Int}(C)$. We let $\pi : \mathbb{R}^2 \to \mathbb{R}^2/\mathbb{Z}^2$ denote the projection and set $\tilde{X} = \pi^{-1}(X)$.

Let $\tilde{U}$ denote the set of vectors tangent to geodesics which are completely contained in $\tilde{X}$. So $\tilde{U}$ is a closed subset of the unit tangent bundle of $(\mathbb{R}^2, \tilde{g})$ which is invariant under the geodesic flow. The geodesic flow is uniformly hyperbolic on $\tilde{U}$ since $\tilde{X}$ has compact quotient $X$ with negative curvature. Since $C$ is contained in the center of $B$ we have $\tilde{T}\mathcal{M} \subseteq \tilde{U}$. So it remains to prove that the geodesic flow is topologically transitive on $\tilde{U}$. To this end we introduce the following symbolism, cf. [9], p. 238. Choose two disjoint geodesic arcs $a$ and $b$ in $X$ such that $X \backslash (a \cup b)$ is simply connected. In $\mathbb{R}^2$ we have the picture shown in Fig. 14.

Let $\gamma : \mathbb{R} \to X$ be a curve which intersects $a \cup b$ transversely in a bi-infinite sequence of points $\gamma(t_i), \ldots < t_i < t_{i+1} < \ldots$, where $t_0$ is determined by $t_0 = \min\{t_i \mid t_i \geq 0\}$. We associate to $\gamma$ its 'symbol' $s = s(\gamma)$ which is the bi-infinite sequence $s = (s_i)_{i \in \mathbb{Z}} \in \{a, b, a^{-1}, b^{-1}\}^{\mathbb{Z}}$ defined as follows:

$$s_i = \begin{cases} a & \text{if } \gamma(t_i) \in a \text{ and } \gamma \text{ crosses } a \text{ from left to right.} \\ a^{-1} & \text{if } \gamma(t_i) \in a \text{ and } \gamma \text{ crosses } a \text{ from right to left.} \\ b & \text{if } \gamma(t_i) \in b \text{ and } \gamma \text{ crosses } b \text{ from left to right.} \\ b^{-1} & \text{if } \gamma(t_i) \in b \text{ and } \gamma \text{ crosses } b \text{ from right to left.} \end{cases}$$

Obviously $s(c)$ is defined for every geodesic $c$ in $X$. Moreover the symbol of a geodesic is reduced, i.e. it does not contain segments of the form $aa^{-1}$, $a^{-1}a$, $bb^{-1}$ or $b^{-1}b$. We consider the set $S$ of all reduced sequences in $\{a, b, a^{-1}, b^{-1}\}^{\mathbb{Z}}$ with the product topology. The shift $T : S \to S$ is defined by $T(s) = \tilde{s}$ where $\tilde{s}_i = s_{i+1}$. We have the following fundamental fact, for details cf. [42].

(9.13)  The $T$-orbits in $S$ are in one-to-one correspondence with the unparametrized oriented geodesics in $X$. This correspondence is a homeomorphism. Explicitly: if the geodesics $c_i$ converge to $c$, i.e. $\lim \dot{c}_i(0) = \dot{c}(0)$, and $c(0) \notin a \cup b$, then $\lim s(c_i) = s(c)$. If $\lim s_i = s$ there exist geodesics $c_i, c$ with $s(c_i) = s_i, s(c) = s$ and $\lim c_i = c$.

Since $T$ is obviously topologically transitive on $S$ we conclude from (9.14) that the geodesic flow of $g$ is topologically transitive on $U = \pi_*(\tilde{U})$. Our claim that the geodesic flow of $\tilde{g}$ is topologically transitive on $\tilde{U}$ is stronger. However,

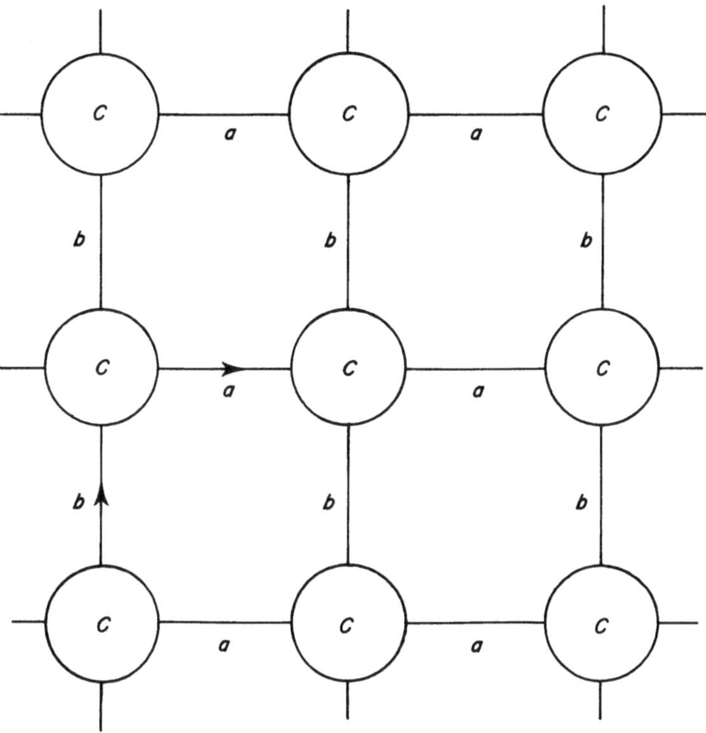

**Fig. 14**

one can easily conclude from (9.13) that for all $v, w \in \tilde{U}$ there exist geodesics $c_i$ in $\tilde{X}$ and a sequence $t_i \to \infty$ such that $\lim \dot{c}_i(0) = v$ and $\lim \dot{c}_i(t_i) = w$.

Next we describe a relation between Mather's work [36] and Morse's work [43]. We present a Riemannian analogue to the main result of [36].

(9.14)   We assume that $(\mathbb{R}^2, \tilde{g})$ covers a torus $(\mathbb{R}^2/\mathbb{Z}^2, g)$ with two disjoint big bumps $B_1$ and $B_2$. Let $C_1$ and $C_2$ be two disks in the centers of $B_1$ and $B_2$. For convenience we assume that $C_1$ and $C_2$ are bounded by closed geodesics. Fix an arbitrary irrational $\alpha \in \mathbb{R}$. We are going to show that there exists a continuum of disjoint closed sets $\mathscr{C}_{\alpha, \varepsilon}$ of geodesics which have exactly the same properties as $\mathscr{M}_\alpha$ except that they do not consist of minimal geodesics. So this is counter-example to the—at first sight tempting—conjecture that $\mathscr{M}_\alpha$ can be characterized by such properties. The precise meaning of 'exactly the same properties as $\mathscr{M}_\alpha$' will become clear below. Similarly to (9.12) we can find three disjoint minimal geodesic arcs $a, b, d$ such that in suitable coordinates a fundamental square of $\mathbb{R}^2/\mathbb{Z}^2$ is as shown in Fig. 15.

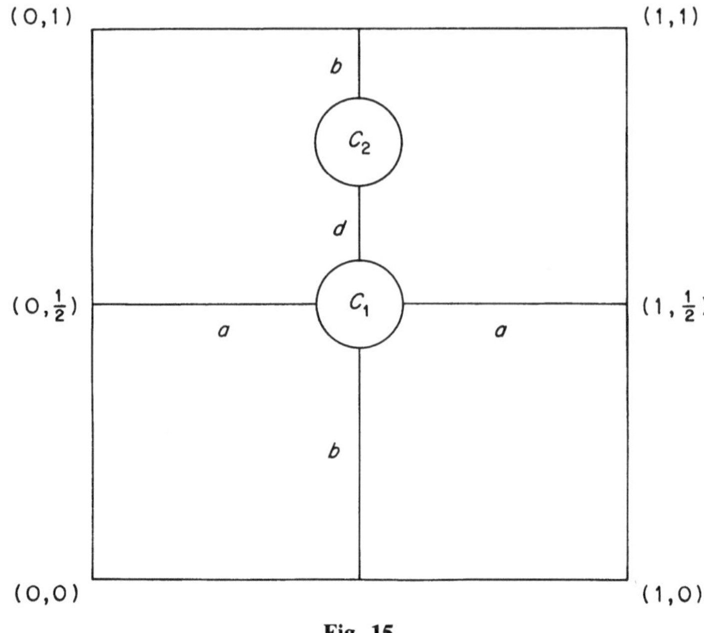


**Fig. 15**

Set $X = \mathbb{R}^2/\mathbb{Z}^2 \backslash \mathrm{Int}(C_1 \cup C_2)$. As in (9.12) we associate a symbol $s = s(\gamma)$ to suitable curves $\gamma : \mathbb{R} \to X$ where in this case $s_i \in \{a, b, d, a^{-1}, b^{-1}, d^{-1}\}$. Since we do not assume that the curvature is negative on $X$ only part of (9.13) is true now:

(9.15)   For every reduced sequence $s \in \{a, b, d, a^{-1}, b^{-1}, d^{-1}\}^{\mathbb{Z}}$ there exists a geodesic $c$ in $X$ with symbol $s(c) = s$ such that the lift of $c$ to the universal Riemannian cover of $X$ is minimal.

If we attach a handle to $\partial X = \partial C_1 \cup \partial C_2$ we see that (9.15) is contained in [43]; however it can as well be proved directly.

Our program is as follows: Since we can change the covering projection by elements of $\mathrm{SL}(2, \mathbb{Z})$ it suffices to treat the case $\alpha \in (0, \frac{1}{2})$. First we construct a family $\gamma_\varepsilon, \frac{1}{2} - \alpha < \varepsilon < \frac{1}{2}$, of curves $\gamma_\varepsilon : \mathbb{R} \to X$ which do not have self-intersections and whose lifts to $\mathbb{R}^2$ have average slope $\alpha$. Let $c_\varepsilon : \mathbb{R} \to X$ denote a geodesic with $s(c_\varepsilon) = s(\gamma_\varepsilon)$ which is minimal when lifted to the universal cover of $X$. Then $c_\varepsilon$ does not have self-intersections and its lifts to $\mathbb{R}^2$ have average slope $\alpha$. Let $\mathscr{C}_{\alpha, \varepsilon}$ denote the set of all geodesics in $\mathbb{R}^2$ which are limits of lifts of $c_\varepsilon$. Since we know $s(c_\varepsilon)$ explicitly we can see that $\mathscr{C}_{\alpha, \varepsilon} \cap \mathscr{C}_{\alpha, \varepsilon'} = \varnothing$ if $\varepsilon \neq \varepsilon'$. The sets $\mathscr{C}_{\alpha, \varepsilon}$ have the same properties as $\mathscr{M}_\alpha$: $\mathscr{C}_{\alpha, \varepsilon}$ is a closed $\mathbb{Z}^2$-invariant set of geodesics of average slope $\alpha$ such that no two geodesics in $\mathscr{C}_{\alpha, \varepsilon}$ intersect. If we choose minimal periodic geodesics as coordinate axes then the geodesics in $\mathscr{C}_{\alpha,\varepsilon}$ can be described by a circle map $f \in \tilde{G}_+$ of rotation number $\alpha$ as in Section 6.

Now we fill in some of the details. We define a lift $\tilde{\gamma}_\varepsilon : \mathbb{R} \to \tilde{X} = \pi^{-1}(X)$ of $\gamma_\varepsilon$: $\tilde{\gamma}_\varepsilon$ is to intersect the coordinate lines $\{i\} \times \mathbb{R}$, $i \in \mathbb{Z}$, in the same points and at the same parameter values as the line $t \to \alpha t$. Since the $\tilde{\gamma}_\varepsilon$ should not intersect $\pi^{-1}(C_1 \cup C_2)$ we have to specify how they move around the holes. This is done by requiring that $\tilde{\gamma}_\varepsilon \,|\, [i, i+1]$ has symbol

$$
\begin{array}{ll}
b & \text{if } i\alpha \in [0, \tfrac{1}{2} - \alpha] \cup (\tfrac{1}{2}, 1) \bmod \mathbb{Z} \\[4pt]
ba & \text{if } i\alpha \in (\tfrac{1}{2} - \alpha, \varepsilon] \bmod \mathbb{Z} \\[4pt]
ad & \text{if } i\alpha \in (\varepsilon, \tfrac{1}{2}] \bmod \mathbb{Z}
\end{array}
$$

Looking at the situation in a fixed fundamental square one sees that $\tilde{\gamma}_\varepsilon$ can be so chosen that $\gamma_\varepsilon = \pi \circ \tilde{\gamma}_\varepsilon$ does not have self-intersections. Obviously $s(\gamma_\varepsilon)$ is a reduced sequence so that there exists a geodesic $c_\varepsilon$ in $X$ with $s(\gamma_\varepsilon) = s(c_\varepsilon)$ and such that $c_\varepsilon$ is minimal when lifted to the universal cover of $(X, g)$. As above we consider $X$ as part of a compact surface of genus 2. Then the universal cover of $X$ becomes part of the unit disk $D$ which covers the compact surface. Because of $s(\gamma_\varepsilon) = s(c_\varepsilon)$ there are lifts of $\gamma_\varepsilon$ and $c_\varepsilon$ which join the same pair of points on $\partial D$. Since $\gamma_\varepsilon$ does not have self-intersections any two lifts of $\gamma_\varepsilon$ to $D$ do not intersect. Since the lifts of $c_\varepsilon$ are minimal they do not intersect either (otherwise they would intersect twice). So $c_\varepsilon$ does not have self-intersections. From $s(c_\varepsilon) = s(\gamma_\varepsilon)$ one easily deduces that the lifts of $c_\varepsilon$ to $\mathbb{R}^2$ have average slope $\alpha$. It remains to prove that the sets $\mathscr{C}_{\alpha,\varepsilon}, \varepsilon \in (\tfrac{1}{2} - \alpha, \tfrac{1}{2})$, are disjoint: for all $\varepsilon \neq \varepsilon'$ there exists $n(\varepsilon, \varepsilon') \in \mathbb{N}$ such that no two segments of $s(\gamma_\varepsilon)$ and $s(\gamma_{\varepsilon'})$ which have length $n(\varepsilon, \varepsilon')$ are identical. This follows from the irrationality of $\alpha$ and the fact that we get the letter $d$ in $s(\gamma_\varepsilon)$ only for those $i \in \mathbb{Z}$ for which $i\alpha \in (\varepsilon, \tfrac{1}{2}] \bmod \mathbb{Z}$. Hence lifts of $c_\varepsilon$ and of $c_{\varepsilon'}$ cannot converge to the same geodesic, i.e. $\mathscr{C}_{\alpha,\varepsilon} \cap \mathscr{C}_{\alpha,\varepsilon'} = \phi$.

Finally we approach the question of the existence of non-recurrent geodesics $c \in \mathscr{M}_\alpha \backslash \mathscr{M}_\alpha^{\mathrm{rec}}$ for irrational $\alpha$. Suppose $(\mathbb{R}^2, \tilde{g}_0)$ covers $(\mathbb{R}^2/\mathbb{Z}^2, g_0)$ and $\mathscr{M}_\alpha^{\mathrm{rec}}(\tilde{g}_0)$ is a Cantor set for some irrational $\alpha \in \mathbb{R}$. Then there exist $c$ and $\bar{c}$ in $\mathscr{M}_\alpha^{\mathrm{rec}}$ bounding an open strip $S$ in $\mathbb{R}^2$ which does not contain elements of $\mathscr{M}_\alpha^{\mathrm{rec}}$. Thus the projection $\pi : \mathbb{R}^2 \to \mathbb{R}^2/\mathbb{Z}^2$ is injective on $S$. Hence any metric on $S$ which coincides with $\tilde{g}_0$ outside a compact subset of $S$ can be extended to a $\mathbb{Z}^2$-periodic metric on $\mathbb{R}^2$ which coincides with $\tilde{g}_0$ outside $S$ and its translates. This allows construction of metrics $\tilde{g}$ on $\mathbb{R}^2$ covering a metric $g$ on $\mathbb{R}^2/\mathbb{Z}^2$ such that $\mathscr{M}_\alpha^{\mathrm{rec}}(\tilde{g}) = \mathscr{M}_\alpha^{\mathrm{rec}}(\tilde{g}_0)$ and such that there exist non-recurrent geodesics $c \in \mathscr{M}(\tilde{g})$ in $S$. However, these examples are exceptional: it is not difficult to prove that for given irrational $\alpha$ there is a generic set of metrics such that $\mathscr{M}_\alpha^{\mathrm{rec}}(\tilde{g}) = \mathscr{M}_\alpha(\tilde{g})$. To see the density of such metrics one simply has to increase a given torus metric $g_0$ on the (open) complement of the union of all geodesics whose lifts are in $\mathscr{M}_\alpha^{\mathrm{rec}}(\tilde{g}_0)$. This change leaves $\mathscr{M}_\alpha^{\mathrm{rec}}(\tilde{g}_0)$ invariant and destroys possible non-recurrent minimal geodesics of slope $\alpha$. Similarly one can show that for given $\alpha \in \mathbb{Q}$ the following is true for generic $\mathbb{Z}^2$-periodic metrics on $\mathbb{R}^2$: $\mathcal{M}_\alpha^{\mathrm{per}}$, $\mathcal{M}_\alpha^+$ and $\mathcal{M}_\alpha^-$ each consist of exactly one $T$-orbit, i.e. the corresponding geodesics on $T^2 = \mathbb{R}^2/\mathbb{Z}^2$ are unique.

## 10   GUIDE TO SOME OF THE RECENT LITERATURE

We briefly review some directions in which this theory has developed recently and which are not treated in this paper: Mather's results were extended by Hall [25], Bernstein [6] and Boyland–Hall [12] to monotone twist maps satisfying topological conditions instead of being area-preserving. In these cases one does not have a variational principle. Mather [40] shows that a Mather set $M_\alpha$ which is a Cantor set can be approximated by invariant Cantor sets with related properties. While the orbits in $M_\alpha$ correspond to minima of the variational principle these new orbits correspond to local minima. Crude examples for this phenomenon are described in [39] and (9.15). For the 1-dimensional problem with $n$ degrees of freedom and $\mathbb{Z}^n$-periodicity ($n \geqslant 3$), e.g. geodesics on a Riemannian $n$-torus, there are perturbation results by Bernstein and Katok [7]. In this case one cannot talk about the order of trajectories and it is known that the results presented here do not generalize in full generality, cf. Hedlund's example in [26]. However, in a different framework the notion of 'order' turns out to be useful also in higher dimensions: Moser [47] considers minimal solutions of $\mathbb{Z}^n$-periodic variational problems for functions $u : \mathbb{R}^{n-1} \to \mathbb{R}$. Moser [47] and Bangert [5] extend many of the ideas presented here to this more general case.

Shorter surveys of parts of the subject of this paper are Chenciner [17], MacKay–Stark [33] and Moser [46], [48].

*Added in proof:* The results on geodesics were obtained independently by M. L. Byalyi and L. V. Polterovich, Geodesic flows on the two-dimensional torus and phase transitions 'commensurability – noncommensurability'. *Functional. Anal. Appl.*, **20** (1986), 260–66. I am indebted to D. V. Anosov for this reference.

## ACKNOWLEDGEMENT

I want to thank Professor J. Moser (ETH Zürich) who introduced me to Aubry's and Mather's work and who pointed out to me that there exists a relation to geodesics on tori. This final version has profited from suggestions by Professor E. Zehnder (Universität Bochum), Dr N. Innami (Nagoya) and by the referee. I am indebted to S. Fischli (Bern) who made the computer draw Figures 10 and 11.

## REFERENCES

[1] V. I. Arnold, *Geometrical Methods in the Theory of Ordinary Differential Equations.* Grundlehren der math. Wissenschaften 250. New York–Heidelberg–Berlin: Springer, 1983.

[2] V. I. Arnold and A. Avez, *Problèmes Ergodiques de la Mécanique Classique.* Paris: Gauthiers-Villars, 1967.

[3] S. Aubry, The twist map, the extended Frenkel–Kontorova model and the devil's staircase. *Physica*, 7D (1983), 240–58.

[4] S. Aubry and P. Y. LeDaeron, The discrete Frenkel–Kontorova model and its extensions I: exact results for the ground states. *Physica*, 8D (1983), 381–422.

[5] V. Bangert, A uniqueness theorem for $\mathbb{Z}^n$-periodic variational problems. *Comment. Math. Helv.*, 62 (1987), 511–31.

[6] D. Bernstein, Birkhoff periodic points for twist maps with the graph intersection property. *Erg. Th. Dynam. Sys.*, 5 (1985), 531–7.

[7] D. Bernstein and A. Katok, Birkhoff periodic orbits for small perturbations of completely integrable Hamiltonian systems with convex Hamiltonians. *Invent. math.*, 88 (1987), 225–41.

[8] G. D. Birkhoff, Surface transformations and their dynamical applications. *Acta Math.*, 43 (1922), 1–119.

[9] G. D. Birkhoff, Dynamical Systems. *Am. Math. Soc. Colloq. Publ. IX.* Providence RI: Am. Math. Soc., 1927.

[10] G. D. Birkhoff, Sur quelques courbes fermées remarquables. *Bull. Soc. Math. France*, 60 (1932), 1–26.

[11] J. S. Birman and C. Series, Geodesics with bounded intersection number on surfaces are sparsely distributed. *Topology*, 24 (1985), 217–25.

[12] P. L. Boyland and G. R. Hall, Invariant circles and the order structure of periodic orbits in monotone twist maps. *Topology*, 26 (1987), 21–35.

[13] H. Busemann, *The Geometry of Geodesics.* New York: Academic Press, 1955.

[14] H. Busemann and F. P. Pedersen, Tori with one-parameter groups of motions. *Math. Scand.*, 3 (1955), 209–20.

[15] P. Buser, Riemannsche Flächen und Längenspektrum vom trigonometrischen Standpunkt aus. *Habilitationsschrift.* Bonn, 1980.

[16] C. Carathéodory, *Calculus of Variations and Partial Differential Equations of the First Order. Part II: Calculus of Variations.* San Francisco: Holden Day, 1967.

[17] A Chenciner, La dynamique au voisinage d'un point fixe elliptic conservatif; de Poincaré et Birkhoff à Aubry et Mather. Sém. Bourbaki, Exposé 622, Vol. 1983/84. *Astérisque*, 121/122 (1985), 147–70.

[18] I. P. Cornfeld, S. V. Fomin and Ya. G. Sinai, *Ergodic Theory.* Grundlehren der math. Wissenschaften 245. New York–Heidelberg–Berlin: Springer, 1982.

[19] A. Denjoy, Sur les courbes définies par les équations différentielles à la surface du tore. *J. de Math. Pure et Appl.*, sér. 9, 11 (1932), 333–75.

[20] R. Douady, Applications du théorème des tores invariants. Thèse 3ème Cycle. Université Paris VII, 1982.

[21] B. A. Dubrovin, A. T. Fomenko and S. P. Novikov, *Modern Geometry— Methods and Applications, Part II.* Graduate Texts in Math. 104. New York: Springer, 1984.

[22] A Duschek and W. Mayer, *Lehrbuch der Differentialgeometrie.* Leipzig–Berlin: Teubner, 1930.

[23] M. Freedman, J. Hass and P. Scott, Closed geodesics on surfaces. *Bull. London Math. Soc.*, 14 (1982), 385–91.

[24] M. Gromov, Structures métriques pour les variétés Riemanniennes. *Rédigé par J. Lafontaine et P. Pansu.* Paris: CEDIC, 1981.

[25] G. R. Hall, A topological version of a theorem of Mather on twist maps. *Erg. Th. Dynam. Sys.*, 4 (1984), 585–603.

[26] G. A. Hedlund, Geodesics on a two-dimensional Riemannian manifold with periodic coefficients. *Ann. of Math.*, 33 (1932), 719–39.

[27] G. A. Hedlund, The dynamics of geodesic flows. *Bull. Am. Math. Soc.*, **45** (1939), 241–60.

[28] R. Herman, Introduction à l'étude des courbes invariantes par les difféomorphismes de l'anneau. *Astérisque*, 103/104 (1983).

[29] E. Hopf, Statistik der Lösungen geodätischer Probleme vom instabilen Typus II. *Math. Ann.*, **116** (1940), 590–608.

[30] E. Hopf, Closed surfaces without conjugate points. *Proc. Nat. Acad. Sci.*, **34** (1948), 47–51.

[31] A. Katok, Some remarks on the Birkhoff and Mather twist theorems. *Erg. Th. Dynam. Sys.*, 2 (1982), 183–94.

[32] B. F. Kimball, Geodesics on a toroid. *Am. J. Math.*, 52 (1932), 29–52.

[33] R. S. MacKay and J. Stark, Lectures on Orbits of Minimal Action for Area-Preserving Maps. Preprint, University of Warwick, May 1985.

[34] R. S. MacKay and I. C. Percival, Converse KAM: theory and practice. *Comm. Math. Phys.*, 98 (1985), 469–512.

[35] J. N. Mather, Existence of quasi-periodic orbits for twist homeomorphisms of the annulus. *Topology*, 21 (1982), 457–67.

[36] J. N. Mather, Non-uniqueness of solutions of Percival's Euler–Lagrange equation. *Comm. Math. Phys.*, **86** (1982), 465–76.

[37] J. N. Mather, Glancing billiards. *Erg. Th. Dynam. Sys.*, 2 (1982), 397–403.

[38] J. N. Mather, A criterion for the non-existence of invariant circles. *Publ. Math. IHES*, **63** (1986), 153–204.

[39] J. N. Mather, Non-existence of invariant circles. *Erg. Th. Dynam. Sys.*, **4** (1984), 301–9.

[40] J. N. Mather, More Denjoy minimal sets for area preserving diffeomorphisms. *Comment. Math. Helv.*, **60** (1985), 508–57.

[41] J. N. Mather, Existence of asymptotic orbits for area-preserving monotone twist diffeomorphisms. Manuscript, 1985.

[42] M. Morse, Recurrent geodesics on a surface of negative curvature. *Trans. Am. Math. Soc.*, **22** (1921), 84–100.

[43] M. Morse, A fundamental class of geodesics on any closed surface of genus greater than one. *Trans. Am. Math. Soc.*, **26** (1924), 25–60.

[44] J. Moser, *Stable and Random Motions in Dynamical Systems*. Ann. of Math. Studies 77. Princeton NJ: Princeton Univ. Press, 1973.

[45] J. Moser, Monotone twist mappings and the calculus of variations. *Erg. Th. Dynam. Sys.*, **6** (1986), 325–33.

[46] J. Moser, Break-down of Stability. In J. M. Jowett, M. Month, S. Turner (eds), *Nonlinear Dynamics Aspects of Particle Accelerators*. Lect. Notes in Physics 247, 492–518. Berlin–New York: Springer, 1986.

[47] J. Moser, Minimal solutions of variational problems on a torus. *Ann. Inst. Henri Poincaré—Analyse non linéaire*, 3 (1986), 229–72.

[48] J. Moser, Recent developments in the theory of Hamiltonian systems. *SIAM Review*, **28** (1986), 459–85.

[49] I. C. Percival, Variational principles for invariant tori and cantori. In M. Month, J. C. Herrara (eds), *Non-Linear Dynamics and the Beam-Beam Interaction. Am. Inst. Phys. Conf. Proc.*, **57** (1980), 310–20.

[50] J. Stillwell, *Classical Topology and Combinatorial Group Theory*. Graduate Texts in Math. 72. New York: Springer, 1980.

[51] E. M. Zaustinsky, Extremals on compact E-surfaces. *Trans. Am. Math. Soc.*, **102** (1962), 433–45.

*Dynamics Reported, Volume 1*
Edited by U. Kirchgraber and H. O. Walther
© 1988 John Wiley & Sons and B. G. Teubner

# 2

# Connecting orbits in scalar reaction diffusion equations

**P. Brunovský**
*Universita Komenského, Bratislava*

and

**B. Fiedler**
*Institut für Angewandte Mathematik, Heidelberg*

## CONTENTS



## 1   INTRODUCTION

We consider the flow of a one-dimensional reaction diffusion equation

$$(1.1) \qquad u_t = u_{xx} + f(u), \ x \in (0, 1)$$

with Dirichlet boundary conditions

$$(1.2) \qquad u(t, 0) = u(t, 1) = 0$$

Let $v$, $w$ denote stationary, i.e. $t$-independent solutions. We say that $v$ connects to $w$, if there exists an orbit $u(t, x)$ of (1.1), (1.2) such that

$$(1.3) \qquad \lim_{t \to -\infty} u(t, \cdot) = v$$

$$\lim_{t \to +\infty} u(t, \cdot) = w$$

i.e. $u(t, \cdot)$ is a heteroclinic orbit connecting $v$ to $w$. In this report we address the following question:

(*)  Given $v$, which stationary solutions $w$ does it connect to?

For a certain class of ordinary differential equations, the question of orbits connecting stationary solutions arose from a study of shock waves via the viscosity method (Gelfand [13], 1959). Later, the main tool to analyze connecting orbits was Conley's topological index ([9], 1978) which found extensive applications to ODE travelling wave problems arising in the thoery of shocks as well as reaction diffusion systems (see [27, §24]). For PDE-flows, Henry [16, §5.3] studied the connection problem of equation (1.1) in its own right, using elementary geometrical arguments and invariant manifold theory. Some other special cases were treated by Conley and Smoller using Conley's index (see [10], [27, §24.D] and the references there). Their approach relied solely on the variational structure of (1.1), and did not exploit maximum principles. A more detailed discussion is postponed to §6.

Apparently it was Hale [14], 1981, who first recognized the importance of maximum principles, notably Matano's result on lap numbers [20], for revealing the Morse–Smale structure of the flow (1.1). Angenent [1] and Henry [17] then showed that stable and unstable manifolds of stationary solutions of (1.1), (1.2) necessarily intersect transversely, if they intersect at all. Note that $v$ connects to $w$, iff the unstable manifold of $v$ does intersect the stable manifold of $w$, provided $v$ and $w$ are hyperbolic stationary solutions. Under the additional assumption

$$(1.4) \qquad f(0) = 0 < f'(0) \text{ and } s \cdot f''(s) < 0 \quad \text{for all } s \neq 0,$$

which was also used in the work of Conley and Smoller, Henry [17] completely solved the connection problem by this transversality approach. His result is contained in our main theorem 1.1 below.

Condition (1.4) appears in the work of Chafee and Infante [8], 1974 on the global bifurcation of stationary solutions of

$$(1.1)_\alpha \qquad u_t = u_{xx} + \alpha^2 f(u),$$

with Dirichlet boundary conditions. In particular, it guarantees that the non-trivial stationary branches bifurcating from zero at $\alpha_k = k\pi/\sqrt{[f'(0)]}$, $k = 1, 2, \ldots$ are globally parametrized over $\alpha \in (\alpha_k, \infty)$ (we assume sublinear

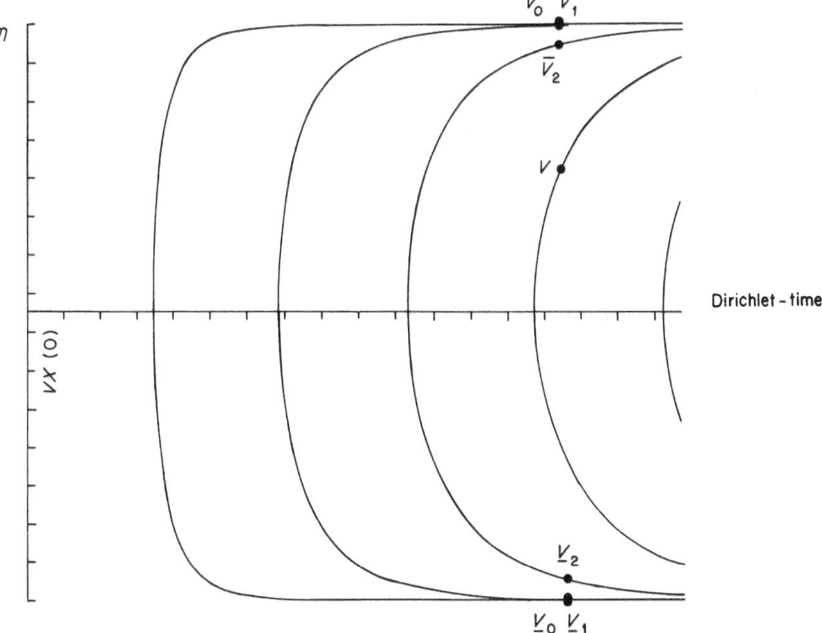



**Fig. 1.** Time map for $f(u) = -(u+1) \cdot u \cdot (u+1)$, Dirichlet problem.

growth of $f$, here) Stated loosely: nontrivial branches have no wiggles (cf. Fig. 1). In slightly different language, these results are contained in [3, ch. VI.10], 1959 already.

In this paper we investigate connecting orbits, even if the stationary branches have wiggles (cf. Fig. 2). We allow $f(0) \neq 0$ and drop assumption (1.4). This greatly increases the complexity of the problem, because it introduces many additional solutions. Besides the Dirichlet case, we also consider Neumann boundary conditions (§6).

Before we state our main result, we fix the technical setting of our investigation and pin-point the precise ingredients to our analysis. For the nonlinearity $f$ we assume only

$$(1.5) \qquad f \in C^2, \ \limsup_{|s| \to \infty} f(s)/s < \pi^2.$$

By $\mathcal{F}$ we denote the set of nonlinearities $f$ satisfying (1.5). We endow $\mathcal{F}$ with the weak Whitney topology (cf. [19]). For $f \in \mathcal{F}$, (1.1), (1.2) define a strongly continuous semiflow on the solution space

$$u(t, \cdot) \in X := H^2 \cap H_0^1$$

(cf. [16]). Let $|\cdot|$ denote the $H^2$-norm on $X$. The growth condition (1.5) on $f$ just ensures that solutions stay bounded in $X$ for all time [16]. The topology

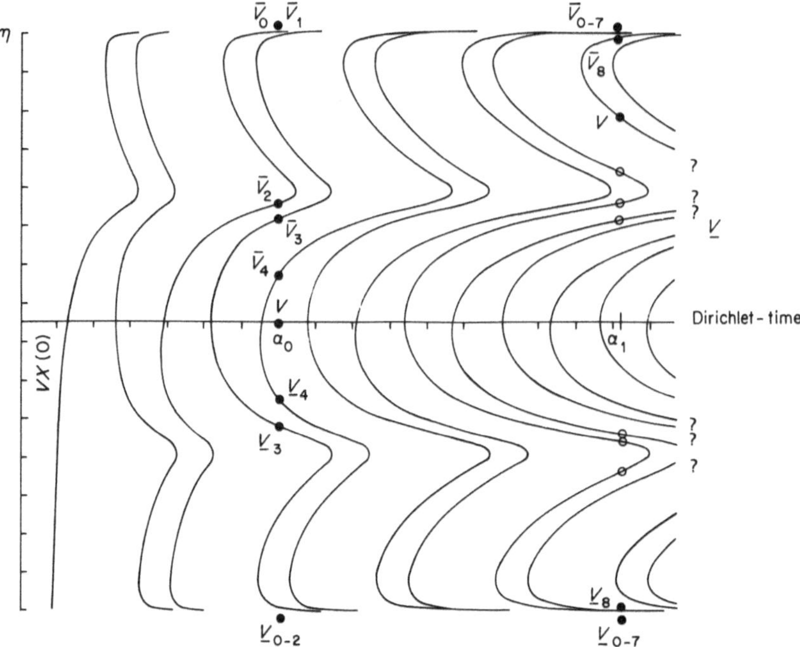


**Fig. 2.** Time  map  for  $f(u) = -(u + 10.2) \cdot u \cdot ((u - 4)^2 + 1.75^2) \cdot (u - 10)$.
Dirichlet problem

on $\mathcal{F}$ is suitable for continuity arguments (see lemmata 2.1 and 3.1 below).

For $f \in \mathcal{F}$ the *gradient structure* of (1.1) guarantees that every orbit tends to some equilibrium via the Ljapunov functional

$$(1.6) \qquad V(u) := \int_0^1 (\tfrac{1}{2} u_x^2 - F(u)) \, \mathrm{d}x, \quad F'(s) := f(s)$$

$$\frac{\mathrm{d}}{\mathrm{d}t} V(u(t, \cdot)) = - \int_0^1 u_t^2 \, \mathrm{d}x$$

[16]. Remember that $V$ was the starting point for the Conley index approach.

Another (discrete) decreasing functional, going back to Nickel [23] essentially, is the *zero number $z$*. For continuous $\Phi : [0, 1] \to \mathbb{R}$, the zero number $z(\Phi)$ for $\Phi \not\equiv 0$ is the maximal integer $n \leqslant \infty$ such that there exist $0 < x_0 < x_1 < \cdots < x_n < 1$ with

$$\Phi(x_i) \cdot \Phi(x_{i+1}) < 0 \quad (0 \leqslant i < n);$$

$z(0) := 0$. By maximum principle arguments (cf. §7 for more details) $t \to z(\tilde{u}(t, \cdot))$ is decreasing along solutions $\tilde{u}(t, \cdot)$ of

$$(1.7) \qquad \tilde{u}_t = \tilde{u}_{xx} + g(x, \tilde{u})$$

with mixed boundary conditions

$$(1.8) \qquad \cos \gamma_j \cdot \tilde{u}(t, j) - \sin \gamma_j \cdot \tilde{u}_x(t, j) = 0, \quad j = 0, 1$$

if $g(x, 0) = 0$. To avoid confusion of solutions $\tilde{u}(t, x)$ of (1.7) with solutions $u(t, x)$ of (1.1) we distinguish solutions as $\tilde{u}$ and $u$. On $g$, we assume

$$(1.9) \qquad g \in C^2, \limsup_{|s| \to \infty} g(x, s)/s < \infty \qquad \text{uniformly in } x \in [0, 1],$$

to guarantee global existence of solutions, and denote the set of those $g$ by $\mathcal{G}$. Note that $\mathcal{F}$ may be viewed as a subset of $\mathcal{G}$. For arguments involving $z(u(t, \cdot))$ we also consider

$$(1.10) \qquad \mathcal{G}_0 := \{ g \in \mathcal{G} \mid g(x, 0) = 0 \quad \text{for all } x \}.$$

Again, $\mathcal{G}$, $\mathcal{G}_0$ are endowed with the weak Whitney topology. Replacing $F$ in (1.6) by $G(x, U)$ with $G_u = g$, equation (1.7) has a gradient structure again.

As a first application of the zero number $z$ we consider a hyperbolic stationary solution $v$ of (1.1), (1.2). By hyperbolic we mean that zero is not an eigenvalue of the linearization $L$ at $v$

$$(1.11) \qquad Lu := u_{xx} + f'(v(x))u$$

$$(1.12) \qquad u(0) = u(1) = 0.$$

In our setting, $L$ is a closed, densely defined linear operator on $X = H^2 \cap H_0^1$ with domain $\mathcal{D}(L) = H^4 \cap H_0^1$, cf. [16]. More specifically, $L$ is self-adjoint on the Hilbert space $X$ and Sturm–Liouville theory applies to $L$ (see e.g. [2, §8; 3, ch. II; 15. ch. XI] for anything on Sturm–Liouville theory). Thus $L$ has discrete spectrum consisting of simple real eigenvalues $\lambda_0 > \lambda_1 > \cdots$ accumulating at $-\infty$. The corresponding eigenfunctions are denotes by $\varphi_k$. By Sturm–Liouville theory, $\varphi_k$ has exactly $k$ sign changes, i.e. $z(\varphi_k) = k$. To investigate the zero number $z$ in a slightly more nonlinear situation, let $W^u(v)$ resp. $W^s(v)$ denote the unstable resp. stable manifold of $v$ (cf. [16]). These manifolds consist of those solutions $u(t, x)$ which tend to $v$ as $t$ tends to $-\infty$ (resp. $+\infty$). Let $i(v) := \dim W^u(v)$ denote the instability index (Morse index) of $v$. Because the tangent space to $W^u(v)$ resp. $W^s(v)$ at $v$ is given by the span of those eigenfunctions $\varphi_k$ for which $\lambda_k$ is positive resp. negative, $i(v)$ is just the number of positive eigenvalues of the linearization $L$ in short:

$$\lambda_k > 0 \text{ implies } k < i(v).$$

If $u$ is a solution of (1.1), (1.2) then $\tilde{u} := u - v$ is a solution of (1.7), (1.2) putting $g(x, \tilde{u}) := f(\tilde{u} + v(x)) - f(v(x))$, and $z(\tilde{u}(t, \cdot))$ is decreasing. Using this fact, it was proved in [5] that

$$(1.13) \qquad z(u_0 - v) < i(v) \text{ for any } u_0 \in W^u(v)$$

and

$$z(u_0 - v) \geqslant i(v) \text{ for any } u_0 \in W^s(v) \backslash \{v\}.$$

Indeed,

$$\lim \frac{\tilde{u}(t, \cdot)}{|\tilde{u}(t, \cdot)|} = \pm \varphi_k$$

as $t \to -\infty$ exists for initial data $u_0 \in W^u(v) \backslash \{v\}$ and equals an eigenfunction $\varphi_k$ of $L$ with positive eigenvalue $\lambda_k$. By Sturm–Liouville theory, $z(\varphi_k) = k$ and we conclude that for $0 > t \to -\infty$

$$z(u_0 - v) = z(\tilde{u}(0, \cdot)) \leqslant z(\tilde{u}(t, \cdot))$$

$$= z\left(\frac{\tilde{u}(t, \cdot)}{|\tilde{u}(t, \cdot)|}\right) \to z(\pm \varphi_k) = k < i(v)$$

for $u_0 \in W^u(v) \backslash \{v\}$. A similar argument can be given for the stable manifold.

As another relation between $i$ and $z$ we mention

(1.14) $$i(v) \in \{z(v), z(v) + 1\}$$

for any stationary solution $v \neq 0$ of (1.1), (1.2). This is proved in §5, lemma 5.1 and serves to distinguish the possible cases in our main theorem below. For example, $i(v) = z(v)$ for $v \neq 0$ and any 'Chafee–Infante–f' satisfying assumption (1.4) above.

For hyperbolic stationary $v$ we define

(1.15) $$\Omega(v) := \{w \mid v \text{ connects to } w \neq v\}$$

and for $0 \leqslant k < i(v)$
$\bar{v}_k$ is the stationary solution $\hat{v}$ with $z(\hat{v}) = k$ such that

$$\hat{v}_x(0) > |v_x(0)| \text{ is minimal,}$$

$\underline{v}_k$ is the stationary solution $\hat{v}$ with $z(\hat{v}) = k$ such that

$$\hat{v}_x(0) < -|v_x(0)| \text{ is maximal.}$$

Note that $v$ enters into the definition of $\bar{v}_k$ and $\underline{v}_k$. Among other things, the proof of our main theorem 1.1 below will also guarantee existence of those $\bar{v}_k$, $\underline{v}_k$ which come up on its various statements. Basically, our growth condition (1.5) on the nonlinearity $f$ is responsible for that. In case (1.5) is violated, some of the $\bar{v}_k$, $\underline{v}_k$ may not exist — see §6 for further discussion At any rate, the $\bar{v}_k$, $\underline{v}_k$ are uniquely defined. With this notation we can state our main result.


## 1.1 Main theorem

Let $f \in \mathscr{F}$ satisfy assumption (1.5) and let $v$ be a hyperbolic stationary solution of (1.1), (1.2). Then $v$ connects to other stationary solutions as follows.

(i)   If $v = 0$, or if $v \neq 0$ and $i(v) = z(v)$, then

$$\Omega(v) = \{\underline{v}_k, \bar{v}_k \mid 0 \leqslant k < i(v)\}.$$

(ii)  If $v_x(0) > 0$ and $i(v) = z(v) + 1$, then

$$\Omega(v) = \Omega_1 \cup \Omega_2 \cup \Omega_3$$

where

$$\Omega_1 = \{\bar{v}_k \mid 0 \leqslant k < i(v)\}$$
$$\Omega_2 = \{\underline{v}_k \mid 0 \leqslant k < i(v) - 1\} \text{ and either}$$
$$\Omega_3 = \{\underline{v}_k \mid k = i(v) - 1\} \text{ or}$$

$\Omega_3$ consists of one or several stationary solutions $w$ with
$$- v_x(0) \leqslant w_x(0) \leqslant v_x(0) \text{ and } i(w) < i(v)$$

(iii) If $v_x(0) < 0$ and $i(v) = z(v) + 1$, then similarly

$$\Omega(v) = \Omega_1 \cup \Omega_2 \cup \Omega_3$$

where

$$\Omega_1 = \{\underline{v}_k \mid 0 \leqslant k < i(v)\}$$
$$\Omega_2 = \{\bar{v}_k \mid 0 \leqslant k < i(v) - 1\} \text{ and either}$$
$$\Omega_3 = \{\bar{v}_k \mid 0 \quad k = i(v) - 1\} \text{ or}$$

$\Omega_3$ consists of one or several stationary solutions $w$ with
$$v_x(0) < w_x(0) \leqslant - v_x(0) \text{ and } i(w) < i(v).$$

For $w$ not necessarily hyperbolic, $i(w)$ denotes the number of strictly positive eigenvalues here.

We illustrate our results, first in a Chafee–Infante situation where $f$ satisfies (1.4) and then for a more general $f$, $f(0) \neq 0$, which exhibits wiggles. For a more in-depth discussion see §6.

Any required information on $v$ as well as the solution $\bar{v}_k$, $\underline{v}_k$ can be read off from the stationary global bifurcation diagram of

$$(1.1)_\alpha \qquad\qquad u_t = u_{xx} + \alpha^2 f(u)$$

with Dirichlet conditions. For a discussion of these difurcation diagrams with nonlinearity $f$ in a generic class see [4, 24, 30]. We represent any stationary solution by a pair $(\alpha, \eta)$ where $\eta = \alpha^{-1} \cdot v_x(0)$ determines $v(\cdot)$. Scaling $\xi := \alpha x$, $v^*(\xi) := v(x)$, any solution $v^*$ of

$$(1.16) \qquad\qquad 0 = v^*_{\xi\xi} + f(v^*)$$

with initial data $v^*(0) = 0$, $v^*_x(0) = \eta$ which satisfies $v^*(\xi) = 0$ at 'time' $\xi = T > 0$ yields a solution ; of $(1.1)_\alpha$ for $\alpha := T$ with $v_x(0) = T\eta = \alpha\eta$. The

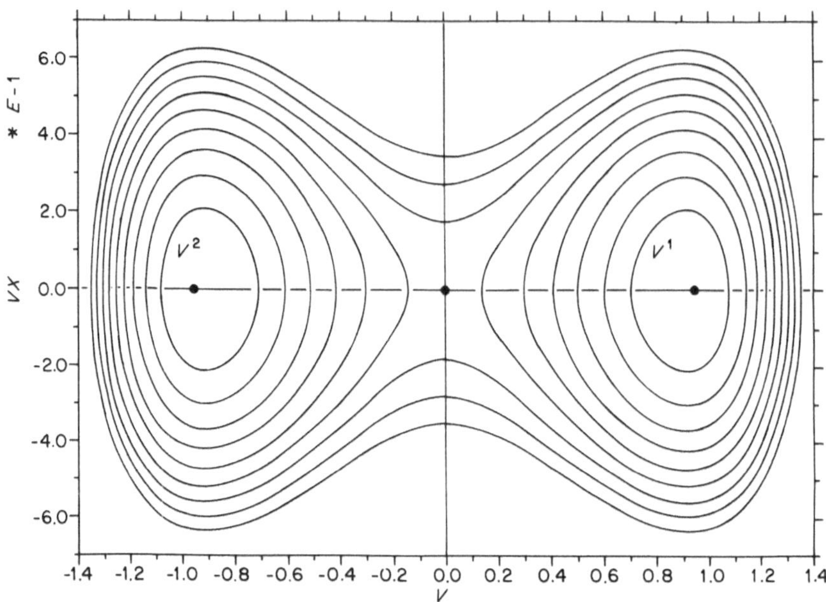

**Fig. 3.** Phase plane for $f(u) = (u + 1)u(u - 1)$

dependence $T = T(\eta)$ of the first zero of $v^*$ on the initial data $\eta = v_x^*(0)$ is usually called the 'time-map' [4, 26, 27, 29–31].

In Figs 3, 4, 5 we present a selection of typical phase portraits of (1.16). Note that the Hamiltonian

$$(1.17) \qquad\qquad H(v^*, v_\xi^*) := \tfrac{1}{2}(v_\xi^*)^2 + F(v^*)$$

is a first integral of (1.16), with $F' = f$ as before. To obtain the phase portrait, we just draw level curves of (1.17). To determine the time-map $T(\eta)$, however, we have to integrate (1.16).

How to obtain the stationary global bifurcation diagram for $(1.1)_\alpha$ once we know $T(\eta)$ for all $\eta$? Given the first positive zero $T(\eta)$ of $v^*$ with $v^*(0) = 0$, $v_\xi^*(0) = \eta$ we may obtain the second zero at

$$\xi = T(\eta) + T(-\eta),$$

because $v^*(T(\eta)) = 0$ by definition, and $v_\xi^*(T(\eta)) = -v_\xi^*(0) = -\eta$ by the first integral (1.17). Here we assume $T(\eta)$ and $T(-\eta)$ are both finite, of course. Proceeding in this manner, we may in fact determine all zeros of $v^*$ from $T(\eta)$ and $T(-\eta)$. Translating this information back to $(1.1)_\alpha$, we can now determine all values $\alpha$ such that the stationary boundary value problem $(1.1)_\alpha$ has a solution $v$ with $v_x(0) = \alpha\eta$. Thus the time-map $T(\eta)$ generates the complete stationary bifurcation diagram of $(1.1)_\alpha$. Note that the ordering of solutions $v(\cdot)$ by $v_x(0)$ which determines $\bar{v}_k$, $\underline{v}_k$ is just the ordering of the corresponding

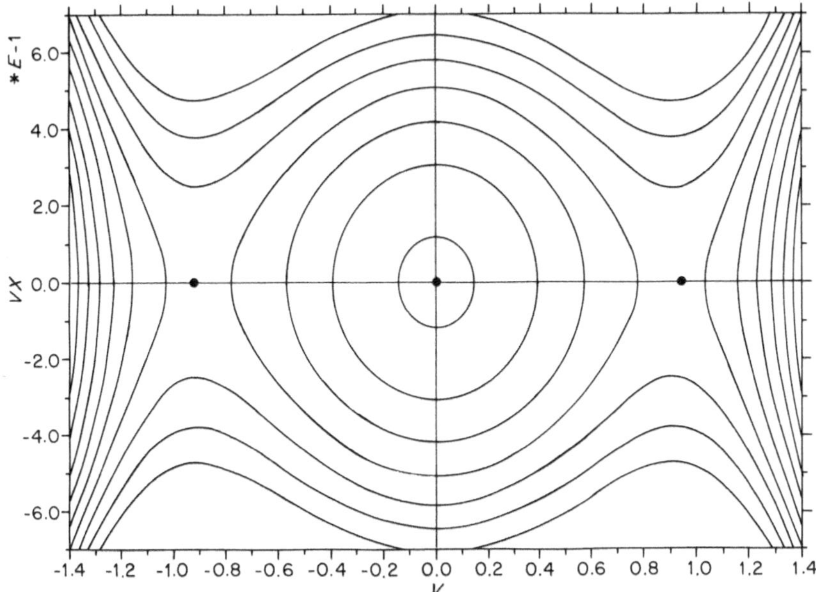

**Fig. 4.** Phase plane for $f(u) = -(u+1)u(u-1)$

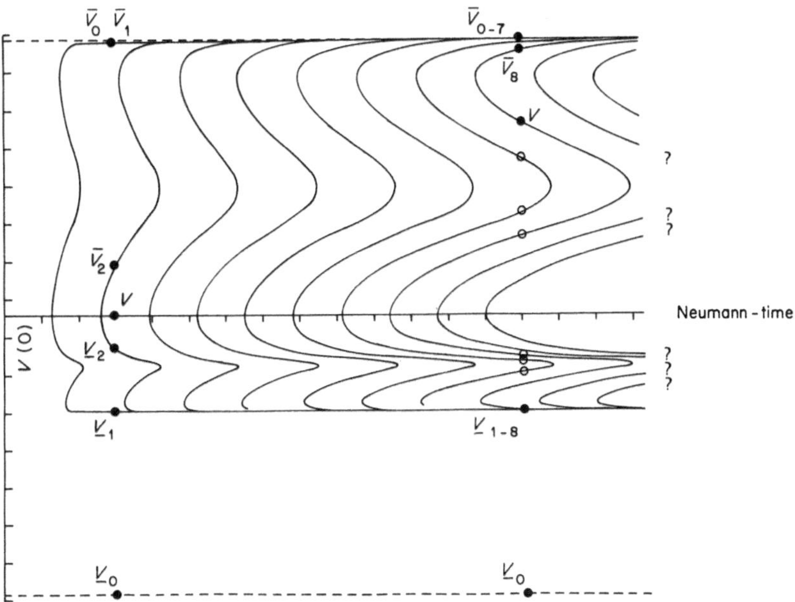

**Fig. 5.** Phase plane for $f(u) = -(u+10.2) \cdot u \cdot ((u-4)^2 + 1.75^2) \cdot (u-10)$

numbers $\eta$. This justifies it to represent any stationary solutions of $(1.1)_\alpha$ by the corresponding pair $(\alpha, \eta)$. For a few typical numerical examples see Figs 1, 2 and 6.

The numerical computations of the time-maps in Figs 1, 2, and 6 used the 27 January 1982 version of the package LSODAR due to L. R. Petzold (Sandia Nat. Lab.) and A. C. Hindmarsh (Lawrence Livermore Nat. Lab.) [18]. It includes automatic method switching between stiff and non-stiff problems, and root-finding. The runs were in double precision on the IBM 3081 D at Universitätsrechenzentrum Heidelberg with a required local relative accuracy of $10^{-4}$. Implementation was done jointly by R. Schaaf.

Let us continue to extract from our bifurcation diagrams the information required by our main theorem 1.1. By uniqueness of solutions of the initial value problem (1.16), branches $(\alpha, \eta)$ with $\eta \neq 0$ are globally parametrized over $\eta$ as $(\alpha(\eta), \eta)$ and do not intersect (the only possible intersections occur at $\eta = 0$) Moreover any stationary solution $v$ with $\eta \neq 0$ has only simple zeros. Thus the zero number $z(v)$ is invariant along each branch. Put differently, the first, second, $k$th intersection point with the stationary diagram $(\alpha, \eta)$ on a line $\eta \equiv$ const. $\neq 0$ starting at $\alpha = 0$ corresponds to a solution $v$ with $z(v) = 0, 1, k - 1$ respectively. With this in mind, for given $v$, the $\bar{v}_k, \underline{v}_k$ are easily determined from diagrams like Figs 1, 2 as indicated.

Next we determine $i(v)$ from the stationary bifurcation diagram. It should

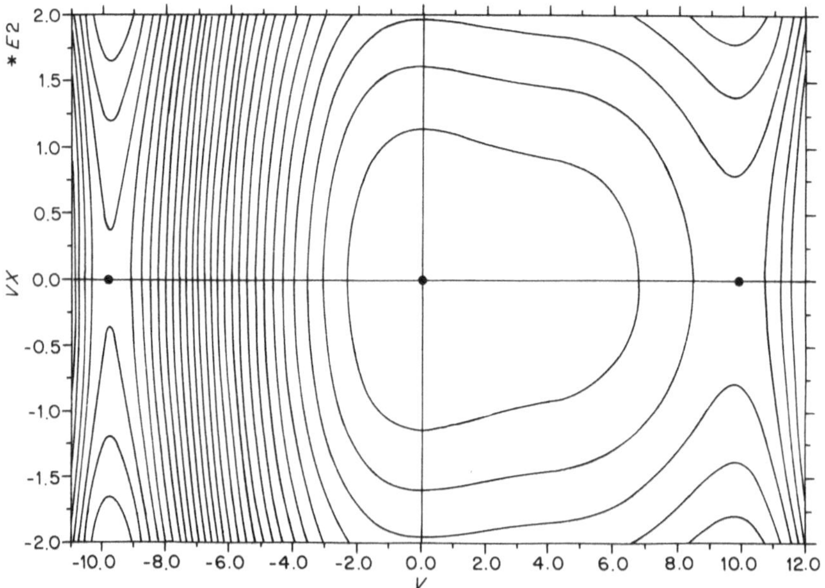

**Fig. 6.** Time map for $f(u) = -(u + 10.2) \cdot u \cdot ((u - 4)^2 + 1.75^2) \cdot (u - 10)$. Neumann problem

be a little bit surprising that this is at all possible because $i(v)$, being the dimension of the unstable manifold at $v$, relates to the dynamics of $(1.1)_\alpha$ rather than to the plain stationary case. In lemma 5.1 below we prove that

$$i(v) \in \{z(v), z(v) + 1\}$$

for hyperbolic stationary $v \equiv 0$, using Sturm–Liouville theory. Moreover it can be shown for $(\alpha_0, v)$ represented by $(\alpha_0, \eta_0)$ on the branch $(\alpha(\eta), \eta)$, $\eta_0 \neq 0$ that $v$ is hyperbolic iff $\alpha'(\eta_0) \neq 0$ (cf. [27, 29] ), and

$$\eta_0 \cdot \alpha'(\eta_0) > 0 \Rightarrow i(v) = z(v)$$
$$\eta_0 \cdot \alpha'(\eta_0) < 0 \Rightarrow i(v) = z(v) + 1.$$

This determines which of the alternatives of the main theorem 1.1 applies to $v \neq 0$. If on the other hand $v \equiv 0$, then the eigenvalues of the linearization $L$ from (1.11), (1.12) of $(1.1)_\alpha$ are determined explicitly as

$$\lambda_k = -k^2 + \alpha^2 f'(0).$$

Denoting the bifurcation points from the trivial solution as $(\alpha_k, 0)$ with $\alpha_k^2 = k^2/f'(0)$ we see that $i(v) = k$ for $v \equiv 0$ and $\alpha$ between $\alpha_k$ and $\alpha_{k+1}$, provided $f'(0) > 0$. If $f'(0) < 0$ then trivially $i(v) = 0$ for all $\alpha$.

As a first example suppose $f$ satisfies (1.4), i.e. $f(0) = 0 < f'(0)$, $sf''(s) < 0$ for $s \neq 0$ in addition to growth condition (1.5). Then Birkhoff and Rota [3] and later Chafee and Infante [8] have proved that $\eta \cdot \alpha'(\eta) > 0$ for each nontrivial branch, and that for $\alpha > \alpha_k = k\pi^2/f'(0)$ there exist exactly two nontrivial solutions $v$ with $z(v) = k - 1$, one with $\eta > 0$ and one with $\eta < 0$. The typical bifurcation diagram is given in Fig. 1. We illustrate a case where $z(v) = 3$. We know $\eta_0 \cdot \alpha'(\eta_0) > 0$, hence $i(v) = z(v) = 3$ and case (i) of the main theorem implies that

$$\Omega(v) = \{\bar{v}_0, \bar{v}_1, \bar{v}_2, \underline{v}_0, \underline{v}_1, \underline{v}_2\}$$

as depicted in Fig. 1. In general, only case (i) occurs — this is the case analyzed by Henry in [17] and we recover his result.

Now we consider an example which exhibits wiggles (Fig. 2) A somewhat simplistic reason for this phenomenon is that $f$ does not satisfy (1.4), this time. First we pick $\alpha = \alpha_0$ and $v = 0$ with $i(v) = 5$. Again case (i) of the theorem applies. This time, however, there is more than one stationary solution of a given zero number $z = 2, 3$ with $\eta > 0$ resp. $\eta < 0$. Thus minimality (maximality) in the definition of $\bar{v}_k (\underline{v}_k)$ comes into effect and we are left with

$$\Omega(v) = \{\bar{v}_k, \underline{v}_k \mid 0 \leqslant k \leqslant 4\}.$$

Next we pick $\alpha = \alpha_1$ and $v$ with $z(v) = 8$ but $i(v) = 9$, $v_x(0) > 0$. Then

$$\Omega_1 = \{\bar{v}_0, ..., \bar{v}_8\}$$
$$\Omega_2 = \{\underline{v}_0, ..., \underline{v}_7\}$$

and either $\Omega_3 = \{v_8\}$ or $\Omega_3$ consists of one or several solutions $w$ with

$$-v_x(0) \leqslant w_x(0) < v_x(0), \qquad i(w) < i(v) = 9.$$

The possible candidates for $w$ are denoted by '?' in Fig. 2. Obviously, theorem 1.1 does not determine completely which other stationary solutions $v$ connects to. For a conjecture how to resolve this problem we refer the reasonably impatient reader to §6. On the other hand, R. Schaaf [26] provides detailed information on the global bifurcation picture, if $f(0) < 0 < f''(0)$ and $f$ is a polynomial with only real zeros. We believe that out theorem can solve the connection problem completely in that case. Note that $f$ in Fig. 2 violates the above condition.

The rest of the paper is organized as follows. In §2 we construct the basic tool to establishing connections: the $y$-map. For given initial datum $\tilde{u}_0$, $y(\tilde{u}_0)$ completely describes the behaviour of $z(\tilde{u}(t, \cdot))$ and of sign $\tilde{u}_x(t, 0)$ along the semi-orbit $\tilde{u}(t, \cdot)$ of $\tilde{u}_0$ under equations (1.7) and (1.8). In §3 we show that $y$ restricted to the unstable manifold $W^u(v)$ induces an essential mapping of spheres. As a corollary, we obtain orbit connections to equilibria $w$ with prescribed $z(w - v)$ and $\text{sign}(w_x(0) - v_x(0))$ (cf (3.4 a,b)). The problem remains to identify $w$. This boils down to two basic lemmata given in §4. They allow us to pass from $z(w - v)$ to $z(w)$ and account for the minimality (maximality) occurring in the definition of $\tilde{v}_k$ ($v_k$). Fitting everything together we prove our main theorem in §5. We devote §6 to a detailed discussion, including a comparison of our approach to those of Henry, and Conley, Smoller, an extension to Neumann boundary conditions (theorem 6.1), a conjecture on the complete answer to the connecting-orbit-problem in the Dirichlet case, and some open questions. In §7 we collect some background material on the behavior of the zero number along solutions $u(t, \cdot)$ of (1.1), (1.2).

## 2  THE $y$-MAP

In this section we construct a continuous mapping

$$y : \{\tilde{u}_0 \in X \mid z(\tilde{u}_0) \leqslant n, \quad \tilde{u}_0 \neq 0\} \to S^n$$

where $S^n$ denotes the standard $n$-sphere in $\mathbb{R}^{n+1}$. Knowing $y(\tilde{u}_0)$ we will know $z(\tilde{u}(t, \cdot))$, $t \geqslant 0$, all along the orbit $\tilde{u}(t, \cdot)$ of (1.7) starting at $\tilde{u}_0$. Moreover, $y$ depends continuously on the nonlinearity $g \in \mathcal{G}_0$ defined in (1.10). Restricting $y$ to an $n$-dimensional sphere $\Sigma^n$ in the unstable manifold of $v \equiv 0$ will provide us with an essential mapping of spheres. With these properties in mind, we will immediately establish existence of connections in §3.

Throughout §2, we consider the equation

$$\begin{aligned}
\tilde{u}_t &= \tilde{u}_{xx} + g(x, \tilde{u}), \qquad 0 < x < 1 \\
\cos \gamma_j \cdot \tilde{u}(t, j) &- \sin \gamma_j \cdot \tilde{u}_x(t, j) = 0, \qquad j = 0, 1
\end{aligned}$$

(2.1)

under the restrictive assumption

$$\cos \gamma_0 \neq 0$$

i.e. we exclude the pure Neumann condition here. For ease of notation, we prefer to write $u$ instead of $\bar{u}$ in this section. Recall from the introduction that $z(u(t, \cdot))$ is non-increasing with $t.$.

We construct the $y$-map. For $u_0 \in X, u_0 \not\equiv 0, z(u_0) \leqslant n$ with orbit $u(t, \cdot)$, define $t_k \in [0, \infty]$ to be the first time that the zero number $z(u(t, \cdot))$ drops below the $k$-level:

$$(2.2a) \qquad t_k := \inf\{t \geqslant 0 \mid z(u(t, \cdot)) \leqslant k\}, \ \tau_k := \tanh t_k \in [0, 1]$$

Note that $0 = \tau_n \leqslant \tau_{n-1} \leqslant \cdots \leqslant \tau_0$. Further we define

$$(2.2b) \qquad \sigma_k := \begin{cases} \operatorname{sign} u_x(t, 0) \text{ for some } t \in (t_k, t_{k-1}), & \text{if } t_k < t_{k-1} \\ 0 & \text{otherwise} \end{cases}$$

The sign $\sigma_k$ is well defined because $u_x(t, 0) \neq 0$ for $t_k < t < t_{k-1}$ by lemma 7.4. The components of the map $y = (y_0, \ldots, y_n)$ are defined as

$$(2.3) \qquad y_0 := \sigma_0 (1 - \tau_0)^{1/2}$$

$$y_k := \sigma_k (\tau_{k-1} - \tau_k)^{1/2}, \qquad 1 \leqslant k \leqslant n.$$

By construction, $y$ maps into $S^n$.

Suppose we know $y(u_0)$. Then we can reconstruct the dropping times $t_k$ above uniquely. Moreover we obtain the signs $\sigma_k$ in case $t_k \neq t_{k-1}$. As an important special case, suppose $y(u_0) = \sigma e_k$ where $e_k$ denotes the $k$th unit vector, $\sigma \in \{-1, 1\}$. This implies $t_0 = \cdots = t_{k-1} = \infty$, $t_k = 0$ and therefore for all $t > 0$:

$$(2.4) \qquad z(u(t, \cdot)) = k$$

$$\sigma \cdot u_x(t, 0) > 0.$$

**2.1 LEMMA** The $y$-map (2.3) depends continuously on $g \in \mathscr{G}_0$ and on $u_0 \in X \backslash \{0\}$ with $z(u_0) \leqslant n$.

*Proof* Throughout the proof we use that the solution $u(t, \cdot)$ of (2.1), viewed as a $C^1$-function of $x$, depends continuously on $g$, $u_0$ and $t$. To be more specific, let this solution be denoted by $u(t, x) = u(t, x; g, u_0)$ emphasizing its actual dependence on $g$ and $u_0$. Then the map

$$\mathscr{G} \times X \times [0, \infty) \to C^1([0, 1], \mathbb{R})$$

$$(g, u_0, t) \mapsto (x \mapsto u(t, x; g, u_0))$$

is continuous, because it is the composition of the analogous map

$$\mathscr{G} \times X \times [0, \infty) \to H^2([0,1], \mathbb{R}) \cap H_0^1([0,1], \mathbb{R})$$

which is continuous by Henry [11], and the continuous Sobolev embedding

$$H^2([0,1], \mathbb{R}) \cap H_0^1([0,1], \mathbb{R}) \to C^1([0,1], \mathbb{R}).$$

Note that we use the weak Whitney topology on $\mathscr{G}$ here. This is sufficient because continuous dependence on initial data is a local property.

We show that $\tau_k \geqslant 0$ depends lower semicontinuously on $(g, u_0) \in \mathscr{G}_0 \times X$. First note that

$$z : C^0([0,1], \mathbb{R}) \to \mathbb{Z}$$

$$\phi \to z(\phi)$$

is lower semicontinuous by definition of $z$. Together with continuity of $u(t, \cdot\,; g, u_0)$ and the definition of $\tau_k$ this implies: for any $\varepsilon > 0$ such that $\tau_k - \varepsilon > 0$, and for $t$ defined by $\tanh t = \tau_k - \varepsilon$ there exists a neighborhood $U$ of $(g, u_0)$ in $\mathscr{G} \times X$ such that for any $(\hat{g}, \hat{u}_0) \in U$ we have

$$z(u(t, \cdot, \hat{g}, \hat{u}_0)) \geqslant z(u(t, \cdot\,; g, u_0)) > k, \text{ and hence}$$

$$\tau_k(\hat{g}, \hat{u}_0) \geqslant \tanh t = \tau_k(g, u_0) - \varepsilon.$$

Thus $\tau_k$ is lower semicontinuous.

We show that $\tau_k \geqslant 0$ is upper semicontinuous if $t_k$ is finite. In that case, lemma 7.3 implies that for any $\varepsilon > 0$ there exists some $t$ such that $\tau_k < \tanh t < \tau_k + \varepsilon$ and all zeros of $x \to u(t, x; g, u_0)$ are simple. Using continuity of $u(t, \cdot\,; g, u_0) \in C^1$ and the definition of $\tau_k$ again, this implies: there exists a neighborhood $U$ of $(g, u_0) \in \mathscr{G} \times X$ such that for any $(\hat{g}, \hat{u}_0) \in U$ we have

$$z(u(t, \cdot\,; \hat{g}, \hat{u}_0)) = z(u(t, \cdot, g, u_0)) \leqslant k, \text{ and hence}$$

$$\tau_k(\hat{g}, \hat{u}_0) \leqslant \tanh t < \tau_k(g, u_0) + \varepsilon.$$

Thus $\tau_k$ is upper semicontinuous and, consequently, continuous.

Finally, we claim that each component $y_k$ of the $y$-map depends continuously on $(g, u_0) \in \mathscr{G}_0 \times X$. We already know that $\tau := (\tau_0, \dots, \tau_{n-1})$ depends continuously on $(g, u_0)$. If $\tau_k < \tau_{k-1}$ at $(g, u_0)$ then lemma 7.4 implies

$$u_x(t, 0; g, u_0) \neq 0$$

for any $t \in (t_k, t_{k-1})$. Fixing any such $t$, there exists a neighborhood $U$ of $(g, u_0)$ in $\mathscr{G} \times X$ such that for any $(\hat{g}, \hat{u}_0) \in U$ we have

$$u_x(t, 0; \hat{g}, \hat{u}_0) \neq 0,$$

by continuous dependence of $u(t, \cdot\,; g, u_0) \in C^1$. Hence $\sigma_k$ is constant on $U$, and $y_k$ is continuous by continuity of $\tau_k, \tau_{k-1}$. If on the other hand $\tau_k = \tau_{k-1}$, then $y_k = 0$ at $(g, u_0)$ and continuity of $\tau$ implies that

$$|y_k| < \varepsilon$$

for all $(\hat{g}, \hat{u}_0)$ in some neighborhood $U$ of $(g, u_0)$ in $\mathcal{G} \times X$, no matter which sign $\sigma_k$ takes. Therefore, $y$ is again continuous and the proof is complete. $\square$

In order to actually find $u_0$ with $y(u_0) = \sigma e_k$, we are interested in surjectivity of the $y$-map (2.3) for nonlinear $g$. As usual, it is much easier to discuss $y$ for linear $g$. But deforming $g$ by a homotopy from the linear to the nonlinear, surjectivity might be destroyed. Fortunately, topology helps us to bridge this gap. In lemma 2.2 below, we prove that for linear $g$

$$y : \Sigma^n \to S^n$$

is an *essential* mapping between spheres $\Sigma^n$ and $S^n$. Essential means that there is no homotopy from $y$ to the constant map. Charmingly, this property is invariant under homotopies to nonlinear $g$ — by definition. Moreover it implies that $y$ remains surjective; or else the image of $y$ would miss some point in $S^n$ and could therefore be contracted to a single point in contradiction to $y$ being essential. Conversely, if $n = 0$ and $y$ is surjective then $y$ is also essential. However, this does not hold for $n > 0$, in general. With this in mind, we turn to the case of linear $g$.

Specialize $g(x, u) = a(x) \cdot u$, $a \in C^2$, and denote the (Sturm–Liouville) eigenvalues and eigenfunctions of

$$(2.5) \qquad\qquad \lambda u = u_{xx} + a(x)u$$

with boundary condition (2.1) by $\lambda_0 > \lambda_1 \ldots$ and $\varphi_0, \varphi_1 \ldots$ as in the introduction. We take $\varphi_k(x)$ normalized to unit length in $X$ with the additional sign convention $\varphi_k'(0) > 0$. Assume that

$$(2.6) \qquad\qquad \lambda_n > 0$$

i.e. $u \equiv 0$ has Morse-index $i(u \equiv 0) \geq n + 1$. Denoting

$$W_n = \text{span}\{\varphi_0, \ldots, \varphi_n\},$$

it is known from Sturm–Liouville theory that $z \leq n$ on $W_n$, see also [5]. Let $\Sigma^n$ denote a sphere centered at 0 in $W_n$.

2.2  LEMMA  Under assumption (2.6), the restriction of the $y$-map

$$y : \Sigma^n \to S^n$$

is essential, i.e. $y$ is not homotopic to a constant. In particular, $y$ is surjective.

*Proof*  The proof is by induction on $n$. For $W_k$, $0 \leq k \leq n$ observe that

$\Sigma^k := \Sigma^n \cap W_k$ is a $k$-dimensional sphere centered at 0. For $n = 0$, $\Sigma^0 = \{ \pm \varphi_0 \}$ for example. In that case $y = y_0 = \sigma_0 = \pm \operatorname{sign} \varphi_0'(0)$ and $y$ is surjective, hence essential. Below, we work our way up from $\Sigma^0$ to $\Sigma^n$ where each $\Sigma^{k-1}$ occurs as an equator in $\Sigma^k$. By a Mayer–Vietoris argument, $y$ is essential from $\Sigma^k$ to $S^k$ if it was essential from $\Sigma^{k-1}$ to $S^{k-1}$. For $k = n$, the lemma will be proved.

Suppose now the lemma is proved for $k - 1$ already. Identify $W_k$ with $\mathbb{R}^{k+1}$ by $(\eta_0, ..., \eta_k) \to \Sigma \eta_j \varphi_j$ and write $\Sigma^k = \{ \Sigma \eta_j^2 = 1 \}$. Note that $W_{k-1}$ with sphere $\Sigma^{k-1}$ is a subspace of $W_k$ and $\Sigma^{k-1}$ becomes an equator of $\Sigma^k$. We denote the closed hemisphere as

$$S_\pm^k := \{ (y_0, ..., y_k) \in S^k \mid \pm y_k \geqslant 0 \}$$
$$\Sigma_\pm^k := \{ (\eta_0, ..., \eta_k) \in \Sigma^k \mid \pm \eta_k \geqslant 0 \},$$

and the equators as

$$S^{k-1} = \{ (y_0, ..., y_k) \in S^k \mid y_k = 0 \}$$
$$\Sigma^{k-1} = \{ (\eta_0, ..., \eta_k) \in \Sigma^k \mid \eta_k = 0 \}.$$

For the restriction we have

$$y^{k-1} := y \mid_{\Sigma^{k-1}} : \Sigma^{k-1} \to S^{k-1},$$

because $\tau_{k-1} = 0$ on $\Sigma^{k-1} \subset W_{k-1}$ implies $y_k = 0$. Moreover

$$y : \Sigma_+^k \to S_+^k$$
$$\Sigma_-^k \to S_-^k .$$

Indeed, $u_0 \in \Sigma_+^k$ with $\eta_k > 0$ and $u(t, \cdot) = \Sigma \exp(\lambda_j t) \eta_j \varphi_j$, $\lambda_0 > \cdots > \lambda_k$ imply

$$\lim_{t \to -\infty} \frac{u(t, \cdot)}{|u(t, \cdot)|} = \operatorname{sign} \eta_k \cdot \phi_k = \varphi_k$$

and, using that $z(u(t, \cdot)$ is decreasing for all real $t$, consequently

either $$\tau_{k-1} = 0,$$

or $$\tau_{k-1} > 0, \qquad \sigma_k = \operatorname{sign} \varphi_k'(0) = +1.$$

Note that for $\tau_{k-1} > 0$ the sign $\sigma_k$ may be evaluated for $t$ near $-\infty$. In the first case $y_k = 0$, whereas in the second case $y_k > 0$. Hence $y$ maps $\Sigma_+^k$ into $S_+^k$; the case of $\Sigma_-^k$ is analogous.

Now $y^{k-1}$ is essential by induction hypothesis, hence its Brouwer degree $\deg y^{k-1}$ is nonzero (cf. [12] for a definition of degree and the topological background used below). Let us consider the Mayer–Vietoris sequence for

$y, k \geqslant 1$ [12]

$$0 \to H_k(\Sigma^k) \to H_{k-1}(\Sigma^{k-1}) \to H_{k-1}(\Sigma^k_+) \oplus H_{k-1}(\Sigma^k_-)$$
$$\quad\;\downarrow \deg y \qquad\qquad \downarrow \deg y^{k-1} \qquad\qquad\qquad \downarrow$$
$$0 \to H_k(S^k) \to H_{k-1}(S^{k-1}) \to H_{k-1}(S^k_+) \oplus H_{k-1}(S^k_-).$$

The homologies of hemispheres are trivial, the other homologies are just $\mathbb{Z}$, hence

$$\deg y = \deg y^{k-1} \neq 0$$

and $y$ is essential [12]. This completes the induction step and the proof of the lemma. $\qquad\qquad\qquad\qquad\qquad\qquad\qquad\qquad\qquad\qquad\qquad\qquad\qquad\quad\;\Box$

A quicker, less explicit proof may be sketched as follows.. Because we consider a linear flow $u(t, \cdot)$, the $y$-map in lemma 2.2 is odd. Thus $y$ is essential by the Borsuk–Ulam Theorem (which can be proved by the Mayer–Vietoris sequence given above).

## 3   ESTABLISHING CONNECTIONS

We use the $y$-map constructed in §2, to establish connections from a stationary hyperbolic solution $v$ to at least $2i(v)$ distinct other stationary solutions; $i(v)$ denotes the Morse index of $v$, as before. Below, we employ homotopy-invariance of the $y$-map to see that the $y$-map induces an essential mapping from a sphere $\Sigma^n$ of dimension $n = i(v) - 1$ around $v$ in the unstable manifold of $v$, and mapping into the standard $n$-sphere $S^n$. Indeed, we investigated the linear case in lemma 2.2 and our result is obtained by standard homotopy to the nonlinear case.

Throughout §3, we again consider the equation

$$(3.1) \qquad\qquad u_t = u_{xx} + g(x, u), \quad 0 < x < 1$$

$$\cos \gamma_j \cdot u(t, j) - \sin \gamma_j \cdot u_x(t, j) = 0, \quad j = 0, 1$$

for $\cos \gamma_0 \neq 0$ and $g \in \mathcal{G}_0$ (in particular $g(x, 0) = 0$, cf. (1.10)). replacing $\tilde{u}$ by $u$ as in §2.

3.1   LEMMA   Suppose $v \equiv 0$ is a hyperbolic stationary solution of (3.1) with unstable manifold $W^u$ of dimension $i(v) > 0$. Let $\Sigma \subset W^u \setminus \{v\}$ be homotopic in $W^u \setminus \{v\}$ to a small sphere centered at $v$ in $W^u$ of dimension $n = i(v) - 1$.

Then for any finite sequence

$$0 = \delta_n \leqslant \delta_{n-1} \leqslant \delta_{n-2} \leqslant \cdots \leqslant \delta_0 \leqslant \infty$$

$$s_k \in \{1, -1\}, \quad 1 \leqslant k \leqslant n$$

there exists an initial datum $u_0 \in \Sigma$ such that the graph $t \to z(u(t, \cdot))$ is characterized by $(\delta_k)$. More precisely for any $0 \leqslant t < \infty$:

(3.2a) $$t \geqslant \delta_k \Leftrightarrow z(u(t, \cdot)) \leqslant k$$

(3.2b) $$\delta_k < t < \delta_{k-1} \Rightarrow \operatorname{sign} u_x(t, 0) = s_k$$

*Proof*   First suppose that the restricted $y$-map

$$y : \Sigma \to S^n$$

is essential (cf. [12] for the topological facts used). Then $y$ is surjective. Now define $\eta$ just as the $y$-map in (2.2–2.3), but replacing $t_k$ by $\delta_k$, and $\sigma_k$ by $s_k$. By surjectivity of $y$, there exists an initial datum $u_0 \in \Sigma$, such that $y(u_0) = \eta$. But knowing $y$, the dropping times $t_k$ and signs $\sigma_k$ associated to the orbit $u(t, \cdot)$ of $u_0$ are uniquely determined as

$$t_k = \delta_k$$

and, in case $\delta_k < \delta_{k-1}$,

$$\sigma_k = s_k.$$

Therefore, it only remains to prove that $y$ is essential. How to achieve this? By a homotopy, of course! We deform $g$ into its linearization, defining

$$g_\beta(x, u) := \beta g(x, u) + (1 - \beta) g_u(x, 0) \cdot u$$

with homotopy parameters $0 \leqslant \beta \leqslant 1$. Simultaneously this deforms the unstable manifold $W^u(g_\beta)$ associated to the stationary solution $v \equiv 0$ of $g_\beta$. Note that our homotopy leaves the linearization at $v \equiv 0$ unchanged. Moreover, $g_\beta \in \mathcal{G}_0$ depends continuously on $\beta$, because $\mathcal{G}_0$ carries the weak Whitney topology Let

$$W^u_{\text{loc}}(g_0) := \operatorname{span}\{\varphi_0, \ldots, \varphi_n\} \cap \{u_0 \in X \mid |u_0| < 2\varepsilon\}$$

denote the (cut-off) tangent space of $W^u(g_\beta)$ at $v \equiv 0$. Then the local unstable manifolds of $g_\beta$ are parametrized by diffeomorphisms

$$p_\beta : W^u_{\text{loc}}(g_0) \to W^u_{\text{loc}}(g_\beta)$$

where $P_\beta^{-1}$ is induced by the orthogonal projection onto $\operatorname{span}\{\varphi_0, \ldots, \varphi_n\}$. Note that $P_\beta$ depends continuously on $\beta$ in the uniform $C^0$-topology. Fix a sphere

$$\Sigma^n := \{u \in W^u_{\text{loc}}(g_0) \mid |u| = \varepsilon\}$$

and let $y^\beta$ denote the restriction to $P_\beta(\Sigma^n)$ of the $y$-map associated to $g_\beta$. After a homotopy, we may assume $\Sigma = P_1(\Sigma^n)$. Finally, define

$$\tilde{y}^\beta := y^\beta \cdot P^\beta : \Sigma^n \to S^n.$$

This mapping is well-defined (recalling from (1.13) that $z \leqslant n$ on $W_{\mathrm{u}}(g_\beta)$), continuous and depends continuously on $\beta$ by lemma 2.1. Lemma 2.2 implies that

$$\tilde{y}^0 = y_0 \circ P_0 = y_0 : \Sigma^n \to S^n$$

is essential. By homotopy-invariance, this implies that $\tilde{y}^1 = y^1 \circ P_1 = y \circ P_1$, and hence $y$, is essential — completing the proof. $\qquad\qquad\square$

As a corollary to lemma 3.1, we obtain connections from $v$ to at least $2i(v)$ different other stationary solutions under a growth restriction on $g$.

**3.2 COROLLARY** Suppose $v$ is a hyperbolic stationary solution of (3.1) with Morse index $i(v) > 0$. In addition, let $g \in \mathscr{G}$ satisfy the growth condition

$$(3.3) \qquad\qquad \varlimsup_{|u| \to \infty} g(x, u)/u \leqslant 0$$

uniformly in $x$ (we do not require $g(x, 0) = 0$, here).

Then for any $0 \leqslant k < i(v)$, $\sigma \in \{1, -1\}$, there exists a stationary solution $w \neq v$ such that $v$ connects to $w$ and

$$(3.4\mathrm{a}) \qquad\qquad z(w - v) = k$$

$$(3.4\mathrm{b}) \qquad\qquad \mathrm{sign}(w_x(0) - v_x(0)) = \sigma.$$

*Proof* Without loss of generality, we may assume $v \equiv 0$. Indeed, let $u$ be a solution of (3.1) with $g \in \mathscr{G}$. Then $\tilde{u} := u - v$ satisfies (3.1), replacing the nonlinearity $g$ there by

$$\tilde{g}(x, \tilde{u}) := g(x, \tilde{u} + v(x)) - g(x, v(x));$$

note that $\tilde{g} \in \mathscr{G}_0$.

Now we apply lemma 3.1 to the solutions of (3.1), picking

$$\delta_j := \begin{cases} 0 \\ \infty \end{cases} \quad \text{for } \begin{matrix} j \geqslant k \\ j < k \end{matrix}$$

$$s_k := \sigma.$$

With initial datum $u_0 \in W^{\mathrm{u}}$ corresponding to this choice, lemma 3.1 asserts for the solution $\tilde{u}(t, \cdot)$ that $z(\tilde{u}(t, \cdot)) = k$, sign $\tilde{u}_x(t, 0) = \sigma$ for all $t > 0$. But $\tilde{u}(t, \cdot)$

converges to a stationary solution $w$, as $t \to \infty$ by assumption (3.3) and the gradient structure of equation (3.1) Because $w \not\equiv 0$ has only simple zeros (it solves the ordinary differential equation $0 = w_{xx} + \tilde{g}(x, w)$), properties (3.4) are immediate from (3.2). This completes the proof of the corollary.

## 4   EXCLUDING CONNECTIONS

Suppose we have constructed a connection from a stationary solution $v$ to a stationary solution $w \neq v$ such that

(4.1a) $$z(w - v) = k$$

(4.1b) $$\mathrm{sign}(w_x(0) - v_x(0)) = \sigma,$$

where $0 \leqslant k < i(v)$, $\sigma \in \{1, -1\}$ are given (this we achieved in corollary 3.2). In this section we try to identify the set of all $w$ such that (4.1) holds for some fixed given $v, k$ and $\sigma$. In general, this set may contain more than just one element. However, for nonlinearities $g(x, u) = f(u)$ independent of $x$ and for Dirichlet boundary conditions, the $w$ in question is determined uniquely in terms of $z(w)$ and $w_x(0)$ — in most cases. The two lemmata below are the crucial tools to determine $w$.

4.1   LEMMA   Suppose $g \in C^2$, and $v, w, \bar{w}$ are three distinct stationary solutions of (1.7), (1.8), $\cos \gamma_0 \neq 0$, such that $w_x(0)$ lies strictly between $v_x(0)$ and $\bar{w}_x(0)$. Then

(4.2) $$z(v - w) \leqslant z(\bar{w} - w)$$

implies that $v$ does not connect to $\bar{w}$.

*Proof*   We prove the lemma by negation. Let $v$ connect to $\bar{w}$ via an orbit $u(t, \cdot)$, $t \in \mathbb{R}$. Then $\tilde{u} := u - w$ satisfies again an equation of the form (1.7), similarly to the proof of corollary 3.2. Hence we may assume without loss of generality that $w = 0$, $g(x, 0) = 0$, $v_x(0) < 0 < \bar{w}_x(0)$. Non-increase of $z(u(t, \cdot))$ then implies

$$z(v - w) = z(v) \geqslant z(\bar{w}) = z(\bar{w} - w).$$

Finally, $z(v) \neq z(\bar{w})$, because $v_x(0)$ and $\bar{w}_x(0)$ have opposite sign (cf. lemma 7.4). Therefore, $z(v - w) > z(\bar{w} - w)$ if $v$ connects to $w$ — and the lemma is proved.   $\square$

The next lemma is based on phase-plane analysis for stationary solutions. This forces us to restrict out attention to autonomous, i.e. $x$-independent

$g(x, u) = f(u)$ and to Dirichlet boundary conditions (the Neumann case is discussed in §6). Going back to §3 and (4.1) we obtained $z(w - v) = k$ for some $w$ that $v$ connects to It is significant, that the lemma below allows us to replace $z(w - v)$ by $z(w)$ itself, if $|w_x(0)| \geq |v_x(0)|$. This enables us to describe connections in terms of $z(w), z(v), i(v)$ alone — rather than $z(w - v)$, which can not be read off from time-map bifurcation diagrams as given in the introduction.

**4.2  LEMMA**  Consider equation (1.1) with Dirichlet boundary conditions (1.2). Let $v^1$ and $v^2$ be two distinct stationary solutions. Then $|v_x^1(0)| \geq |v_x^2(0)|$ implies

$$(4.3) \qquad\qquad z(v^1 - v^2) = z(v^1)$$

*Proof*  Recall that any stationary solution $v$ of (1.1), (1.2) has only simple zeros, or else $v \equiv 0$. If $v^1 \equiv 0$, then $v_x^2 = 0$ implies $v^2 \equiv 0$, and $v^2$ cannot be distinct from $v^1$. Hence we consider $v^1 \not\equiv 0$, only. We partition the non-empty set

$$\{x \in [0, 1] \mid v_x^1(x) \neq 0\} = I_0 \cup \cdots \cup I_{n+1}$$

into its disjoint connected components (intervals) $I_j, 0 \leq j \leq n + 1, n = z(v^1)$. Note that $0 \in I_0, 1 \in I_{n+1}$. It is sufficient to show that each $I_j$ contains exactly one zero of $v^1 - v^2$, this zero is automatically simple.

Each interval $I_j$ contains at least one zero of $v^1 - v^2$. Indeed, for $j = 0, n + 1$ these are given by $x = 0, 1$, respectively. For $0 < j < n + 1$ this follows because

$$\operatorname{sign}(v^1 - v^2) = \operatorname{sign} v^1$$

at the endpoints of $I_j$. To see this, just note that $v^1$ attains both its extreme values at the endpoints of $I_j$ and that the orbit of $(v^1(x), v_x^1(x))$ does not lie inside the orbit of $(v^2(x), v_x^2(x))$ for the Hamiltonian system

$$(4.4) \qquad\qquad 0 = v_{xx} + f(v)$$

in the $(v, v_x)$-plane, by assumption (cf. e.g. Fig. 4).

Each interval $I_j$ contains at most one zero of $v^1 - v^2$. This is again immediate from the fact that the orbit of $(v^1(x), v_x^1(x))$ does not lie inside the orbit of $(v^2(x), v_x^2(x))$, which implies that

$$\operatorname{sign}(v_x^1(x) - v_x^2(x)) = \operatorname{sign} v_x^1(x)$$

whenever $v^1(x) = v^2(x)$ for some $x \in I_j$ (cf. Fig. 4).

Hence each interval $I_j$ contains exactly one (simple) zero of $v^1 - v^2$. This implies $z(v^1 - v^2) = n = z(v^1)$ and the proof is finished.  $\square$

## 5  PROOF OF THEOREM 1.1

We combine the results of Sections 3 and 4 to prove theorem 1.1. First, we use corollary 3.2 to establish connections from $v$, stationary and hyperbolic with positive Morse index $i(v)$, to some stationary $w \neq v$ such that

(5.1a) $$z(w - v) = k$$

(5.1b) $$\mathrm{sign}(w_x(0) - v_x(0)) = \sigma,$$

where $0 \leqslant k < i(v)$, $\sigma \in \{1, -1\}$ can be prescribed arbitrarily. Vice versa, any stationary solution $w$ that $v$ connects to has to satisfy (5.1) for some appropriately chosen $0 \leqslant k < i(v)$ and $\sigma \in \{1, -1\}$ because, by (1.13), $z(u_0 - v) < i(v)$ for any initial data $u_0$ in the unstable manifold $W^u$ of $v$. Below we employ lemmata 4.1 and 4.2 to identify $w$ as described in theorem 1.1. Remember that lemma 4.2 required autonomous phase plane analysis; therefore we consider equation (1.1) with Dirichlet boundary condition (1.2) throughout this section.

As a final preparation to the proof of theorem 1.1, we show that its cases (i)–(iii) are the only possible ones.


5.1  LEMMA  Let $v \neq 0$ be a hyperbolic stationary solution of (1.1) with Dirichlet boundary condition (1.2). Then Morse index $i(v)$ and zero number $z(v)$ are related by
$$i(v) \in \{z(v), z(v) + 1\}.$$

*Proof*  By Rolle's theorem and phase-plane analysis of
$$0 = v_{xx} + f(v)$$

we have $z(v_x) = z(v) + 1$ for Dirichlet boundary conditions on $v$.

Now consider the linearization (1.11), (1.12) and its eigenvalues $\lambda_n$, eigenfunctions $\varphi_k$ as in (2.5) with $a(x) := f'(v(x))$. For $n := i(v) - 1$ we have
$$\lambda_n > 0 > \lambda_{n+1},$$
$$z(\varphi_n) = n, \ z(\varphi_{n+1}) = n + 1$$

On the other hand, $u := v_x$ also satisfies (2.5) with $\lambda = 0$. By the Sturm–Liouville comparison theorem [3, 15], between any two consecutive zeros of $\varphi_n$ there is a zero of $v_x$, and between any two consecutive zeros of $v_x$ there is a zero of $\varphi_{n+1}$— all these zeros being simple. This respectively implies
$$n + 1 \leqslant z(v_x), \text{ and } z(v_x) - 1 \leqslant n + 1, \text{ i.e.}$$
$$i(v) = n + 1 \leqslant z(v_x) = z(v) + 1, \text{ and}$$
$$i(v) = n + 1 \geqslant z(v_x) - 1 = z(v),$$

completing the proof of the lemma. A similar idea can be found in [27, Lemma 24.16]. $\qquad\square$

**Proof of theorem 1.1:**

Using corollary 3.2 as above, it remains to identify those stationary $w$ satisfying (5.1) which $v$ does connect to. We have to consider three cases. First we address $v \neq 0$ and $0 \leqslant k < z(v)$. Then we analyze $v \equiv 0$ (in case $f(0) = 0$) for $0 \leqslant k < i(v)$ These two cases result in part (i) of theorem 1.1 (note that $z(v) = i(v)$ is assumed there for $v \neq 0$). Finally, we consider $v \neq 0$, $k = z(v) = i(v) - 1$ as the only remaining case. Replacing $f(s)$ by $-f(-s)$ we may in fact assume $v_x(0) > 0$ in that case without loss of generality.

**Case 1: $v \neq 0$, $0 \leqslant k < z(v)$**

By lemma 4.2, $k < z(v)$ implies

(5.2a) $$k = z(w - v) = z(w),$$

(5.2b) $$|w_x(0)| > |v_x(0)|.$$

If $\sigma = +1$ (resp. $-1$) then $w_x(0)$ is above (resp. below) $v_x(0)$ and, by (5.2b), also above (resp. below) $\pm\,|v_x(0)|$. On the other hand $w$ is the minimal (resp. maximal) stationary solution with that property by lemma 4.1 (recall that $v$ does not connect to $w$ by assumption). Therefore $w = \bar{v}_k$ (resp. $\underline{v}_k$).

**Case 2:   $v \equiv 0$, $0 \leqslant k < i(v)$**

Rereading case 1, (5.2a–b) are automatic for $v \equiv 0$. Leaving the remaining arguments of case 1 unchanged we conclude again $w = \bar{v}_k$ (resp. $\underline{v}_k$).

**Case 3:   $v_x(0) > 0$, $k = z(v) = i(v) - 1$**

If $\sigma = +1$, then $|w_x(0)| = w_x(0) > v_x(0) = |v_x(0)|$, hence (5.2a–b) hold and case 1 applies identifying $w$ as $\bar{v}_k$. Likewise, if $\sigma = -1$ and $w_x(0) < -v_x(0)$, we conclude that $w = \underline{v}_k$. However, if $\sigma = -1$ and $w_x(0) \geqslant -v_x(0)$ (i.e. $-v_x(0) \leqslant w_x(0) < v_x(0)$) complications arise. In this one remaining case, we first claim that $i(w) < i(v)$. From (1.13) it is immediate that

$$z(v - w) < \dim W^{\mathrm{u}}(v) = i(v).$$

Now follow $\tilde{u}(t, \cdot) := u(t, \cdot) - w$ for $t \to \pm\infty$ along an orbit $u(t, \cdot)$ connecting $v$ to $w$. Similarly to the argument given in the introduction, $\tilde{u}(t, \cdot)/|\tilde{u}(t\,\cdot)|$ converges to an eigenfunction $\varphi_k$ with eigenvalue $\lambda_k \leqslant 0$ of the linearization at

$w$ as $t \to +\infty$ and

$$z(\tilde{u}(t, \cdot)) = z(\tilde{u}(t, \cdot)/|\tilde{u}(t, \cdot)|) \geqslant z(\varphi_k) = k \geqslant i(w)$$

for all real $t$. In the limit $t \to -\infty$ this yields

$$z(v - w) \geqslant \dim W^{\mathrm{u}}(w) = i(w).$$

Finally we claim that there is no additional connection to any stationary $\bar{w}$ with

$$z(\bar{w}) = k, \quad \bar{w}_x(0) < -v_x(0).$$

Indeed, $w_x(0)$ is between $\bar{w}_x(0)$ and $v_x(0)$. Further, by lemma 4.2

$$z(v - w) = z(v) = k = z(\bar{w}) = z(\bar{w} - w).$$

Now lemma 4.1 implies that $v$ does not connect to $\bar{w}$. This completes the proof of theorem 1.1. $\qquad\qquad\square$

# 6   DISCUSSION

We compare our result with previous results by Henry [17] for Dirichlet boundary conditions. Then we present a theorem for the Neumann case, sketching the necessary modifications Finally we indicate some open problems.

Under the additional assumption (1.4): $f(0) = 0 < f'(0)$, $s \cdot f''(s) \geqslant 0$ for $s \neq 0$, Henry [17, §4] works out all the connecting orbits for the Dirichlet problem (1.1), (1.2). As outlined in the introduction, (1.4) implies $i(v) = z(v)$ for any nonzero stationary solution $v$ — and Henry's result is recovered by case (i) of theorem 1.1

The method of Henry [17] is in some sense complementary to ours. One establishes, for a general class of scalar parabolic equations including nonlinearities $f(x, u, u_x)$ and nonlinear boundary conditions that stable and unstable manifolds of hyperbolic equilibria always intersect transversely (cf. [17, theorem 7], degenerate equilibria are treated as well, and [1]). Consequently, if $\bar{v}_k$, $\underline{v}_k$ are hyperbolic, too, the sets $C(v, \bar{v}_k)$ and $C(v, \underline{v}_k)$ of orbits connecting $v$ to $\bar{v}_k$ resp. $\underline{v}_k$ are seen to be immersed submanifolds of $X$ of dimension $i(v) - k$, provided these sets are guaranteed to be non-empty by theorem 1.1. On the other hand, our approach and in particular lemma 3.1 shows, that the graph of $z(u(t, \cdot) - v)$ can be prescribed arbitrarily on orbits in $C(v, \bar{v}_k)$ resp $C(v, \underline{v}_k)$, along with the signs of $u_x(t, 0)$ where $z(u(t, \cdot) - v)$ is locally constant. Of course, the obvious limitation $i(v) > z \geqslant k$ has to be observed. Even under restrictive assumptions on $f$, this was previously unknown.

The first results on connecting orbits were obtained by Conley and Smoller (cf [10, 27] and references there) by an entirely different method. Their approach relied solely on the Ljapunov functional $V$ from (1.6), rather than the discrete 'Ljapunov' functional $z$. Given $V$, Conley's index theory could be applied to establish connections from $v$ to $w$, typically in cases where $i(v) = i(w) + 1$. For an exposition on Conley's index see e.g. [9], [27, §§22–24] and in particular [27, lemma 24.12 and theorem 24.14]. A very rudimentary account would run as follows. Consider Fig. 1 and take a parameter $\alpha$ between the first and second bifurcation point. We have three stationary solutions: $v \equiv 0$, $\bar{v}_0$ and $\underline{v}_0$ in our notation. The Conley index may be defined as follows. Let $S$ be an isolated invariant set with isolating neighborhood $N_1$ (i.e. $S$ is the maximal invariant subset of $N_1$ and is contained in the interior of $N_1$). Let $N_2$ denote the exit set of $N_1$, i.e. those points of $\partial N_1$ which leave $N_1$ in forward time. Then $h(S)$, the Conley-index of $S$, is the homotopy type of $N_1/N_2$, i.e. of $N_1$ with the exit set collapsed to a (distinguished) point. In our example, take $S$ to be successively $v$, $\bar{v}_0$, $\underline{v}_0$, and finally the maximal compact invariant set $\mathcal{A}$. Then

$$h(v) = \Sigma^1$$
$$h(\bar{v}_0) = h(\underline{v}_0) = \Sigma^0,$$
$$h(\mathcal{A}) = \Sigma^0,$$

where $\Sigma^k$ denotes the $k$-sphere with some distinguished point, and $h(\mathcal{A})$ is computed by homotopy to parameters $\alpha$ below the first bifurcation point. It is a result of Conley that

$$h(\mathcal{A}) = h(v) \vee h(\bar{v}_0) \vee h(\underline{v}_0),$$

if $\mathcal{A} = \{v, \bar{v}_0, \underline{v}_0\}$; here $v$ denotes the wedge product. But

$$\Sigma^0 \neq \Sigma^1 \vee \Sigma^0 \vee \Sigma^0,$$

hence $\mathcal{A}$ cannot consist of stationary solutions only — it must also contain connecting orbits. By symmetry and because $\bar{v}_0$, $\underline{v}_0$ are both stable, these orbits have to connect $v$ to both $v_0$ and $\underline{v}_0$.

Henry's result, as well as ours, contain all information that was originally gained by Conley's index. But curiously enough, these later approaches are both based on the discrete 'Ljapunov' functional given by the zero number $z$. In other words, maximum principles are emphasized, rather than the variational structure which plays only a marginal role in the results of Henry and ourselves. The Ljapunov functional $V$ was used only to guarantee convergence of $u(t, \cdot)$ to a critical point.

Indeed, it appears that connecting orbits can be worked out even without any variational structure. As a concrete exmaple, we mention the scalar delay equation

$$(6.1) \qquad \dot{x}(t) = -x(t) + f(x(t-1))$$

with negative feedback: $f(0) = 0$, $f'(0) < 0$, $x \cdot f(x) < 0$ for $x \neq 0$. Mallet-Paret [22] uses a discrete 'Ljapunov' functional with properties quite similar to our zero number — we call it $z$, again — in order to work out connecting orbits for (6.1). But (6.1) is not a variational problems solutions $x(t)$ cannot be expected to converge to equilibrium for $t \to \infty$, in general. Thus, the role of 'stationary solutions' has to be replaced by the 'maximal compact invariant subset of $\{z = k\}$' for the various integers $k$. Such sets with different $k$ may be connected by solutions of (6.1). In fact, the 'Ljapunov' functional $z$ alone still enabled Mallet-Paret [22] to detect some connecting orbits via Conley's index method (the analogous approach, based on $z$, was never tried for our reaction diffusion problem). Again, Conley's method seems to be limited to establishing connections between maximal invariant subsets with adjacent $k$. Recently, this difficulty has been circumvented by topological considerations which use only $z$, but are not explicitly related to Conley's index. Summarizing, both equation (6.1) and equations (1.1), (12) exhibit a discrete 'Ljapunov' functional $z$ with the enticing property that it takes on many different values near $v \equiv 0$, e.g. on $W^{u}(v)$, which can be studied by homotopy to the linear case.

Returning to the variational setting once more, we notice that $V(v) > V(w)$ if $v$ connects to $w$. Knowing which $w$ our $v$ connects to, it should be possible to obtain this relation directly from the phase portrait of the Hamiltonian system (4.4). Except — we do not know, how. In this context it should be noted that R. Schaaf [26] has proved monotonicity of $V$ along those parts of the stationary bifurcation diagram which consist of nondegenerate hyperbolic solutions.

For scalar reaction diffusion equations with Neumann boundary condition

$$(1.1) \qquad u_t = u_{xx} + f(u), \quad 0 < x < 1$$

$$(6.2) \qquad u_x(t, 0) = u_x(t, 1) = 0,$$

and growth condition

$$(6.3) \qquad \overline{\lim_{|s| \to \infty}} \; f(s)/s < 0, \quad f \in C^2$$

we present an analogue to theorem 1.1. For any $C^1$-function $v: [0, 1] \to \mathbb{R}$, we define the lap number $l(v)$ (cf. [20]) by

$$(6.4) \qquad l(v) = \begin{cases} z(v_x) + 1, & \text{if } v_x \not\equiv 0 \\ 0, & \text{if } v_x \equiv 0. \end{cases}$$

Given a stationary solution $v$, let $\bar{v}_k$ (resp. $\underline{v}_k$) denote the stationary solution $\tilde{v}$ with minimal $\tilde{v}(0) > $ range $(v)$ (resp. maximal $\tilde{v}(0) < $ range $(v)$) As in the Dirichlet case, the proof of theorem 6.1 below in particular proves existence of all $\bar{v}_k$, $\underline{v}_k$ which occur in its statements. Uniqueness of the $\bar{v}_k$, $\underline{v}_k$ is obvious viewing the stationary boundary value problem as an initial value problem.

Given $v$, the solutions $\bar{v}_k$, $\underset{\sim}{v}_k$ and $l(v)$, $i(v)$ can be identified from global bifurcation pictures as given, e.g. in [24]. As in the Dirichlet-case, numerical bifurcation diagrams can be obtained by rescaling (1.1) to (1.16) and appropriately defining the Neumann-time $\hat{T} = \hat{T}(v^*(0))$ as the first positive zero of $v_x^*(\xi)$. Implementation details were the same as before, and a concrete example is shown in Fig. 6.

**6.1 THEOREM** Under assumption (6.1) above, let $v$ be a hyperbolic stationary solution with lap-number $l(v)$ and Morse-index $i(v) > 0$ of equation (1.1) with Neumann boundary condition.

Then $v$ connects to other stationary solutions (denoted by $\Omega(v)$ again) as follows.

(i)   If $v \equiv$ constant, or if $i(v) = l(v)$, then

$$\Omega(v) = \{\bar{v}_k, \underset{\sim}{v}_k \mid 0 \leqslant k < i(v)\}.$$

(ii)   If $v(0) = \max v \neq \min v$ and $i(v) = l(v) + 1$, then

$$\Omega(v) = \Omega_1 \cup \Omega_2 \cup \Omega_3,$$

where

$$\Omega_1 = \{\bar{v}_k \mid 0 \leqslant k < i(v)\}$$

$$\Omega_2 = \{\underset{\sim}{v}_k \mid 0 \leqslant k < i(v) - 1\} \quad \text{and}$$

either  $\quad\Omega_3 = \{\underset{\sim}{v}_k \mid k = i(v) - 1\}$

or  $\quad\Omega_3$ consists of one or several stationary solutions $w$ with range$(w) \subset$ range$(v)$ and $i(w) < i(v)$.

(iii)  An analogous statement holds for $v(0) = \min(v) \neq \max(v)$

*Sketch of proof*   First we adapt the $y$-map to Neumann boundary conditions. In fact we only replace $u_x(t, 0)$ by $u(t, 0)$ in the definition (2.2b) of $\sigma_k$. As before, $y$ is continuous (cf. also [20]) and essential with this definition and §3 remains valid. In particular, for any $0 \leqslant k < i(v)$, $\sigma \in \{1, -1\}$ there exists an initial datum $u_0 \in W^u(v)$ such that $u(t, \cdot) \underset{t \to \infty}{\longrightarrow} w$ and

$$z(w - v) \equiv k,$$
$$\text{sign}(w(0) - v(0)) = \sigma.$$

Analogously to lemma 4.1, $v$ does not connect to $\bar{w}$ if there exists a stationary $w$ with $w(0)$ between $v(0)$ and $\bar{w}(0)$ satisfying

$$z(v - w) \leqslant z(\bar{w} - w).$$

This accounts for the minimality (maximality) in the definitions of $\tilde{v}_k$ $(\underline{v}_k)$. The appropriate modification of lemma 4.2 to Neumann conditions tells that

$$(6.5) \qquad z(v^1 - v^2) = \begin{cases} l(v^1) \geqslant 1 \\ 0 \end{cases} \quad \text{if} \quad \begin{array}{l} \text{range}(v^2) \subset \text{range}(v^1) \\ \text{range}(v^2) \cap \text{range}(v^1) = \varnothing \end{array}$$

for $v^1 \not\equiv v^2$ and is proved as before. Note that up to an interchange of $v^1$ and $v^2$ one of these alternatives has to occur. Relation (6.5) translates $z(v^1 - v^2)$ to the lap numbers $l(v^1)$ or $l(v^2)$, except when the ranges of $v^1$ and $v^2$ do not intersect. Taking $v^2 = v$, $v^1 = \tilde{v}$, with $z(\tilde{v} - v) = 0$ and *minimal* $\tilde{v}(0) > \text{range } v$, however, it is immediate that $\tilde{v} \equiv \text{const.}$, i.e $l(\tilde{v}) = 0$, from $(v, v_x)$ phase-plane analysis. Just observe that any two closed trajectories either are nested or their intersections with the $v$-axis are separated by a stationary (saddle type) solution (cf. Fig. 3).

By the final Sturm–Liouville ingredient that

$$i(v) \in \{l(v), l(v) + 1\}$$

for non-constant $v$, we observe that theorem 6.2 (i)–(iii) cover all possible cases. Replacing $z(v)$, $z(w)$ by $l(v)$, $l(w)$ the proof of theorem 6.1 is completed analogously to §5. $\qquad\qquad\qquad\qquad\qquad\qquad\qquad\qquad\qquad\qquad\qquad\square$

For mixed boundary conditions we are lacking a bridge to cross the gap separating $z(v - w)$ from $z(w)$ itself. Of course, corollary 3.2 still establishes connections. But we are unable to identify the target $w$ intrinsically, e.g. by looking at the global bifurcation diagram.

Note that our growth condition on $f$ can be weakened for both the Dirichlet and the Neumann case to include arbitrary linear growth at infinity. If we just assume

$$\varlimsup_{|s| \to \infty} f(s)/s < \infty,$$

some of our stationary solutions $v_k$ (bar or tilde) may not exist any more. In the bifurcation diagram of

$$0 = v_{xx} + \alpha^2 f(v),$$

$v_x(0)$ or $v(0)$ may escape to infinity at some finite value $\alpha$. This is best seen in case $f(v)$ is linear, $f(v) = f'_\infty \cdot v$, $f'_\infty > 0$, where vertical stationary bifurcation occurs for any $\alpha_j$ such that $-\alpha_j^2 \cdot f'_\infty$ is an eigenvalue of $v_{xx}$ For nonlinear $f$, this picture is somewhat perturbed but essentially correct. Still, by our $y$-map, there remains a trajectory $u(t, \cdot)$ in $W^u(v)$ with

$$z(u(t, \cdot) - v) \equiv k$$

$$\text{sign}(u_x(t, 0) - v_x(0)) = \sigma$$

for all $t \geqslant 0$ (here we consider the Dirichlet case, for simplicity) But $u(t, \cdot)$ cannot remain bounded, unless it converges to some equilibrium $v_k$. If $\alpha$ is such that $v_k$ does not exist, then the $y$-map leads to an unbounded solution $u(t, \cdot)$ — we have established a connection from $v$ to infinity. For more information on stationary solutions in such problems with $f(x, u) = f(u) + \sin x$ and 'jumping nonlinearities' $0 < f'(-\infty) < f'(+\infty)$ we recommend [21].

Penultimately, we propose a conjecture giving, for the Dirichlet case, a complete description of the set $\Omega(v)$ of all stationary solutions $w$ that $v$ connects to. Recall from theorem 1.1 that $\Omega(v)$ consisted of three disjoint subset

$$\Omega(v) = \Omega_1 \cup \Omega_2 \cup \Omega_3,$$

in case $v_x(0) \neq 0$, $i(v) = z(v) + 1$, and there was some unsettled alternative for $\Omega_3$. In order to describe $\Omega_3$ precisely we define:

$\underline{v}$ is the stationary solution $\hat{v}$ with $\hat{v} \equiv 0$ or $z(\hat{v}) = z(v)$, such that

$$\hat{v}_x(0) < |v_x(0)| \text{ is maximal}$$

$\bar{v}$ is the stationary solution $\hat{v}$ with $\hat{v} \equiv 0$ or $z(\hat{v}) = z(v)$, such that

$$\hat{v}_x(0) > -|v_x(0)| \text{ is minimal.}$$

These stationary solutions are uniquely defined. They are claimed to exist wherever they figure in the conjecture below.

**6.2  CONJECTURE**  Let the assumptions of theorem 1.1 be satisfied..

(i)  If $v_x(0) > 0$ and $i(v) = z(v) + 1$, then
$\Omega_3$ consists of $\underline{v}$ and all stationary solutions $w$ such that

$$\underline{v}_x(0) < |w_x(0)| < v_x(0) \text{ and } z(w) < z(v).$$

(ii)  If $v_x(0) < 0$ and $i(v) = z(v) + 1$, then similarly
$\Omega_3$ consists of $\bar{v}$ and all stationary solutions $w$ such that

$$v_x(0) < -|w_x(0)| < \bar{v}_x(0) \text{ and } z(w) < z(v).$$

The proof of this conjecture is work in progress. Restricting attention to $v_x(0) > 0$, the alternative for $\Omega_3$ in theorem 1.1 is resolved by the conjecture as follows. If $v_k = \underline{v}$ for $k = i(v) - 1 = z(v)$, then there are no stationary solutions $\hat{v}$ with

$$z(\hat{v}) = z(v), \quad |\hat{v}_x(0)| < v_x(0)$$

by definition of $\underline{v}_k$. Similarly, $\hat{v} \equiv 0$ is not a stationary solution. Using a more detailed analysis of the stationary bifurcation diagram than given in this chapter, this implies that the set of stationary $w$ considered in the conjecture is empty. If, on the other hand, $\underline{v}_k \neq \underline{v}$, then

$$\underline{v}_{k,x}(0) < \underline{v}_x(0) < v_x(0)$$

and lemma 4.1 implies that $v$ does not connect to $\underline{v}_k$. Then the second alternative of theorem 1.1 applies, and the set of $w$ considered there contains the set of $w$ from conjecture 6.2, because any $w$ from conjecture 6.2 satisfies

$$|w_x(0)| < v_x(0), \text{ and}$$
$$i(w) \leqslant z(w) + 1 < z(v) + 1 = i(v)$$

These remarks reconcile conjecture 6.2 and theorem 1.1.

Finally, nothing global is known for higher dimensions of the space variable $x$. We are lacking an analogue of the zero number $z(\phi)$. Within the class of rotationally symmetric solutions in a ball, the problem seems tractable. But introducing polar coordinates, this is essentially the one-dimentional case again. For Conley's index, using only the variational structure, these problems do not matter — at least in principle. To our knowledge, no attempt has been made so far to push this advantage to its limits. For special systems, i.e. higher dimensional $u$, but one space dimension for $x$, Smoller and Shi [28] and Conley and Smoller [11] obtained information on the flow, again by Conley's index. But in general, we suffer from the lack of a zero number.


## 7  APPENDIX ON $z$

We collect a few useful facts on the behaviour of the zero number $z$ along solutions of the equation

$$(7.1) \qquad u_t = u_{xx} + g(x, u), \quad x \in (0, 1)$$

$$(7.2) \qquad u(t, 0) = u(t, 1) = 0.$$

Throughout this section we assume $g \in \mathscr{G}_0$, i.e. $g(x, 0) = 0$ and $g \in C^2$ with the linear growth condition (1.9). We try to convince the reader of most of these facts by an example — rigorous proofs are given in the references as indicated.


### 7.1  LEMMA [5]  For $g \in \mathscr{G}_0$, the map

$$[0, \infty) \to N_0$$
$$t \to z(u(t, \cdot))$$

is non-increasing with $t$ along solutions $u(t, x)$ of (7.1), (7.2).


### 7.2  EXAMPLE  Consider a solution $u$ such that $x \to u(t_0, x)$ has only simple

zeros except for a double zero at $x_0$, i.e.

$$u(t,x) = \tfrac{1}{2}(x - x_0)^2$$

for $x$ near $x_0$. Then locally near $(t_0, x_0)$

$$u_t = u_{xx} + g(x, u) = 1 + g(x, u) > 0,$$

because $g(x_0, u(t_0, x_0)) = 0$. Therefore

$$z(u(t_0 + \varepsilon, \cdot)) = z(u(t_0, \cdot)) = z(u(t_0 - \varepsilon, \cdot)) - 2$$

for small $\varepsilon > 0$, and indeed $z$ decreases by 2 at $t = t_0$. A rigorous proof uses maximum principles like [5, 20, 23, 27, 32], instead. Historically, Nickel [23] first used such arguments and the essential idea of proof is due to him.

7.3  LEMMA [5, 20]  Assume $g \in \mathscr{G}_0$ and $z(u(0, \cdot)) < \infty$. Then the set of times $t > 0$, such that $x \to u(t, x)$ has only simple zeros, is open and dense in $\mathbb{R}^+$.

Indeed, openness is obvious by continuity of the flow $t \to u(t, \cdot) \in C^1$. Density again uses maximum principles The benevolent reader may reconsider example 7.2 to get convinced that multiple isolated zeros of $u(t_0, \cdot)$ become simple or disappear immediately for $t \ne t_0$.

7.4  LEMMA  Assume $g \in \mathscr{G}_0$ and $z(u(0, \cdot)) < \infty$. Define the dropping times $t_k$ of $z(u(t, \cdot))$ as in (2.2a), and assume $t_k < t_{k-1}$. Then

$$u_x(t, 0) \ne 0 \qquad \text{for all } t \in (t_k, t_{k-1})$$

and, in particular, sign $u_x(t, 0)$ does not depend on $t \in (t_k, t_{k-1})$.

For lack of reference, a rigorous proof of lemma 7.4 is given below. The idea, however, is again basic. As in example 7.2, the occurrence of a multiple isolated zero of $u(t_0, \cdot)$ at $x = 0$, $t_0 \in (t_k, t_{k-1})$ would force $z(u(t, \cdot))$ to decrease at $t = t_0$ in contradiction to $z(u(t, \cdot))$ being locally constant near $t$ by definition of $t_k$ and $t_{k-1}$.

*Proof of lemma 7.4:*  We already noted that

$$z(u(t, \cdot)) = k \qquad \textit{for all } t \in (t_k, t_{k-1}).$$

Now [6, lemma 2] implies that for any $t \in (t_k, t_{k-1})$ there exists an $\varepsilon > 0$ such that $u(x', t')$ has one definite sign independent of $x', t'$, provided that

$$0 < x' < \varepsilon$$
$$|t - t'| < \varepsilon.$$

This fact just uses the strong maximum principle [6, 27, 32]. But then

$$u_x(t', 0) \neq 0,$$

provided that $|\, t - t' \,| < \varepsilon$, again by the strong maximum principle [6, 27, 32]. This completes the proof.   □


## ACKNOWLEDGEMENT

The authors wish to express their gratitude to Renate Schaaf for her interest and her valuable suggestions, and to Peter Kunkel for his numerical advice.


## REFERENCES


[1]  S. Angenent, The Morse–Smale property for a semilinear parabolic equation, to appear in *J. Diff. Eq.* **62** (1986), 427–442.

[2]  F. V. Atkinson, *Discrete and Continuous Boundary Problems*. Academic Press, London, 1964.

[3]  G. Birkhoff and G.-C. Rota, *Ordinary Differential Equations*. Ginn & Co, Boston, 1959.

[4]  P. Brunovský and S.-N. Chow, Generic properties of stationary state solutions of reaction diffusion equations. *J. Diff. Eq.*, **53** (1984), 1–23.

[5]  P. Brunovský and B. Fiedler, Numbers of zeros on invariant manifolds in scalar reaction diffusion equations, *Nonl. Analysis* **10** (1986), 179–193.

[6]  P. Brunovský and B. Fiedler, Simplicity of zeros in scalar parabolic equations, *J. Diff. Eq.* **62** (1986), 237–241.

[7]  P. Brunovský and B Fiedler, Heteroclinic connections of stationary solutions of scalar reaction diffusion equations. *Proc. Int. Stefan Banach Inst.*, B. Bojarski (ed.), Warszawa, 1984, to appear.

[8]  N. Chafee and E. Infante, A bifurcation problem for a nonlinear parabolic equation. *J. Appl. Anal.*, **4** (1974), 17–37.

[9]  C. C. Conley, Isolated invariant sets and the Morse index. *Conf. Board Math. Sci.*, 38, AMS, Providence, 1978.

[10] C. C. Conley and J. Smoller, Topological techniques in reaction diffusion equations, in *Biologal Growth and Spread*, Proc., Heidelberg, 1979, Jäger, Rost, Tautu (eds), Springer Lect. Notes Biomath., 38, 473–83.

[11] C. C. Conley and J. Smoller, Birfucation and stability of stationary solutions of the Fitz–Hugh–Nagumo equations, preprint, 1984.

[12] A. Dold, *Lectures on Algebraic Topology*. Springer, Heidelberg, 1972.

[13] I. Gelfand, Some problems in the theory of quasilinear equations. *AMS Transl.*, ser. 2, **29** (1963), 295–381.

[14] J. K. Hale, Infinite dimensional dynamical systems, in *Geometric Dynamics*, Proc. Rio De Janeiro 1981, J. Palis Jr. (ed.), Springer Lect. Notes Math. 1007, 379–400.

[15] Ph. Hartman, *Ordinary Differential Equations* (2nd edn). Birkhäuser, Boston, 1982.

[16] D. Henry, *Geometric Theory of Semilinear Parabolic Equations*. Springer Lect. Notes Math., New York, 1981.

[17] D. Henry, Some infinite dimensional Morse–Smale systems defined by parabolic differential equations. *J. Diff. Eq.*, **59** (1985), 165–205.

[18] A. C. Hindmarsh, LSODE and LSODI, two new initial value ordinary differential equation solvers. *ACM-Signum Newsletter 15* (1980), 10–11.

[19] M. W. Hirsch, *Differential Topology*. Springer, New York, 1976.

[20] H. Matano, Nonincrease of the lap-number of a solution for a one-dimensional semi-linear parabolic equation. *J. Fac. Sci. Univ. Tokyo Sec.* IA, 29 (1982), 401–41.

[21] D. Hart, A. C. Lazer and P. J. McKenna, Multiple solutions of two point boundary value problems with jumping nonlinearities, to appear in *J. Diff. Eq.*

[22] J. Mallet-Paret, Morse decompositions and global continuation of periodic solutions for singularly perturbed delay equations, in *Systems of Nonlinear Partial Differential Equations*, J. M. Ball (ed.), D. Reidel, Dordrecht, 1983, 351–66.

[23] K. Nickel, Gestaltaussagen über Lösungen parabolischer Differentialgleichungen. *Crelles J. Reine Angew. Math.*, **211** (1962), 78–94.

[24] P. Poláčik, Generic bifurcation of stationary solutions of the Neumann problem for reaction diffusion equations. Thesis, Bratislava, 1984.

[25] M. Protter and H. Weinberger, *Maximum Principles in Differential Equations*, Prentice-Hall, 1967.

[26] R. Schaaf, Global solution branches via time-maps, preprint 1986.

[27] J Smoller, *Shock Waves and Reaction Diffusion Equations*. Springer, New York, 1983.

[28] J. A. Smoller and J. S. Shi, Analytical and topological methods for reaction diffusion equations, in *Modelling of Patterns in Space and Time*, Proc. Heidelberg, 1983, Jäger, Murray (eds), Springer Lect. Notes Biomath., 55, 350–63.

[29] J. Smoller, A. Tromba and A. Wasserman, Nondegenerate solutions of boundary value problems. *Nonl. Analysis*, **4** (1980), 207–215.

[30] J. Smoller and A. Wasserman, Generic bifurcations of steady state solutions. *J. Diff. Eq.*, **52** (1984), 432–8.

[31] J. Smoller and A. Wasserman, Global bifurcation of steady state solutions. *J. Diff. Eq.*, **39** (1981), 269–90.

[32] W. Walter, *Differential and Integral Inequalities*. Springer, Heidelberg, 1970.

*Dynamics Reported, Volume 1*
Edited by U. Kirchgraber and H. O. Walther
© 1988 John Wiley & Sons and B. G. Teubner

# 3

# Qualitative Theory of Nonlinear Resonance by Averaging and Dynamical Systems Methods

**James Murdock**
*Department of Mathematics, Iowa State University*

## CONTENTS



## INTRODUCTION

This paper is a self-contained exposition of the theory of averaging for periodic and quasiperiodic systems, with the emphasis being on the author's research (part of it joint work with Clark Robinson) on qualitative aspects of nonlinear resonance. Many topics in averaging theory are not covered, among them: averaging for systems more general than quasiperiodic; relations between averaging and multiple time-scale methods; Eckhaus's approach to averaging; combinations of averaging with matching of asymptotic expansions. The principal question which is addressed is: when does averaging (to first or higher order) lead to an accurate qualitative description of the solutions of the original (unaveraged) equation? By qualitative description we mean both locally (existence and stability of certain invariant sets such as periodic orbits, or almost invariant 'lingering' sets) and globally (connecting orbits between invariant sets, or claims that in certain large regions all orbits drift in a certain direction).

The mathematical background assumed is roughly: advanced calculus, including Fourier series; linear algebra through Jordan canonical forms; and a graduate level introduction to differential equations including Floquet theory. No knowledge of averaging is assumed, but some familiarity with standard problems in nonlinear oscillations or celestial mechanics will make the material more meaningful. (No concrete examples are introduced until the last section.) Occasionally a result from modern dynamical systems theory will be used without proof, but a complete explanation of the meaning of the result and the conditions under which it holds will be given.

No references are given in the main text. Each section concludes with a subsection called 'Notes' which contains historical comments, references, and remarks to the specialist which are not confined to the list of prerequisites mentioned above.

Particular thanks are due to Jürgen Moser for introducing me to the problems of nonlinear resonance; to Clark Robinson for his expertise in dynamical systems theory, which, combined with my knowledge of averaging, made many of these results possible; and to Urs Kirchgraber for asking me to write a unified exposition of this work incorporating improvements which have become clear since the original papers were published.


## 1  NONLINEAR OSCILLATORS

The simplest oscillator is $\ddot{x} + \omega^2 x = 0$, which takes the form $\dot{x} = -\omega y$, $\dot{y} = \omega x$ as a system of first-order equations in Cartesian coordinates, or $\dot{r} = 0$, $\dot{\theta} = \omega$ in polar coordinates. The Cartesian form generalizes to $\dot{u} = Au$, $u \in \mathbb{R}^n$, where $A$ is a constant real matrix with pure imaginary eigenvalues (hence $n$ is even) and a full set of $n$ eigenvectors; these conditions guarantee that all solutions are quasiperiodic. (Quasiperiodicity is defined in Section 2.) This is the general *linear oscillator*, and it reduces (by diagonalizing $A$) to $n/2$ uncoupled linear oscillators of the simple type with which this paragraph began. If the linear oscillator is perturbed to $\dot{u} = Au + \varepsilon g(u, t, \varepsilon)$, where $g$ is periodic or quasiperiodic in $t$, it is in general no longer possible to uncouple the equations. The substituion $u = e^{At}x$ reduces the problem to

$$(1.1) \qquad\qquad \dot{x} = \varepsilon f(x, t, \varepsilon), \qquad x \in \mathbb{R}^n$$

where $f(x, t, \varepsilon) = e^{-At} g(e^{At}x, t, \varepsilon)$ is quasiperiodic, having for frequencies those of $g$ together with those of $e^{At}$. System (1.1) is one of our two primary objects of study. In studying (1.1) for its own sake, there is no need to assume $n$ is even. The unperturbed ($\varepsilon = 0$) solutions of (1.1) are constant.

Our second primary object of study is obtained similarly by generalizing, and then perturbing, the polar coordinate representation $\dot{r} = 0$, $\dot{\theta} = \omega$ of the simplest oscillator. This leads naturally to

$$(1.2) \qquad \begin{aligned} \dot{r} &= \varepsilon f(r, \theta, \varepsilon) & r \in \mathbb{R}^n \quad \text{or} \quad (\mathbb{R}^+)^n \\ \dot{\theta} &= \Omega(r) + \varepsilon g(r, \theta, \varepsilon) & \theta \in \mathbb{T}^m. \end{aligned}$$

The statement '$\theta \in \mathbb{T}^m$' in (1.2) is equivalent to saying that $\theta \in \mathbb{R}^m$ and $f$ and $g$ are periodic with period $2\pi$ in each component of $\theta$. $\mathbb{T}^m$ denotes the $m$-dimensional torus, that is, $\mathbb{R}^m$ with points identified when their components differ by multiples of $2\pi$. (Thus, for instance, a solution of (1.2) will be called *periodic* if $r$ returns to its original value and $\theta$ returns to an equivalent, but possibly unequal, value after a certain interval of time.) The $n$ variables comprising $r$ are called *amplitudes* (or, in some cases, *action variables*, a term arising from Hamilton–Jacobi theory) and these are constant for the unperturbed equations ($\varepsilon = 0$). The $m$ variables in $\theta$ are called *angles*, and when $\varepsilon = 0$ each $\theta_i$ rotates at angular frequency $\Omega_i(r)$ which may depend on the amplitudes. Thus, in contrast to the first case, even the unperturbed form of (1.2) may be nonlinear, although in a very simple way. Since (1.2) is easily integrated when $\varepsilon = 0$ (although nonlinear), we refer to (1.2) as a *nearly integrable oscillator*; by contrast (1.1) is called *nearly linear*. (This use of the term 'integrable' is also motivated by its technical meaning in Hamilton–Jacobi theory, except that there one always has $m = n$.)

Systems (1.1) and (1.2) are not unrelated. In fact (1.1) can be regarded as a special case of (1.2) with constant $\Omega$, while (1.2) can 'locally' be put into the form (1.1). We now discuss each of these relations in greater detail.

By the definition of quasiperiodicity (Section 2) thee exist frequencies $\omega = (\omega_1, \ldots, \omega_m)$ and a function $F : \mathbb{R}^n \times \mathbb{T}^m \times \mathbb{R} \to \mathbb{R}^n$ such that $f(x, t, \varepsilon) = F(x, \omega t, \varepsilon)$. Now the system

$$\begin{aligned} \dot{x} &= \varepsilon F(x, \theta, \varepsilon) & x \in \mathbb{R}^n \\ \dot{\theta} &= \omega & \theta \in \mathbb{T}^m \end{aligned}$$

is of the form (1.2) with constant frequencies, and is equivalent to $\dot{x} = \varepsilon F(x, \theta_0 + \omega t, \varepsilon)$, which contains (1.1) by setting $\theta_0 = 0$. Thus (1.1) can be put in the form (1.2).

In the other direction, choose a fixed value $r_*$ of $r$ and introduce 'dilated amplitudes' $R$ and 'spinning angles' $\Theta$ by $r = r_* + \sqrt{(\varepsilon)}R$, $\theta = \Omega(r_*)t + \Theta$. With $\mu = \sqrt{\varepsilon}$ and $\omega_* = \Omega(r_*)$, equation (1.2) becomes

$$(1.3) \qquad \begin{aligned} \dot{R} &= \mu f(r_* + \mu R, \omega_* t + \Theta, \mu^2) \\ \dot{\Theta} &= \mu \Lambda(R, \mu) + \mu^2 g(r_* + \mu R, \omega_* t + \Theta, \mu^2) \end{aligned}$$

where

$$\Lambda(r, \mu) = \begin{cases} [\Omega(r_* + \mu R) - \omega_*]/\mu & \text{for } \mu \neq 0 \\ \Omega'(r_*)R & \text{for } \mu = 0. \end{cases}$$

System (1.3) is of the form (1.1) with $x = (R, \Theta)$ and with $\mu$ in place of $\varepsilon$. (1.3) is called a *localization* of (1.2) at $r = r_*$ for the following reason. The perturbation methods for (1.1) to be studied in later sections require that $\varepsilon$ be small and $x$ be confined to a compact set. For (1.3) this means, for instance, studying solutions in a region $\| R \| \leqslant C$; no restriction needs to be placed on $\Theta$ because of periodicity ($\mathbb{T}^m$ is compact). Now $\| R \| \leqslant C$ corresponds to $\| r - r_* \| \leqslant C\sqrt{\varepsilon}$, a 'band' around the torus $r = r_*$ which is narrow because $\sqrt{\varepsilon}$ is small. Thus, although (1.3) is formally equivalent to (1.2), it cannot be used to deduce results about (1.2) except for solutions having $r$ close to $r_*$. This explains our remark that (1.2) is locally equivalent to (1.1).

In subsequent sections we will study (1.1) by the method of averaging and construct approximate solutions. Then methods of dynamical systems theory will be used to deduce qualitative features of the exact solutions from a knowledge of the approximations. These results are not general but will require additional hypotheses: sometimes that $f$ be periodic (and not merely quasiperiodic), or quasiperiodic of restricted types; sometimes that the system be non-conservative (non-Hamiltonian) in specific ways. Finally the results for (1.1) will be applied to (1.2) via localizations such as (1.3). In order to do this it is necessary to decide which localizations are qualitatively significant, which requires making a distinction between 'passive' and 'active' resonances and showing that under certain conditions (again of a non-conservative character) only finitely many resonances are active. We conclude with the example to which these methods have been applied in greatest detail, a model of spin/orbit resonance (or more or less equivalently, a perturbation of a strongly nonlinear spring—a form of Duffing equation).


## Notes

A general introduction to nonlinear oscillations, still of value although dated, is [35]. The concept of action/angle variables for integrable Hamiltonian systems, which is a primary motivation for the system (1.2), is explained in [2]. These are two personal favorites out of a large number of books on nonlinear oscillations, Hamiltonian systems, celestial mechanics, and perturbation theory which provide examples and background for this section. Unfortunately almost every book in the field, even some of those containing the best insights, also have serious errors, and it would almost be irresponsible to give a general bibliography without going into detail about the strong and weak points of each book. The best advice is to read critically, to accept only what you can prove yourself, and to be a little unsure of even what you think you can prove. Authors of world-renowned stature in this field have made amazing blunders, and several of my own have also reached print. (Of course the present version is perfect in every respect, so feel free to throw caution to the winds and read on.)

## 2  QUASIPERIODIC FUNCTIONS

A function $f: \mathbb{R} \to \mathbb{R}^n$ of class $C^\infty$ is called *quasiperiodic* if there is an integer $m$, a $C^\infty$ function $F: \mathbb{T}^m \to \mathbb{R}^n$, and a frequency vector $\omega = \text{col}(\omega_1, \ldots, \omega_m)$ such that $f(t) = F(\omega t) = F(\omega_1 t, \ldots, \omega_m t)$. The letters 'col' indicate that $\omega$ is regarded as a column vector. Note that $F$ itself is assumed to be of class $C^\infty$; this does not follow from the smoothness of $f$. (To avoid technicalities we assume once and for all that all functions are $C^\infty$.)

If $f$ is quasiperiodic, the triple $(m, \omega, F)$ will be called a *representation* of $f$. There are many representations for any $f$. As a trivial example, $m$ can be increased by adding zero components to $\omega$. It is more interesting to ask for the smallest $m$ with which a given $f$ can be represented. For example, if $f(t) = F(\omega_1 t, \omega_2 t)$ with $\omega_1 = \sqrt{2}$ and $\omega_2 = -2\sqrt{2}$ is a representation with $m = 2$, the same $f(t)$ can be represented with $m = 1$ by writing $f(t) = G(\lambda t)$, where $G(\phi) = F(\phi, -2\phi)$ and $\lambda = \sqrt{2}$. This reduction is possible (regardless of the function $F$) because $\omega_1$ and $\omega_2$ satisfy the commensurability relation $2\omega_1 + \omega_2 = 0$, a linear relation with integer coefficients. (Warning: it is not permissible to write $f(t) = G(\omega_2 t)$ with $G(\phi) = F(-\frac{1}{2}\phi, \phi)$ since this $G$ is not $2\pi$-periodic, i.e., is not a map of $\mathbb{T}^1 \to \mathbb{R}^n$ and so does not qualify as a representation.)

To explore the reduction of $m$ in general, it is useful to introduce the additive group $\mathbb{Z}^m$ of integer row vectors $\nu = (\nu_1, \ldots, \nu_m)$. For any frequency vector $\omega \in \mathbb{R}^m$, the *resonance group* or *commensurability group* of $\omega$ is the subgroup $\omega^\perp = \{\nu \in \mathbb{Z}^m : \nu\omega = 0\}$; here $\nu\omega = \nu_1\omega_1 + \cdots + \nu_m\omega_m$, since $\nu$ is a row and $\omega$ is a column. $\mathbb{Z}^m$ is similar to a vector space in that every subgroup of $\mathbb{Z}^m$ has a basis, in terms of which each element can be expressed uniquely as a linear combination with integer coefficients. Given an $\omega \in \mathbb{R}^m$ with $m$ not too large, a basis for $\omega^\perp$ can usually be found by inspection. The rank or dimension of $\omega^\perp$ (the number of elements in a basis) is well defined (independent of the basis chosen) and will be called $s$. Given a quasiperiodic function $f$ with representation $(m, \omega, F)$, and given a basis for $\omega^\perp$, we will show how to construct a representation $(r, \lambda, G)$ with $r = m - s$ for which $\lambda^\perp = \{(0, \ldots, 0)\}$ is the zero subgroup of $\mathbb{Z}^r$; such a set of frequencies is called *rationally independent*. No further reduction is possible on the basis of the frequency structure alone. (For a given $f$, $G$ may turn out to be independent of some of the variables in $\mathbb{T}^r$, making further reduction possible.)

The heart of the reduction procedure is to construct a basis $v^1, \ldots, v^m$ of $\mathbb{Z}^m$ (each $v^i$ being an integer vector $v^i = (v^i_1, \ldots, v^i_m)$) such that the last $s$ of the basis vectors, $v^{r+1}, \ldots, v^m$, form a basis for the subgroup $\omega^\perp$. If $\mathbb{Z}^m$ were a vector space, such a basis would exist automatically: any basis for $\omega^\perp$ could be completed to a basis for the full space by adjoining $m - s$ linearly independent vectors. Such a construction does not work in $\mathbb{Z}^m$, and in fact the desired basis exists only because $\omega^\perp$ is a special type of subgroup of $\mathbb{Z}^m$ called a *pure subgroup*. But before turning to these technicalities, we will show how the existence of such a basis solves the reduction problem.

Let $A$ be the $m \times m$ integer matrix whose rows are the basis elements $v^1, \ldots, v^m$. The matrix $A$ is invertible and its inverse, whose rows express the standard basis $(1, 0, \ldots, 0), \ldots, (0, \ldots, 0, 1)$ of $\mathbb{Z}^m$ in terms of $v^1, \ldots, v^m$, will have integer entries; such a matrix (an integer matrix with integer inverse) is called *unimodular*. Furthermore $A\omega = (\lambda_1, \ldots, \lambda_r, 0, \ldots, 0) = (\lambda, 0)$ with $\lambda^\perp = \{0\}$: the last $s$ components of $A\omega$ will be zero since the last $s$ rows of $A$ belong to $\omega^\perp$, while the first $r$ components of $A\omega$ will be rationally independent (will have a trivial resonance group) because any linear relation with integer coefficients holding among $\lambda_1, \ldots, \lambda_r$ would imply a similar commensurability relation among $\omega_1, \ldots, \omega_m$ not belonging to $\omega^\perp$. The matrix $A$ defines an invertible map $\mathbb{T}^m \to \mathbb{T}^m$ by $\phi = A\theta, \theta = A^{-1}\phi$: since $A$ and $A^{-1}$ are integer matrices, if the components of $\theta$ are augmented by multiples of $2\pi$ the same happens to $\phi$, and vice versa. Given a representation $(m, \omega, F)$ of a quasiperiodic function $f(t)$, let $F_A : \mathbb{T}^m \to \mathbb{R}^n$ and $G : \mathbb{T}^r \to \mathbb{R}^n$ be defined by $F_A(\phi) = F(A^{-1}\phi)$ and $G(\phi_1, \ldots, \phi_r) = F_A(\phi_1, \ldots, \phi_r, 0, \ldots, 0)$. Then $G(\lambda t) = F_A(\lambda t, 0) = F(\omega t) = f(t)$ and so $(r, \lambda, G)$ is the desired minimal representation of $f$.

It remains only to construct the basis $v^1, \ldots, v^m$ (equivalently, the matrix $A$) given a basis for $\omega^\perp$. We begin with the $s \times m$ integer matrix $B$ whose rows are the given basis elements of $\omega^\perp$. Integer row operations on $B$ (interchanging rows, adding to one row an integer multiple of another row, or multiplying a row by $-1$; only $-1$ is allowed because other integers do not have multiplicative inverses in $\mathbb{Z}$) give other matrices whose rows form alternative bases for $\omega^\perp$. Column operations on $B$ correspond to expressing the same basis for $\omega^\perp$ in terms of a nonstandard basis for $\mathbb{Z}^m$. To see this, let $\alpha^1, \ldots, \alpha^m$ be the standard basis for $\mathbb{Z}^m$, that is $\alpha^1 = (1, 0, \ldots, 0)$ etc. Then the $i$th row $[b^i_1 \ldots b^i_m]$ of $B$ denotes the element $b^i_1\alpha^1 + \cdots + b^i_m\alpha^m$ of $\mathbb{Z}^m$. If a new basis $\beta^1, \ldots, \beta^m$ for $\mathbb{Z}^m$ is defined by $\alpha^j = \beta^j$ for $j \neq k$, $\alpha^k = \beta^k + t\beta^l$, the same element of $\mathbb{Z}^m$ may be expressed $b^i_1\beta^1 + \cdots + (b^i_l + tb^i_k)\beta^l + \cdots + b^i_m\beta^m$. So if the $k$th column of $B$ is multiplied by $t$ and added to the $l$th column, the resulting matrix expresses the same basis for $\omega^\perp$ as before, but in terms of the new basis $\beta^1, \ldots, \beta^m$ for $\mathbb{Z}^m$. The other column operations correspond to interchanging two basis elements or reversing one of them. Now we appeal to the fact that any integer matrix such as $B$ may be reduced by integer row and column operations to the *Smith normal form*

$$\begin{bmatrix} \delta_1 & 0 & & \cdots & & \cdots & \\ 0 & \delta_2 & 0 & \cdots & & \cdots & \\ & & \cdot & & & & \\ & & & \cdot & & & \\ 0 & \cdots & 0 & \delta_s & 0 & \cdots & 0 \end{bmatrix}.$$

(This is not obvious over the integers as it is over a field, but depends on facts about greatest common divisors.) None of the $\delta_i$ will be zero, since the rows of $B$ are linearly independent (being basis elements for $\omega^\perp$). We claim that in fact every $\delta_i = \pm 1$. This results from the fact that $\omega^\perp$ is a *pure subgroup*: if the components of any $\nu \in \omega^\perp$ have a common factor $t$, so that $\nu/t$ is an integer vector, then $\nu/t \in \omega^\perp$. (This is easy to see from the definition of $\omega^\perp$.) If $\delta_1$ (say) is not equal to 1 or $-1$, then $(\delta_1, 0, ..., 0)$ is divisible by $\delta_1$ and $(1, 0, ..., 0) \in \omega^\perp$; but then $(\delta_1, 0, ..., 0)$ could not be a basis element for $\omega^\perp$. So each $\delta_i$ is $\pm 1$, and all of them can be made $+ 1$ (by a few more row operations). Only the column operations needed in reducing $B$ to this diagonal form are important; these correspond to successive changes in the basis for $\mathbb{Z}^m$, beginning with the standard basis. Let the final basis be $\gamma^1, ..., \gamma^m$, in terms of which $B$ is diagonal with ones. Then $\gamma^1, ..., \gamma^s$ is a basis for $\omega^\perp$ which is part of a basis for $\mathbb{Z}^m$. Moving these elements to the end gives the desired basis $\nu^1, ..., \nu^m = \gamma^{s+1}, ..., \gamma^m, \gamma^1, ..., \gamma^s$. Observe that the sequence of row and column operations leading to diagonal form is not unique, and neither is the basis $\nu^1, ..., \nu^m$.

For future reference we summarize the previous discussion in a theorem.

2.1 THEOREM   Given $\omega \in \mathbb{R}^m$ there exists an integer $r \leqslant m$ and an $m \times m$ unimodular matrix $A$ such that $A\omega = \mathrm{col}(\lambda_1, ..., \lambda_r, 0, ..., 0)$ with $\lambda_1, ..., \lambda_r$ being rationally independent ($\lambda^\perp = 0$).

Next we turn to Fourier series and averages. Since $F: \mathbb{T}^m \to \mathbb{R}^n$ is $C^\infty$, its Fourier series $F(\theta) = \Sigma a_\nu e^{i\nu\theta} = \Sigma a_{\nu_1...\nu_m} e^{i(\nu_1\theta_1 + \cdots + \nu_m\theta_m)}$ converges uniformly and absolutely. The summation is over all $\nu \in \mathbb{Z}^m$ and the coefficients $a_\nu$ are $n$-dimensional column vectors with complex entries satisfying $a_{-\nu} = \bar{a}_\nu$. The constant term $a_0 = a_{00...0}$ is a real $n$-vector equal to the average

$$(1/2\pi)^m \int_0^{2\pi} \cdots \int_0^{2\pi} F(\theta) \, d\theta_1 ... \, d\theta_m$$

of $F$ over $\mathbb{T}^m$. Passing now to $f(t) = F(\omega t) = \Sigma a_\nu e^{i\nu\omega t} = \Sigma a_\nu e^{i(\nu_1\omega_1 + \cdots + \nu_m\omega_m)t}$ we have a nicely convergent series which could be misleading: the 'constant term' is not $a_0$ only (if $\omega$ satisfies commensurabilities) but $\Sigma_{\nu \in \omega^\perp} a_\nu$, since $\nu\omega = 0$ for $\nu \in \omega^\perp$. We will see that this equals the time average of $f$. In fact,

2.2 THEOREM   For any quasiperiodic function $f(t) = F(\omega t)$ the following are equal:

(i) $\displaystyle\sum_{\nu \in \omega} a_\nu;$

(ii) the time average $\displaystyle\lim_{T \to \infty} \frac{1}{T} \int_0^T f(t)\, dt;$

(iii) the 'space average' of $F_A$ over the $r$-dimensional torus consisting of $\phi \in \mathbb{T}^m$ with $\phi_{r+1} = \cdots = \phi_m = 0$ (in other words, the average of $G$ over $\mathbb{T}^r$);

(iv) a 'space average' of $F$ over the $r$-torus which is the closure of the curve $\theta = \omega t$ in $\mathbb{T}^m$, the averages being taken with respect to a suitable measure on the $r$-torus.

The equivalence of (i) and (iii) is the most immediate. First of all $F_A(\phi) = F(A^{-1}\phi) = \sum_\nu a_\nu e^{i\nu A^{-1}\phi} = \sum_\mu a_{\mu A} e^{i\mu\phi}$. (Note that as $\mu$ ranges over all integer row vectors, so does $\mu A$. So $\nu = \mu A$ is a legitimate substitution.) Therefore

$$F_A(\phi_1, \ldots, \phi_r, \ldots, 0) = \sum_{\mu_1, \ldots, \mu_r} \left( \sum_{\mu_{r+1}, \ldots, \mu_m} a_{\mu A} \right) e^{i(\mu_1\phi_1 + \cdots + \mu_r\phi_r)},$$

a Fourier series in $r$ variables whose average (its constant term) is

$$\sum_{\mu_{r+1}, \ldots, \mu_m} a_{(0, \ldots, 0, \mu_{r+1}, \ldots, \mu_m)A} = \sum_{\nu \in \omega^-} a_\nu.$$

Next, and most importantly, we prove the equivalence of (ii) and (iii), at the same time proving the *existence* of (ii), which is not exactly obvious. Begin with $f(t) = G(\lambda t)$, $G(\phi) = \sum_\nu b_\nu e^{i\nu\phi}$, $\nu \in \mathbb{Z}^r$, $\lambda$ rationally independent. For any integer $N$, define $f_N(t) = \sum_{\|\nu\| \leqslant N} b_\nu e^{i\nu\lambda t}$, where $\|\nu\| = |\nu_1| + \cdots + |\nu_r|$. Since $f_N$ is a finite sum, it can be seen by calculation that

$$\lim_{T \to \infty} \frac{1}{T} \int_0^T f_N(t)\, dt$$

exists and equals $b_0$, independently of $N$. On the other hand

$$\left| \frac{1}{T} \int_0^T (f(t) - f_N(t))\, dt \right| \leqslant \sup_{0 \leqslant t \leqslant T} |f(t) - f_N(t)| \leqslant \sup_{0 \leqslant t < \infty} |f(t) - f_N(t)|,$$

and since $f_N$ converges to $f$ uniformly on the whole real line, this approaches zero as $N$ increases. Let $\varepsilon > 0$ be given. First choose $N$ so that $\sup_{0 \leqslant t < \infty} |f(t) - f_N(t)| < \varepsilon$. Then

$$\left| \frac{1}{T} \int_0^T (f(t) - f_N(t))\, dt \right| < \varepsilon \text{ for all } T > 0.$$

Next choose $T_0$ so that (for the already chosen value of $N$)

$$\left| \frac{1}{T} \int_0^T f_N(t)\, dt - b_0 \right| < \varepsilon \text{ for all } T > T_0.$$

It follows from the triangle inequality that

$$\left| \frac{1}{T} \int_0^T f(t)\, dt - b_0 \right| < 2\varepsilon \text{ for all } T > T_0.$$

Since $\varepsilon$ is arbitrary, this establishes that

$$\lim_{T \to \infty} \frac{1}{T} \int_0^T f(t)\, dt$$

exists and equals $b_0$, the average value of $G$. The equivalence of (iii) and (iv) depends on the fact that the torus in (iii) is the closure of the curve $\phi = (\lambda, 0)t$ and that the map $\theta = A^{-1}\phi$ carries this curve and torus differentiably onto the curve and torus in (iv). (To see that $\phi = \lambda t$ is dense in the torus $\phi_{r+1} = \cdots = \phi_m = 0$, suppose there is an open set $U$ in this torus that is not intersected by the curve. Then there is a function $F$ such that the restriction of $F_A$ to this torus is zero outside $U$ and one on an open set contained in $U$. But then $F$ has zero time average and non-zero space average, violating the equality of (ii) and (iii).) The average in (iii) is taken with respect to the 'ordinary' $r$-measure $d\phi_1 \ldots d\phi_r$. But the map $\theta = A^{-1}\phi$ is not an isometry, so the correct measure on the $\theta$ torus is an induced one: first the 'flat torus' Riemannian metric $ds^2 = d\phi_1^2 + \cdots + d\phi_m^2$ is pulled back (via $\phi = A\theta$) to the $\theta$ torus and then restricted to the embedded $\mathbb{T}^r$, where it gives rise to an $r$-measure.

It is worthwhile to note how we did *not* prove Theorem 1.2. We could have integrated $f(t) = \Sigma b_\nu e^{i\nu\lambda t}$ to obtain

$$\int_0^T f(t)\, dt = b_0 T + \sum_{\nu \neq 0} b_\nu \frac{e^{i\nu\lambda T} - 1}{i\nu\lambda}.$$

This is valid, and the series converges, because definite integrals of uniformly convergent series may be taken termwise. (Note that the integration depends on the rational independence of $\lambda$; if some $\nu\lambda = 0$, the corresponding term would integrate to $b_\nu T$.) But after dividing by $T$, there is nothing obvious which justifies taking the limit termwise as $T \to \infty$. So we did the same calculation for the finite sum $f_N(t)$ instead.

Often in dealing with a quasiperiodic function $f(t)$ we assume that the original representation $(m, \omega, F)$ is already reduced (so that $r = m, \lambda = \omega, G = F$). Therefore we repeat the definite integration formula in the form

$$(2.1) \qquad \int_0^T f(t)\, dt = a_0 T + \sum_{\nu \neq 0} a_\nu \frac{e^{i\nu\omega t} - 1}{i\nu\omega} \text{ if } \omega^\perp = 0.$$

Although this termwise definite integral is valid, the following indefinite integral is usually not:

$$\int f(t)\, dt = a_0 t + \sum_{\nu \neq 0} \frac{a_\nu}{i\nu\omega}\, e^{i\nu\omega t}.$$

Since the exponential function is never zero, there is no lower limit of integration for which the definite integral coincides with this indefinite integral, and one cannot use the usual theorem to justify the termwise integration. Neither is it possible in general to prove convergence by estimating the coefficients $a_\nu/i\nu\omega$, since the denominators $\nu\omega$, while not equalling zero (because of the assumed rational independence), can become arbitrarily close to zero, making the corresponding coefficients $a_\nu/i\nu\omega$ large unless the $a_\nu$ are sufficiently small. (This constitutes what might be called the 'easy' small divisor problem, in contrast to the much harder one which arises out of this one in the Kolmogoroff–Arnol'd–Moser theory.) But there are certain important cases in which the termwise indefinite integral is valid:

2.3 THEOREM    If $f(t)$ is quasiperiodic with rationally independent frequencies $\omega_1, \ldots, \omega_m$ and Fourier series $f(t) = \Sigma\, a_\nu e^{i\nu\omega t}$, the formula

$$\int f(t)\, dt = a_0 t + \sum_{\nu \neq 0} \frac{a_\nu}{i\nu\omega}\, e^{i\nu\omega t}$$

is valid in the following special cases:

(1) If $f$ is periodic ($m = 1$).
(2) If $f$ has finitely many harmonics (only finitely many $a_\nu \neq 0$).
(3) If only one of the frequencies of $f$ has infinitely many harmonics. (If this frequency is $\omega_m$, this means that the set of $\nu$ for which $a_\nu \neq 0$ contains only finitely many values of $\nu_1, \ldots, \nu_{m-1}$.)
(4) If $f$ is real analytic and $\omega_1, \ldots, \omega_m$ are 'badly incommensurable'.

In all cases, once convergence is proved, equality with $\int f(t)\, dt$ follows since the differentiated series is uniformly convergent. In case 1 there are no small divisors and the inequality $\| a_\nu/i\nu\omega \| \leqslant \| a_\nu \|/|\omega|$ establishes convergence by the comparison test. In case 2 the series is finite and convergence questions do not arise. In case 3, the summation can be written

$$\sum_{\nu_1 = -N}^{N} \sum_{\nu_2 = -N}^{N} \cdots \sum_{\nu_{m-1} = -N}^{N} \sum_{\nu_m = -\infty}^{\infty} \frac{a_\nu}{i\nu\omega}\, e^{i\nu\omega t}$$

for some $N$. For each value of $\nu_1, \ldots, \nu_{m-1}$ the summation over $\nu_m$ has a single smallest divisor, occurring when $\nu_m\omega_m$ is closest to $-(\nu_1\omega_1 + \cdots + \nu_{m-1}\omega_{m-1})$, and so converges by an estimate similar to case 1, the remaining summation is finite. Case 4 is technical and is mentioned here for completeness although it will not be used. If $f$ is real analytic the coefficients $a_\nu$ decay exponentially with $\| \nu \| = | \nu_1 | + \cdots + | \nu_m |$. 'Badly incommensurable' means that there is a certain lower bound on $| \nu\omega |$ in terms of $\| \nu \|$ which together with exponential decay implies convergence. Most $\omega$ (in the sense of measure theory) are badly incommensurable, and yet the set of $\omega$ which are incommensurable but not badly incommensurable is dense.

A function $f$ of $t$ and other variables is called quasiperiodic in $t$ if there is a smooth function $F$ of $\theta \in \mathbb{T}^m$ and the other variables such that $f = F$ when $\theta = \omega t$; the values of $\omega_i, i = 1, \ldots, m$ are not allowed to depend on the other variables, but must be constant. Everything we have said holds in this larger context, with the Fourier coefficients $a_\nu$ now being smooth functions of the additional variables.

Finally, a very elementary theorem is needed occasionally to guarantee that a certain construction produces a $C^\infty$ function. If $f(\varepsilon) = \sum_{n=0}^\infty a_n \varepsilon^n$ is analytic and $f(0) = 0$, then $a_0 = 0$ and so $f(\varepsilon) = \varepsilon g(\varepsilon)$ with $g(\varepsilon) = \sum_{n=1}^\infty a_n \varepsilon^{n-1}$. If $f$ is $C^\infty$, its Taylor series is not necessarily convergent (but is of course asymptotic), so this series argument does not suffice to prove the existence of $g(\varepsilon)$. However, if we define $g(\varepsilon) = \int_0^1 f'(\varepsilon\eta)\, d\eta$ then $g(\varepsilon)$ is clearly $C^\infty$, and by the fundamental theorem of calculus $g(\varepsilon) = f(\varepsilon)/\varepsilon$ for $\varepsilon \neq 0$ (since $f(0) = 0$ by hypothesis), while $g(0) = f'(0)$. The equation $f(\varepsilon) = \varepsilon g(\varepsilon)$ holds for all $\varepsilon$.

**2.4 THEOREM** Suppose $f(x, t, \varepsilon)$ is defined and of class $C^\infty$ with values in $\mathbb{R}^k$ for all $x \in \mathbb{R}^n$, $t \in \mathbb{R}$, $\varepsilon \in \mathbb{R}$, and $f(x, t, 0) \equiv 0$. Then the function defined by

$$g(x, t, \varepsilon) = \begin{cases} f(x, t, \varepsilon)/\varepsilon & \text{for } \varepsilon \neq 0 \\ f_\varepsilon(x, t, 0) & \text{for } \varepsilon = 0 \end{cases}$$

is of class $C^\infty$. If $f$ is periodic or quasiperiodic in $t$, $g$ is periodic or quasiperiodic with the same frequencies.

The proof is simply to observe, as before, that $g(x, t, \varepsilon) = \int_0^1 f(x, t, \varepsilon\eta)\, d\eta$. In the quasiperiodic case, let $f(x, t, \varepsilon) = F(x, \omega t, \varepsilon)$, $G(x, \theta, \varepsilon) = \int_0^1 F(x, \theta, \varepsilon\eta)\, d\eta$, and $g(x, t, \varepsilon) = G(x, \omega t, \varepsilon)$.

An application of this theorem has already appeared in equation (1.3), with $\mu$ for $\varepsilon$, $R$ for $x$, and no $t$. The conclusion is that $\Lambda(R, \mu)$ is of class $C^\infty$.

## Notes

The theory of quasiperiodic functions was introduced by Bohl and later generalized by Bohr [4] into the theory of almost periodic functions. Among almost periodic functions, quasiperiodic functions are those with a finite basis for the module of frequencies. The present exposition is based upon part of my paper [18], but contains no fundamental ideas which are original, except perhaps for the observation that resonance groups are pure subgroups and that this is the basic reason for the existence of the matrix $A$. The theory of pure subgroups is given in [13]. Theorem 2.2 is essentially the Kronecker–Weyl theorem, proved with considerable discussion in [34]. Theorem 2.3 has a long history. Parts 1 and 2 are very old, part 3 is an easy observation of my own, and part 4 is the elementary starting point for the difficult work on small divisor series done by Siegel, Kolmogoroff, Arnol'd, and Moser, expounded in [17]. (This theory deals with power series whose coefficients are series of the form dealt with in Theorem 2.3 part 4.) Theorem 2.4 I learned from [14] p. 15. It has the following consequence: a smooth function $f(x, \varepsilon)$ satisfying an asymptotic estimate $f(x, \varepsilon) = 0(\varepsilon^k)$ uniformly for $x$ in a fixed set $K$ can be written $f(x, \varepsilon) = \varepsilon^k g(x, \varepsilon)$, where $g$ is smooth and hence bounded on $K$. This principle must be used with caution, because frequently the symbol $0(\varepsilon^k)$ denotes a bound on an $\varepsilon$-dependent domain.


## 3   AVERAGING TRANSFORMATIONS

An *averaging transformation* for (1.1)—which may or may not exist—is a near-identity transformation carrying (1.1) into an equation which is autonomous up to some order $k$ in $\varepsilon$; $k$ is called the *order* of the averaging transformation. In detail: the *original system* is

$$(3.1) \qquad\qquad \dot{x} = \varepsilon f(x, t, \varepsilon), \qquad x \in \mathbb{R}^n;$$

the *averaging transformation* is

$$(3.2) \qquad\qquad x = y + \varepsilon u(y, t, \varepsilon);$$

and the *averaged system* (or *full averaged system*) is

$$(3.3) \qquad\qquad \dot{y} = \varepsilon g(y, \varepsilon) + \varepsilon^{k+1} \hat{g}(y, t, \varepsilon).$$

The *truncated averaged system* is

$$(3.4) \qquad\qquad \dot{z} = \varepsilon g(z, \varepsilon).$$

It is assumed that $f$ is periodic or quasiperiodic in $t$, and $u$ is required to be periodic or quasiperiodic with the same frequencies; the same will then be true of $\hat{g}$. The present section focuses on the existence of an averaging transformation for various classes of functions $f$. The ultimate goal is to use (3.4) to obtain approximate solutions to (3.3) and hence (by using (3.2)) to (3.1). The question of the quantitative and qualitative accuracy of such approximations is addressed in later chapters.

    The term 'averaging' refers to the fact (which will emerge in due course) that $g(y, 0)$, the term of order $\varepsilon$ in the averaged system, is the average over time of $f(y, t, 0)$, the term of order $\varepsilon$ in the original system (with $x$ replaced by $y$).

In other words, denoting this average by $\bar{f}$, the first order ($k = 1$) truncated averaged system will always turn out to be $\dot{z} = \varepsilon\bar{f}(z)$. There are some heuristic reasons for expecting that this equation may give useful approximate solutions. Namely, for small $\varepsilon$, $x$ varies slowly in (3.1) while $f$ oscillates rapidly by comparison; so $x$ may be expected to respond primarily to the average value of $f$. But this argument is of no value in establishing the degree of accuracy of the approximations or in finding approximations of higher order ($k > 1$).

It is best to approach the question of averaging transformations from a reversed perspective. Namely, we consider the effect of an arbitrary transformation of the form (3.2) on equation (3.1) and determine the transformed equation, which will look like (3.3) except that $g$ will depend on $t$. Then it will be possible to write down the condition that $g$ be independent of $t$, and try to solve for $u$.

If (3.2) is to be a valid coordinate transformation it must have a smooth inverse. Our first task is to show that this is true, for suitable domains and small $\varepsilon$. Let $u(y, t, \varepsilon)$ be an arbitrary $C^\infty$ function, quasiperiodic in $t$, so that $u(y, t, \varepsilon) = U(y, \omega_1 t, \ldots, \omega_m t, \varepsilon)$ for some $U: \mathbb{R}^n \times \mathbb{T}^m \times \mathbb{R} \to \mathbb{R}^n$. Let $A$ be any set in $R^n$ (regarded as a set of values of $y$), and let $\mathbf{A}(A, \varepsilon_0) = \mathbf{A}(\varepsilon_0)$ denote $A \times R \times [0, \varepsilon_0]$. Then (3.2) defines a map carrying points $(y, t, \varepsilon) \in \mathbb{R}^n \times \mathbb{R} \times \mathbb{R}$ into points $(x, t, \varepsilon)$, with the values of $t$ and $\varepsilon$ unchanged. As $(y, t, \varepsilon)$ varies over $\mathbf{A}(\varepsilon_0)$, $(x, t, \varepsilon)$ will vary over a set $\mathbf{B}(A, \varepsilon_0) = \mathbf{B}(\varepsilon_0)$ which need not be a Cartesian product: each 'slice' of $\mathbf{B}(\varepsilon_0)$ for fixed $t$ and $\varepsilon$ will be the image of $A$ under (3.2) with the given value of $t$ and $\varepsilon$. Thus (3.2) defines a map $\mathbf{A}(A, \varepsilon_0) \to \mathbf{B}(A, \varepsilon_0)$ which we claim has a smooth inverse for sufficienty small $\varepsilon_0$, if $A$ is bounded. The proof requires that we define sets of $(y, \theta, \varepsilon)$ and $(x, \theta, \varepsilon)$ corresponding to the sets $\mathbf{A}$ and $\mathbf{B}$ of $(y, t, \varepsilon)$ and $(x, t, \varepsilon)$; namely define $\mathbf{C}(A, \varepsilon_0) = A \times \mathbb{T}^m \times [0, \varepsilon_0]$ and let $\mathbf{D}(A, \varepsilon_0) \subset \mathbb{R}^n \times \mathbb{T}^m \times [0, \varepsilon_0]$ be the image of $\mathbf{C}(A, \varepsilon_0)$ under $x = y + \varepsilon U(y, \theta, \varepsilon)$, with $\theta$ and $\varepsilon$ preserved.

3.1 THEOREM   For any bounded open set $A \subset \mathbb{R}^n$ there exists an $\varepsilon_0$ such that the map $\mathbf{A}(A, \varepsilon_0) \to \mathbf{B}(A, \varepsilon_0)$ defined by (3.2) is one to one and has a $C^\infty$ inverse expressible in the form $y = x + \varepsilon v(x, t, \varepsilon)$.

The proof is rather technical. First choose any $\varepsilon_1 > 0$, let $\|\ \ \|$ be any matrix norm, and let $M$ be the maximum of $\| U_y(y, \theta, \varepsilon) \|$ on $\mathbf{C}(\bar{A}, \varepsilon_1)$ a compact set. ($\bar{A}$ is the closure of $A$.) Choose $\varepsilon_2 \leqslant \min(\varepsilon_1, 1/M)$. We claim first that the map $\mathbf{C}(\bar{A}, \varepsilon_2) \to \mathbf{D}(\bar{A}, \varepsilon_2)$ is locally invertible with a $C^\infty$ inverse. This follows from the inverse function theorem, since at each point $(y, \theta, \varepsilon) \in \mathbf{C}(\bar{A}, \varepsilon_2)$ the Jacobian matrix $I + \varepsilon U_2$ is invertible, the inverse being expressible as the convergent geometric series

$$\sum_{l=0}^{\infty} (-1)^l \varepsilon^l [U_y(y, t, \varepsilon)]^l.$$

Next we claim that there is some $\varepsilon_0, 0 < \varepsilon_0 \leq \varepsilon_2$, for which the mapping $C(\bar{A}, \varepsilon_0) \to D(\bar{A}, \varepsilon_0)$ is (globally) one to one. If not, select a sequence $\varepsilon_n \to 0$ and for each $\varepsilon_n$ choose two points $(y_n, \theta_n, \varepsilon_n)$ and $(y_n', \theta_n, \varepsilon_n)$ with $y_n \neq y_n'$ which map to the same point $(x_n, \theta_n, \varepsilon_m)$. (Note that $\theta_n$ must be equal for both points—we did not say $\theta_n'$—because $\theta$ is preserved by the mapping.) Since $C(\bar{A}, \varepsilon_0)$ is compact there exists a subsequence of the indices $n$ for which $(y_n, \theta_n, \varepsilon_n)$ converges to some $(y_*, \theta_*, 0)$, and some further subsequence of the indices such that $(y_n', \theta_n, \varepsilon_n)$ converges to $(t_*', \theta_*, 0)$. (The second subsequence of indices must be a subset of the first so that $\theta_*$ is the same for both.) By continuity, $(y_*, \theta_*, 0)$ and $(y_*', \theta_*, 0)$ must map to the same point; but for $\varepsilon = 0$ the mapping is the identity, so $y_*' = y_*$. Now in a neighborhood of $(y_*, \theta_*, 0)$ the mapping is one to one (by the argument from the inverse function theorem); but for some $n$, both $(y_n, \theta_n, \varepsilon_n)$ and $(y_n', \theta_n, \varepsilon_n)$ belong to this neighborhood, so they cannot map to the same point. This contradiction establishes that there is an $\varepsilon_0 > 0$ such that $C(\bar{A}, \varepsilon_0) \to D(\bar{A}, \varepsilon_0)$ is one to one, and therefore invertible. The inverse coincides at each point with the local inverse, which is $C^\infty$, and therefore the global inverse is $C^\infty$. Setting $\theta = \omega t$ returns us to the case of $A(\varepsilon_0) \to B(\varepsilon_0)$. Finally the existence of $v$ follows from the $C^\infty$ division theorem (theorem 2.4).

We stated the theorem for bounded open sets since open sets are usually wanted, although the proof requires passing to the compact closure; of course, to be $C^\infty$ on a compact set means to be $C^\infty$ on a slightly larger open set, and this is true since $u$ is $C^\infty$ everywhere and the inverse function theorem gives local inverses at each point of the boundary of $A$ which 'slosh over' the boundary. It follows that if the boundary of $A$ is a smooth manifold, the same is true of the boundary of each slice (for fixed $t$ and $\varepsilon$) of $\mathbf{A}$ and $\mathbf{B}$. These boundary considerations will be important when we consider topological conjugacy.

Having clarified the manner in which (3.2) is a valid coordinate transformation, we turn to the question of using it to transform (3.1). By differentiating (3.2) with respect to time and using (3.1), the transformed equation is seen to be

$$(3.5) \qquad \dot{y} = \varepsilon (I + \varepsilon u_y(y, t, \varepsilon))^{-1} (f(y + \varepsilon u(y, t, \varepsilon), t, \varepsilon) - u_t(y, t, \varepsilon)).$$

This ungainly expression is not easy to work with, but it is important to write it down because it shows that the equation for $y$ has the same form $\dot{y} = \varepsilon g(y, t, \varepsilon)$ as the equation for $x$, and is quasiperiodic in $t$ with the same frequencies (assuming that $u$ has the same frequencies as $f$). Therefore the following expanded forms of (3.1), (3.2), and (3.5) are possible:

$$(3.6) \qquad \dot{x} = \varepsilon f_1(x, t) + \varepsilon^2 f_2(x, t) + \cdots + \varepsilon^k f_k(x, t) + \varepsilon^{k+1} \hat{f}(x, t, \varepsilon)$$

$$(3.7) \qquad x = y + \varepsilon u_1(y, t) + \cdots + \varepsilon^k u_k(y, t)$$

$$(3.8) \qquad \dot{y} = \varepsilon g_1(y, t) + \cdots + \varepsilon^k g_k(y, t) + \varepsilon^{k+1} \hat{g}(y, t, \varepsilon).$$

On the basis of (3.5) we can assert that the transformation (3.7) carries (3.6) into (3.8), where the functions $g_j$ remain to be determined from the $f_j$ and $u_j$. We have chosen a fixed order $k$ for explicit study, and have included only terms up to that order in (3.7) because further terms will affect only $\hat{g}$, not $g_1$ through $g_k$.

The $g_i$ are calculated as follows: First, substitute (3.7) into (3.6) and expand fully in powers of $\varepsilon$ through order $k$. Second, differentiate (3.7) with respect to $t$ and use (3.8) to eliminate $\dot{y}$. Third, equate coefficients of $\varepsilon^i$ in the previous two series and solve for $g_i$. If $k = 2$, the first step gives $\dot{x} = \varepsilon f_1(y, t) + \varepsilon^2 \{ f_2(y, t) + f_{1x}(y, t)u_1(y, t) \} + \cdots$, the second gives $\dot{x} = \varepsilon \{ g_1 + u_{1t} \} + \varepsilon^2 \{ g_2 + u_{1y}g_1 + u_{2t} \}$ with all functions evaluated at $(y, t)$, and the third gives

$$
\begin{aligned}
(3.9) \quad & g_1(y, t) = f_1(y, t) - u_{1t}(y, t) \\
& g_2(y, t) = f_2(y, t) + f_{1x}(y, t)u_1(y, t) - u_{1y}(y, t)g_1(y, t) - u_{2t}(y, t).
\end{aligned}
$$

We have somewhat awkwardly written $f_{1x}(y, t)$ for $(\partial/\partial x)f_1(x, t)$ evaluated at $x = y$; this is equivalent to $(\partial/\partial y)f_1(y, t) = f_{1y}(y, t)$ but it is perhaps better to use $f_{1x}(y, t)$ to remember that $f_1$ is originally a function of $x$. (In any case, the use of letter subscripts to denote partial derivatives is logically incorrect, although a useful concession to human psychology. Ideally a partial derivative should always be denoted by the number of the argument with respect to which the derivative is taken. For example, suppose the function $h$: $\mathbb{R}^2 \to \mathbb{R}$ has the property that its derivative with respect to the first argument is symmetric in its arguments. This may be expressed clearly as $h_1(x, y) = h_1(y, x)$, but neither $h_x(x, y) = h_x(y, x)$ nor $h_x(x, y) = h_y(y, x)$ is unambiguous.)

For computational applications it is necessary to carry out the full calculation of $g_i$ to the desired order $k$. This is best done by the algorithms of the Lie transform method, in which the so-called generator of the transformation figures in place of the transformation itself. But for present theoretical purposes the Lie method offers no advantages and would require a long digression. The crucial point for us is the following theorem, illustrated by equations (3.9).

**3.2 THEOREM** The transformation (3.7) carries (3.6) into (3.8), with $g_i(y, t) = f_i(y, t) - u_{it}(y, t) + K_i(y, t)$, where $K_i$ is expressible in terms of $f_j$ and $u_j$ for $j < i$, and their derivatives.

The proof is simply to examine the steps in computing $g_i$, described just before equation (3.9). When (3.7) is substituted into (3.6) each term $\varepsilon^j f_j(x, t)$ becomes $\varepsilon^j f_j(y, t)$ plus higher order terms. Therefore the coefficient of $\varepsilon^i$ consists of $f_i(y, t)$ plus terms arising from $f_j(x, t)$ for $j < i$. When (3.7) is differentiated, the derivative of $y$ is $\dot{y}$ which contributes $g_j$ to the $\varepsilon^j$ term; the derivative of $\varepsilon^j u_j$ is $\varepsilon^j \{ u_{jy}\dot{y} + u_{jt} \} = \varepsilon^j u_{jt}$ plus higher order. Therefore the coefficient of

$\varepsilon^i$ contains $g_i + u_{it}$ plus terms deriving from $u_j$ with $j < i$. Equating coefficients in the two equations gives the theorem. Each $K_i$ is given by a universal formula in terms of $f_j, u_j$ and their derivatives for $j < i$; for instance (3.9) shows that $K_1 = 0$ and $K_2 = f_{1x}u_1 - u_{1y}g_1$. The formula for each order $K_i$ is independent of the ultimate order $k$ to which computations are being carried. This has the practical significance that further terms can be found at any time without recalculating earlier terms.

At this point we are able to change perspective from 'given $u$ find $g$' to 'given $g$, find $u$ if possible'. The key here is to express the equation of Theorem 3.2 as a sequence of differential equations for $u_i$ to be solved recursively:

$$(3.10) \quad \begin{aligned} u_{1t}(y, t) &= f_1(y, t) - g_1(y, t) \\ u_{it}(y, t) &= \{f_i(y, t) + K_i(y, t)\} - g_i(y, t), \qquad i = 2, \dots, k. \end{aligned}$$

Assuming that $g_1, \dots, g_k$ are given and $u_j$ has been found for $j = 1, \dots, i - 1$, the term $K_i$ is computable and one may attempt to solve the equation for $u_i$. We will see that these equations are not solvable (remembering that only quasiperiodic solutions are acceptable) unless the $g_i$ are chosen properly. Therefore the actual strategy is that at the $i$th stage only $g_1, \dots, g_{i-1}$ and $u_1, \dots, u_{i-1}$ (and all of $f_1, \dots, f_k$) are known. This is sufficient to determine $K_i$. Next one must choose a suitable $g_i$ (based on knowledge of $f_i + K_i$) and then solve for $u_i$, making both $g_i$ and $u_i$ available so that $K_{i+1}$ may be found and the next stage attempted. (If the result is to be an averaging transformation, the $g_i$ must be chosen to be independent of $t$.) Therefore the existence of averaging transformations hinges on the existence of quasiperiodic solutions to differential equations of the form $u_t(y, t) = h(y, t)$. It is this question to which we now turn.

Assuming that $h(y, t)$ is quasiperiodic with rationally independent frequencies $\omega = \mathrm{col}(\omega_1, \dots, \omega_m)$, there exists a Fourier series $h(y, t) = \Sigma a_\nu(y) e^{i\nu\omega t}$, the summation being over integer vectors $\nu = (\nu_1, \dots, \nu_m)$. The function

$$(3.11) \qquad u(y, t) = \bar{u}(y) + a_0(y)t + \sum_{\nu \neq 0} \frac{a_\nu(y)}{i\nu\omega} e^{i\nu\omega t},$$

where $\bar{u}(y)$ is arbitrary, will clearly provide the general solution of $u_t = h$ if the series on the right-hand side converges and is differentiable term by term. This solution is obtained by termwise indefinite integration of $h$, and we have seen in Theorem 2.3 that such series need not converge except in special circumstances. Assuming that such a circumstance holds, $u(y, t)$ will still not give a quasiperiodic solution unless $a_0(y) = 0$, that is, unless the average of $h(y, t)$ over $t$ equals zero. In this case $u(y, t)$ will be quasiperiodic and $\bar{u}(y)$, which is arbitrary, will be its average. (One could attempt to solve $u_t = h$ by the termwise definite integration formula (2.1) and thereby avoid all convergence difficulties, but although this series always converges, it need not always be quasiperiodic: if $\omega t$ is replaced by $\theta$, the series need not converge for all $\theta$.)

Returning to the solution of (3.10), it is now clear how the $g_i$ must be chosen if they are to be independent of $t$: since the right-hand side must have zero average, $g_1(y)$ must be the average of $f_1(y, t)$, while $g_i$ must equal the average of $f_i + K_i$ for $i \geq 2$. Then (3.10) may be solved recursively for $u_i$ and $g_i$, $i = 1, ..., k$, provided that one of the special circumstances in Theorem 2.3 holds. Pulling the details together gives Theorem 3.3 below, our main existence theorem for averaging transformations.

3.3 THEOREM Let (3.1) be quasiperiodic with rationally independent frequencies $\omega_1, ..., \omega_m$. There exists an averaging transformation (3.2), quasiperiodic with the same frequencies, carrying (3.1) to (3.3), provided one of the following cases holds:

(1) $f$ is periodic
(2) $f$ has finitely many harmonics
(3) Only one of the frequencies of $f$ has infinitely many harmonics
(4) $f$ is real analytic and $\omega_1, ..., \omega_m$ are badly incommensurable.

For every bounded open set $A$ there is an $\varepsilon_0$ such that the transformation (3.2) is valid for $(y, t, \varepsilon) \in \mathbf{A}(A, \varepsilon_0)$. The averaging transformation may be chosen so that $x = y$ at $t = 0$, that is, $u(y, 0, \varepsilon) = 0$; in this case, if $f$ is periodic, $x = y$ at each period $t = nT$.

In the theorem the hypotheses (1) through (4) apply to $f$, and hence to each $f_i$. It is necessary to check that they apply to $K_i$ in order that Theorem 2.3 can be used for (3.10). This is done inductively. Case 1 is immediate. In case 2, suppose $f_j$ and $u_j$ have finitely many harmonics for $j < i$. In the formation of $K_i$ the differentiations do not introduce new harmonics, but the multiplications may. (Think of $K_2 = f_{1x}u_1 - u_{1y}g_1$.) However, only finitely many new harmonics can arise at each step, and $f_i + K_i$ satisfies case 2 of Theorem 2.3. Then $u_i$ will have finitely many harmonics, allowing the induction to continue. Case 3 is similar. In case 4, each $K_i$ is real analytic. The one point remaining to be checked is the possibility of achieving $u(y, 0, \varepsilon) = 0$. This results from the freedom to choose $\bar{u}(y)$ in (3.11): in solving (3.10), each $\bar{u}_i(y)$ must be chosen so that $u_i(y, 0) = 0$. The advantage of doing so is that an initial value problem (at time 0) for (3.1) will correspond to a problem for (3.3) having the same initial conditions. In addition, if $f$ is periodic, solutions of (3.1) and (3.3) with the same initial conditions will be equal at 'stroboscopic times' $t = nT$. (In another terminology, the systems have the same period map.) When the $\bar{u}_i$ are chosen this way and $f$ is periodic, one speaks of the *stroboscopic method*. (Otherwise, $\bar{u}_i$ is usually taken to be 0.)

In cases not covered by Theorem 3.3 there is still what might be called *pseudo-averaging*. The idea is to choose $g_i(y, t)$ recursively in such a way that

(3.10) is solvable and $g_i$ is simpler than $f_i$. One way to do this, which always reduces the solution of (3.10) to case 2 of Theorem 2.3, is to choose $g_i$ to contain all the 'high harmonics' of $f_i + K_i$, so that $f_i + K_i - g_i$ consists of finitely many 'low' harmonics. In this case $g_i$ is not independent of $t$, but since (by the decay of Fourier coefficients) high harmonics have small amplitudes, $g_i$ can be made nearly constant. This is not good enough for obtaining computable approximate solutions with error estimates, but is sufficient for some qualitative arguments in later sections. We formalize it as

3.4 THEOREM Let (3.1) be quasiperiodic with rationally independent frequencies $\omega_1, \ldots, \omega_m$ and let $N_1, \ldots, N_k$ be positive integers. Then there exists a pseudo-averaging transformation of the form (3.7), quasiperiodic with the same frequencies, carrying (3.1) into (3.8), such that each $g_i$ consists of a constant term and harmonics of order $\| \nu \| > N_i$ only.

At the $i$th stage in the construction the $g_j$ and $u_j$ for $j < i$ depend only on the previously used cutoffs $N_j$, $j < i$. This is true therefore of $K_i$ also. Now $g_i$ is taken to consist of the constant term of $f_i + K_i$, plus the harmonics of order $> N_i$. Then $u_i$ can be found from (3.10). Since $N_i$ does not enter until $g_i$ is defined, $N_i$ need not be prescribed in advance but may be determined after $K_i$ is found, so as to make $g_i$ as nearly constant as needed. (We will use this later in the following form: if the average of $f_i + K_i$ is not zero, then $N_i$ can be taken large enough that $g_i$ is bounded away from zero.)

**Notes**

The method of averaging has its roots in celestial mechanics, but under this name is usually attributed to Krylov, Bogoliubov, and Mitropolskii, standard references being [15] and [3]. These books present the $k$th order method of averaging for periodic systems. A more modern reference, covering also the quasiperiodic badly irrational case, is [28]. The method of averaging has been extended far beyond the periodic and quasiperiodic cases, but usually only to first order. For almost periodic systems this appears already in [3], a brief exposition appearing in [9]. (The latter author, Fink, is skeptical of some of the claims made by Hale [11] concerning averaging of almost periodic systems.)

It is worth mentioning the relation between the first-order method of averaging for almost periodic systems, and the methods given in this section. Consider the quasiperiodic case. At each stage of averaging, we have seen that it is necessary to solve equation $u_y = h$ with $h(y, t) = \Sigma a_\nu(y) e^{i\nu\omega t}$, where $a_0(y) = 0$. Let $u(y, t, \eta) = \Sigma_{\nu \neq 0} [ a_\nu(y) / i\nu\omega + \eta ]$. For small $\eta \neq 0$, this is an approximate solution to $u_y = h$; the presence of $\eta$ in the denominator causes the series to converge regardless of the smallness of $\nu\omega$, much as the series

(3.11) can be forced to converge by a frequency cut-off. The immediate result of this idea is a form of pseudo-averaging similar to Theorem 3.4, introducing parameters $\eta_1, ..., \eta_k$ at successive stages of (approximate) averaging in place of $N_1, ..., N_k$. If this procedure is used to first order only, and $\eta_1$ is then set equal to $\varepsilon$, the result (after some careful estimating) is a first-order averaging in which (3.3) is replaced by $\dot{y} = \varepsilon g + 0(\varepsilon)$ instead of the stronger $\dot{y} = \varepsilon g + 0(\varepsilon^2)$ which our methods give. Thus it is possible in this weakened sense to average any quasiperiodic system, without regard to the restrictions in Theorem 3.3, to first order. The method generalizes to almost periodic systems, where $u(y, t, \eta)$ must be defined as

$$\int_{-\infty}^{t} e^{-\eta(t-s)} h(y, s) \, ds,$$

which reduces to the series above in the quasiperiodic case.

In section 3 of [19] I attempted to generalize this idea to higher order averaging, but it seems now that my method does not work. The idea was to use pseudo-averaging with parameters $\eta_1, ..., \eta_k$ and then to choose functions $\eta_i(\varepsilon)$ which approach zero rapidly as $\varepsilon \to 0$. (Similarly, one could make the cutoffs $N_i$ in Theorem 3.4 into functions of $\varepsilon$ which approach infinity so rapidly as $\varepsilon \to 0$ that the high harmonic nonautonomous parts of $g_1, ..., g_k$ are of order greater than $\varepsilon^k$.) In so doing, the coefficients $u_i$ and $g_i$ in (3.7) and (3.8) become functions of $\varepsilon$, so that these are no longer straightforward Taylor expansions. What I failed to notice is that one loses control over the magnitude of the terms in (3.7) and can no longer even guarantee that it is a near-identity transformation. It is doubtful whether anything can be salvaged from this idea. (The rest of the paper [19] is correct.)

Further extensions of the method of averaging exist, including even some second-order results for systems that are not even almost periodic. The most complete treatment of the current state of such results, with many references, is [32]. A quite different sort of encyclopedic survey of averaging, with emphasis on computational procedures and examples, is Chapter 5 of [25]. This reference includes the use of Lie transforms (see the remarks prior to Theorem 3.2 above).


## 4  ASYMPTOTIC ESTIMATES


Under the assumption that an averaging transformation exists for (1.1), various estimates can be derived which compare solutions of the truncated averaged system, the full averaged system, and the original system. For reference, we have:

(4.1)     original     $\dot{x} = \varepsilon f_1(x, t) + \cdots + \varepsilon^k f_k(x, t) + \varepsilon^{k+1} \hat{f}(x, t, \varepsilon)$

(4.2)     transformation  $x = y + \varepsilon u_1(y, t) + \cdots + \varepsilon^k u_k(y, t)$

(4.3)     full averaged   $\dot{y} = \varepsilon g_1(y) + \cdots + \varepsilon^k g_k(y) + \varepsilon^{k+1} \hat{g}(y, t, \varepsilon)$

(4.4)     truncated      $\dot{z} = \varepsilon g_1(z) + \cdots + \varepsilon^k g_k(z)$

The solutions of these systems with initial condition $x$ (or $y$ or $z$) $= a$ at $t = 0$ will be denoted $x(t, a, \varepsilon)$, $y(t, a, \varepsilon)$, and $z(t, a, \varepsilon)$. Considering $z$ as the most easily solvable system, one wants to construct approximations to $y$ and $x$. It is reasonable to think that $z(t, a, \varepsilon)$ approximates $y(t, a, \varepsilon)$ since the $y$ and $z$ equations agree up to order $\varepsilon^k$. Therefore it is reasonable to try to approximate $x$ by feeding $z$ (instead of $y$) into the transformation (4.2). In fact two approximations to $x$ are defined, the *kth approximation* $X$ and the *improved kth approximation* $X_{\text{imp}}$. These are defined by

(4.5)     $X_{\text{imp}}(t, a, \varepsilon) = z(t, a, \varepsilon) + \varepsilon u_1(z(t, a, \varepsilon), t) + \cdots + \varepsilon^k u_k(z(t, a, \varepsilon), t)$

(4.6)     $X(t, a, \varepsilon) = z(t, a, \varepsilon) + \varepsilon u_1(z(t, a, \varepsilon), t) + \cdots + \varepsilon^{k-1} u_{k-1}(z(t, a, \varepsilon), t)$

The difference is that $X$ omits and $X_{\text{imp}}$ includes the $k$th order term in the transformation. It may seem unnatural to use $X$ instead of $X_{\text{imp}}$, but omitting the last term in the transformation does not affect the asymptotic accuracy of the solution as $\varepsilon \to 0$. Furthermore, the sequence of steps in finding the averaged equations is such that $g_k$ is determined (as the average of $f_k + K_k$) before $u_k$ is found (by solving $u_{kt} = f_k + K_k - g_k$). Therefore the truncated averaged equations can be fully determined without finding $u_k$, and since $u_k$ also does not improve the asymptotic accuracy of the approximate solution, a good deal of computation can be saved by using $X$ instead of $X_{\text{imp}}$. One warning is in order whether using $X$ or $X_{\text{imp}}$: unless the stroboscopic method has been adapted (so that $u_i = 0$ at $t = 0$), the initial condition satisfied by $X(t, a, \varepsilon)$ is not $a$ but rather $X(0, a, \varepsilon) = A(a, \varepsilon) = a + \varepsilon u_1(a, 0) + \cdots + \varepsilon^{k-1} u_{k-1}(a, 0)$, and for $X_{\text{imp}}$ it is $A_{\text{imp}}(a, \varepsilon) = A(a, \varepsilon) + \varepsilon^k u_k(a, 0)$. So $X(t, a, \varepsilon)$ approximates $x(t, A(a, \varepsilon), \varepsilon)$ and not necessarily $x(t, a, \varepsilon)$.

Most treatments of averaging emphasize the approximation of $x$ by $X$. The first step in the argument is to study the approximation of $y$ by $z$. Since the present study is chiefly aimed at qualitative results, and since the qualitative behavior of the total family of solutions is the same for $x$ and $y$ (since these are mapped one to another by a homeomorphism), there is little need for us to study $x$ in preference to $y$. Therefore we will give more attention than usual to the approximation of $y$ by $z$, although for completeness the approximation of $x$ by $X$ will be presented in the manner of an afterthought.

The relationship of $y$ to $z$ does not involve the averaging transformation, but is strictly a matter of truncation. The fundamental result, briefly stated, is that $\| y(t, a, \varepsilon) - z(t, a, \varepsilon) \|$ is small of order $\varepsilon^k$ for long time intervals with length of order $1/\varepsilon$. Thus as $\varepsilon$ decreases, the approximation improves in two ways: the error decreases, and the interval of validity of the error estimate increases. For later use we need to be precise about the constants that appear in these estimates, and want them to be uniform for $a$ in compact sets. Let $K$ be any compact set in $\mathbb{R}^n$ and let $K_r$ for $r > 0$ be its 'compact $r$-halo', that is, the union of all closed $r$-balls centered at points of $K$.

**4.1 LEMMA**  Given $K, r > 0$, and $\varepsilon_0 > 0$, there exists $\mathbf{T} = \mathbf{T}(K, r, \varepsilon) > 0$ such that $y(t, a, \varepsilon)$ and $z(t, a, \varepsilon)$ belong to $K_r$ for all $a \in K$, $|t| \leqslant \mathbf{T}/\varepsilon$, and $0 \leqslant \varepsilon \leqslant \varepsilon_0$.

For proof, let $M = M(K, r, \varepsilon_0)$ denote the maximum of $\| g_1(y) \| + \varepsilon \| g_2(y) \| + \cdots + \varepsilon^{k-1} \| g_k(y) \| + \varepsilon^k \| \hat{g}(y, t, \varepsilon) \|$ for $y \in K_r$, $0 \leqslant \varepsilon \leqslant \varepsilon_0$, and all $t$. Such a maximum exists because $\hat{g}$ is quasiperiodic and so can be replaced by a function on $K \times \mathbb{T}^m \times [0, \varepsilon_0]$, which is compact. Using the inequality $\mathrm{d} \| y \|/\mathrm{d}t \leqslant \| \mathrm{d}y/\mathrm{d}t \|$, which can be expressed 'you can't *run away* (from the origin) faster than you can *run*', it follows that both $\mathrm{d} \| y \|/\mathrm{d}t$ and $\mathrm{d} \| z \|/\mathrm{d}t$ are less than $\varepsilon M$ for any solutions $y(t)$ and $z(t)$ as long as these solutions are in $K_r$. Since any solution starting in $K$ must travel at least a distance $r$ before leaving $K_r$, and can do so at a rate no faster than $\varepsilon M$, such solutions must remain in $K_r$, both forwards and backwards in time, at least until $\varepsilon M |t| = r$. This proves the lemma with $\mathbf{T} = r/M$.

It would be more convenient if it were possible to choose $\mathbf{T} > 0$ arbitrarily instead of $r$. One might attempt to argue as follows. Given $K$, $\mathbf{T}$, and $\varepsilon_0$, let the $\mathbf{T}$-halo $K_\mathbf{T}$ of $K$ be the set of all points reached by solutions $y(t, a, \varepsilon)$ or $z(t, a, \varepsilon)$ for $a \in K$, $|t| \leqslant \mathbf{T}/\varepsilon$, and $0 \leqslant \varepsilon \leqslant \varepsilon_0$. Then Lemma 4.1 is true by definiton (with $K_r$ replaced by $K_\mathbf{T}$). The difficulty is that $K_\mathbf{T}$ need not be compact and so maxima such as $M$ (which are needed again) need not exist. To see that $K_\mathbf{T}$ need not be compact, consider the case $k = 1$, $\dot{z} = \varepsilon g_1(z)$. With $\tau = \varepsilon t$ this becomes $\mathrm{d}z/\mathrm{d}\tau = g_1(z)$, and such a system in $\mathbb{R}^n$ can have solutions which approach infinity in finite time. If $\tau = \mathbf{T}$ is such a time, $K_\mathbf{T}$ will be unbounded.

Theorem 4.2 below is the fundamental asymptotic estimate for the method of averaging. Lemma 4.1 guarantees that the hypotheses of Theorem 4.2(a) hold if $V$ is taken to be $K_r$. Other instances will occur later. The variant 4.2(b) is included to meet a need in Section 6.

**4.2 THEOREM**  (a) Let $K$ and $V$ be compact sets in $\mathbb{R}^n$ with $K$ contained in the interior of $V$. Let $\varepsilon_0 > 0$ and $\mathbf{T} > 0$ be any constants such that $y(t, a, \varepsilon)$ and $z(t, a, \varepsilon)$ belong to $V$ for all $a \in K$, $|t| \leqslant \mathbf{T}/\varepsilon$, $0 \leqslant \varepsilon \leqslant \varepsilon_0$. Then there exists a constant $c > 0$ such that

$$(4.7) \qquad\qquad \| y(t, a, \varepsilon) - z(t, a, \varepsilon) \| < c\varepsilon^k$$

for all $a \in K$, $|t| \leqslant \mathbf{T}/\varepsilon$, $0 \leqslant \varepsilon \leqslant \varepsilon_0$.

(b) Let $K$ and $V$ be compact sets in $\mathbb{R}^n$ with $K$ contained in the interior of $V$. Let $\varepsilon_0 > 0$ and $\mathbf{T} > 0$ be arbitrary. Then there exists a constant $c$ such that (4.7) holds for all $a \in K$, $0 \leqslant \varepsilon \leqslant \varepsilon_0$, and for all $t$ satisfying the following conditions: $|t| \leqslant \mathbf{T}/\varepsilon$ and both $y(s, a, \varepsilon)$ and $z(s, a, \varepsilon)$ belong to $V$ for all $s$ between 0 and $t$ ($0 \leqslant s \leqslant t$ if $t > 0$, $t \leqslant s \leqslant 0$ if $t < 0$).

To prove (a), let $B$ be the maximum of $\|\hat{g}(y, t, \varepsilon)\|$ on $V \times \mathbb{R} \times [0, \varepsilon_0]$ (which exists by quasiperiodicity) and let $L_i$ be a Lipschitz constant for $g_i(y)$ on $V$ (for instance $L_i = \max \|\partial g_i/\partial y\|$ in a suitable matrix norm). Then for $a \in K$, $0 \leqslant \varepsilon \leqslant \varepsilon_0$, and $|t| \leqslant \mathbf{T}/\varepsilon$, the distance $\rho(t, a, \varepsilon) = \|y(t, a, \varepsilon) - z(t, a, \varepsilon)\|$ satisfies the differential inequality $d\rho/dt \leqslant (\varepsilon L_1 + \cdots + \varepsilon^k L_k)\rho + \varepsilon^{k+1}B$ with initial conditions $\rho = 0$ at $t = 0$; this follows from (4.3), (4.4), and the 'running away' inequality. The differential inequality can be solved exactly as a linear differential equation would be, resulting in $\rho \leqslant \varepsilon^{k+1}B\, \delta^{-1}(e^{\delta t} - 1)$ where $\delta = \varepsilon L_1 + \cdots + \varepsilon^k L_k \geqslant \varepsilon L_1$. Now $\delta^{-1} \leqslant 1/\varepsilon L_1$, so $\varepsilon^{k+1}\delta^{-1}$ is of order $\varepsilon^k$. Furthermore $\delta t$ is bounded for $0 \leqslant \varepsilon \leqslant \varepsilon_0$ and $|t| \leqslant \mathbf{T}/\varepsilon$; for such $\varepsilon$ and $t$, $|\delta t| \leqslant (L_1 + \varepsilon_0 L_2 + \cdots + \varepsilon_0^{k-1}L_k)\mathbf{T}$. Therefore $\rho$ is of order $\varepsilon^k$ for such $a$, $t$, and $\varepsilon$. (Technical point: if we want the same result for $-\varepsilon \leqslant \varepsilon \leqslant \varepsilon_0$ the claim $\delta \geqslant \varepsilon L_1$ fails, but $\delta^{-1}$ is still of order $1/\varepsilon$ if $\varepsilon_0$ is decreased, if necessary, until $[-\varepsilon_0, \varepsilon_0]$ contains no zeros of $\delta$.) The proof of (b) is exactly the same, except that the estimates cease to apply if either $y$ or $z$ leaves $V$ before $|t|$ reaches $\mathbf{T}/\varepsilon$.

The following corollary states the situation in the important special case that $K$ is a single point.

**4.3 COROLLARY**  Let $V$ be a compact set in $\mathbb{R}^n$ and let $\varepsilon_0 > 0$ be given. Then for each $a \in V$ there exists $\mathbf{T} > 0$ such that $y(t, a, \varepsilon)$ and $z(t, a, \varepsilon)$ belong to $V$ for $|t| \leqslant \mathbf{T}/\varepsilon$, $0 \leqslant \varepsilon \leqslant \varepsilon_0$. For any such $\mathbf{T}$, there exists $c > 0$ such that $\|y(t, a, \varepsilon) - z(t, a, \varepsilon)\| < c\varepsilon^k$ for $|t| \leqslant \mathbf{T}/\varepsilon$, $0 \leqslant \varepsilon \leqslant \varepsilon_0$.

Let $r$ be the distance from $a$ to $\partial V$ (the boundary of $V$) and let $K = \{a\}$. Then $K_r \subseteq V$, and the existence of $\mathbf{T}$ follows from Lemma 4.1. The rest of the corollary then follows from Theorem 4.2.

For certain purposes it is necessary to estimate not only the distance between $y$ and $z$, but also the distance between their derivatives with respect to the initial point $a$.

**4.4 COROLLARY**  Given a compact set $K$ and $\varepsilon_0 > 0$ there exists $\mathbf{T} > 0$ and $c > 0$ such that both (4.7) and

$$(4.8) \qquad \|y_a(t, a, \varepsilon) - z_a(t, a, \varepsilon)\| < c\varepsilon^k$$

hold for $a \in K$, $|t| \leqslant \mathbf{T}/\varepsilon$, $0 \leqslant \varepsilon \leqslant \varepsilon_0$.

To prove (4.8) does not require a new argument. We shall show that the matrices $y_a$ and $z_a$ satisfy differential equations of the same form as (4.3) and (4.4), so that the argument leading to (4.7) yields (4.8) as well. Let $G_i(t, a, \varepsilon) = \partial g_i(y)/\partial y$ evaluated at $y(t, a, \varepsilon)$, $\hat{G}(t, a, \varepsilon) = \partial \hat{g}(y, t, \varepsilon)/\partial y$ evaluated at $(y(t, a, \varepsilon), t, \varepsilon)$. Then differentiating $\dot{y}(t, a, \varepsilon) = \varepsilon g_1(y(t, a, \varepsilon)) + \cdots$ with respect to $a$ yields $\dot{y}_a(t, a, \varepsilon) = \varepsilon G_1(t, a, \varepsilon)y_a(t, a, \varepsilon) + \cdots + \varepsilon^{k+1}\hat{G}(t, a, \varepsilon)y_a(t, a, \varepsilon)$ and similarly for $\dot{z}_a$ with $\hat{G}$ omitted. This does not quite fit the pattern of (4.3) and (4.4) because the $G_i$ depend on $t$ (and are not even quasiperiodic). This can be remedied by forming the following coupled systems of $n + n^2$ differential equations for $y, Y$ and $z, Z$, the $Y$ and $Z$ being $n \times n$ matrices:

$$(4.9) \qquad \dot{y} = \varepsilon g_1(y) + \cdots + \varepsilon^k g_k(y) + \varepsilon^{k+1}\hat{g}(y, t, \varepsilon)$$

$$\dot{Y} = \varepsilon \frac{\partial g_1}{\partial y}(y)Y + \cdots + \varepsilon^k \frac{\partial g_k}{\partial y}(y)Y + \varepsilon^{k+1}\frac{\partial \hat{g}}{\partial y}(y, t, \varepsilon)Y$$

$$(4.10) \qquad \dot{z} = \varepsilon g_1(z) + \cdots + \varepsilon^k g_k(z)$$

$$\dot{Z} = \varepsilon \frac{\partial g_1}{\partial z}(z)Z + \cdots + \varepsilon^k \frac{\partial g_k}{\partial z}(z)Z.$$

The solution of (4.9) with initial conditions $y = a$, $Y = I$ at $t = 0$ will be $y = y(t, a, \varepsilon)$, $Y = y_a(t, a, \varepsilon)$, because along this solution $\partial g_i/\partial y = G_i$. Then the application of (4.7) to (4.9) and (4.10) implies (4.8), by way of the triangle inequality. (One first gets $\| (y, Y) - (z, Z) \| \leq c\varepsilon^k$ for a norm on $\mathbb{R}^{n+n^2}$. But this is $\leq \| (y, 0) - (z, 0) \| + \| (0, Y) - (0, Z) \|$.) The only point to notice is that given a compact set $K$ of initial conditions $a$ for $y$, the set $K \times \{I\}$ of initial conditions $(a, I)$ for $(y, Y)$ is also compact, and it is this set whose $r$-halo must be taken in Lemma 4.1, and hence determines the values of **T** and $c$ for which (4.7) and (4.8) together are valid.

In the case that (4.1)–(4.3) are periodic in $t$ with period $T$, it will be important (in Section 6) to consider the 'augmented' version of (4.3), an autonomous system in one higher dimension,

$$(4.11) \qquad \begin{aligned} \dot{y} &= \varepsilon g_1(y) + \cdots + \varepsilon^k g_k(y) + \varepsilon^{k+1}\hat{g}(y, T\theta/2\pi, \varepsilon) \\ \dot{\theta} &= 2\pi/T, \end{aligned}$$

and the similar augmented version of (4.4), consisting of (4.4) together with $\dot{\theta} = 2\pi/T$. Here $\theta$ is to be regarded as belonging to the circle $S^1 = \mathbb{R} \bmod 2\pi$, so (4.11) is a vector field on $\mathbb{R}^n \times S^1$. (System (4.11) includes, in addition to (4.3), all similar systems with $t$ shifted. The symbol $S^n$ is a standard notation for the $n$-sphere, as $T^n$ is for the $n$-torus. The circle can be denoted either $S^1$ or $T^1$.) The advantage of (4.11) over (4.3) is that its solutions form a *flow* (or more precisely, a partial flow, since solutions need not exist for all time). There is a unique orbit of (4.11) through each point of $\mathbb{R}^n \times S^1$; each 'orbit' carries infinitely many 'solutions' differing only by a shift of $t$. Let $\phi^t_\varepsilon(y_0, \theta_0)$ denote the solution of (4.11) with initial conditions $y = y_0$, $\theta = \theta_0$ at $t = 0$. For each $t$, $\phi_\varepsilon^t$ is defined for a subset of $\mathbb{R}^n \times S^1$ (the points whose orbits have not become unbounded by time $t$), and $\phi_\varepsilon^t$ is a smooth map with smooth inverse (a diffeomorphism). Let $\psi_\varepsilon^t$ be the similar map for (4.4) augmented. In this context, the previous results take the following form.

4.5 COROLLARY   Given compact sets $K \subset V \subset \mathbb{R}^n$, there exists constants $\varepsilon_0 > 0$, $\mathbf{T} > 0$ such that $\phi^t$ and $\psi^t$ are defined on $K \times S^1 \to V \times S^1$ for $0 < \varepsilon < \varepsilon_0$, $|t| \leqslant \mathbf{T}/\varepsilon$. For any such $\varepsilon_0$ and $\mathbf{T}$, $\phi_\varepsilon^t$ and $\psi_\varepsilon^t$ are $\varepsilon^k - C^1$-close; that is, there exists $c > 0$ such that $\| \phi_\varepsilon^t - \psi_\varepsilon^t \|$ and $\| D\phi_\varepsilon^t - D\psi_\varepsilon^t \|$ are less than $c\varepsilon^k$ on $K \times S^1$ for $0 < \varepsilon < \varepsilon_0$, $|t| \leqslant \mathbf{T}/\varepsilon$. (Here $D$ denotes the Jacobian matrix with respect to the variables in $\mathbb{R}^n \times S^1$.)

The only new feature of this corollary is that $D\phi_\varepsilon^t$ includes a derivative with respect to the initial angle $\theta_0$. If we write $y(t, a, b, \varepsilon)$, $\theta(t, a, b, \varepsilon)$ for the solution of (4.11) with initial conditions $y = a$, $\theta = b$ at $t = 0$, then $\theta(t, a, b, \varepsilon) = 2\pi t/T + b$ and

$$D\phi_\varepsilon^t(a, b) = \begin{bmatrix} y_a & \theta_a \\ 0 & 1 \end{bmatrix} = \begin{bmatrix} Y & w \\ 0 & 1 \end{bmatrix}$$

where $Y$ is an $n \times n$ matrix and $w$ is an $n$-vector. Then $y, Y, w$ satisfies the system

$$\dot{y} = \varepsilon g_1(y) + \cdots + \varepsilon^k g_k(y) + \varepsilon^{k+1} \hat{g}(y, t + bT/2\pi, \varepsilon)$$

$$\dot{Y} = \varepsilon \frac{\partial g_1}{\partial y}(y)Y + \cdots + \varepsilon^k \frac{\partial g_k}{\partial y}(y)Y + \varepsilon^{k+1} \frac{\partial \hat{g}}{\partial y}(y, t + bT/2\pi, \varepsilon)Y$$

$$\dot{w} = \frac{\partial g_1}{\partial y}(y)w + \cdots + \varepsilon^k \frac{\partial g_k}{\partial y}(y)w$$

$$+ \varepsilon^{k+1} \left\{ \frac{\partial \hat{g}}{\partial y}(y, t + bT/2\pi, \varepsilon)Y + \frac{T}{2\pi} \frac{\partial \hat{g}}{\partial t}(y, t + bT/2\pi, \varepsilon) \right\},$$

with initial conditions $y = a$, $Y = I$, $w = 0$. This system is the same as (4.9) except that $b$ enters as a parameter in the $\hat{g}$ terms and the $w$ equation is adjoined. There is a similar system analogous to (4.10), and the difference between solutions of these systems is estimated in the usual way since the terms involving $\hat{g}$ have bounds independent of $b$.

Finally we turn to the comparison of $x$, $X_{\mathrm{imp}}$, and $S$, defined in (4.5) and (4.6).

4.6 COROLLARY   Let $K$ be a compact subset of $\mathbb{R}^n$, and let $\varepsilon_0 > 0$ be given. There exist constants $\mathbf{T} > 0$, $c > 0$ such that $\| x(t, A(a, \varepsilon), \varepsilon) - X(t, a, \varepsilon) \| <$

$c\varepsilon^k$ and $\| x(t, A_{\mathrm{imp}}(a, \varepsilon), \varepsilon) - X_{\mathrm{imp}}(t, a, \varepsilon) \| < c\varepsilon^k$ for $a \in K$, $|t| \leqslant \mathbf{T}/\varepsilon$, $0 \leqslant \varepsilon \leqslant \varepsilon_0$.

Choose $r > 0$, form $K_r$, and obtain $\mathbf{T}$ from Lemma 4.1. Then apply Theorem 4.2 with $V = K_r$. Write the transformation (4.2) as $x = U(y, t, \varepsilon)$. Then since $U$ is Lipschitz on $K_r$, (4.7) implies that $\| U(y(t, a, \varepsilon), t, \varepsilon) - U(z(t, a, \varepsilon), t, \varepsilon) \|$ is of order $\varepsilon^k$ for $a \in K$, $0 \leqslant \varepsilon \leqslant \varepsilon_0$, and $|t| \leqslant \mathbf{T}/\varepsilon$. But this is just $\| x(t, A_{\mathrm{imp}}(a, \varepsilon), \varepsilon) - X_{\mathrm{imp}}(t, a, \varepsilon) \|$, so $X_{\mathrm{imp}}$ approximates $x$ to order $\varepsilon^k$. Since $X_{\mathrm{imp}}$ and $X$ agree to this order it cannot hurt to replace $X_{\mathrm{imp}}$ by $X$; since $A_{\mathrm{imp}}$ and $A$ agree to this order and $x(t, b, \varepsilon)$ is Lipschitz in $b$ (on a suitable compact set containing $A(a, \varepsilon)$ and $A_{\mathrm{imp}}(a, \varepsilon)$ for $a \in K$, $0 \leqslant \varepsilon \leqslant \varepsilon_0$) it cannot hurt to replace $A_{\mathrm{imp}}$ by $A$. So we conclude $\| x(t, A(a, \varepsilon), \varepsilon) - X(t, a, \varepsilon) \|$ is of order $\varepsilon^k$ for the usual set of $(t, a, \varepsilon)$. This is the standard asymptotic estimate for averaging.

## Notes

The basic asymptotic estimates for averaging are derived in [3] and [28]. The specific details in the present chapter, such as the hypotheses of Theorem 4.2 (a) and (b), and the estimate for the derivatives in Corollary 4.4, are new (to my knowledge) and are especially formulated to meet needs in Section 6.

The estimates in this section are all for expanding time intervals of order $1/\varepsilon$. In some cases there exist estimates valid on longer intervals. If the first $r$ of the averaged terms $g_1, \ldots, g_r$ are equal to zero, the $k$th order average is accurate to $O(\varepsilon^{k-j})$ on a time interval of order $1/\varepsilon^{j+1}$ for any $j = 0, 1, \ldots, r$ ([21]). (Part of this result is given in [32].) There are also estimates valid for all time for solutions in the basin of an attractor [32]; some remarks on this are given in the notes to Section 5 below.

## 5   EXISTENCE AND HYPERBOLICITY OF PERIODIC SOLUTIONS

This section and the next concern systems periodic in time. Such systems always admit averaging (case 1 of Theorem 3.3), which can be done in the stroboscopic manner. Thus the original system is

$$(5.1) \qquad \dot{x} = \varepsilon f(x, t, \varepsilon)$$

where $f(x, t + T, \varepsilon) = f(x, t, \varepsilon)$ for all $(x, t, \varepsilon)$. The transformation

$$(5.2) \qquad x = y + \varepsilon u_1(y, t) + \cdots + \varepsilon^k u_k(y, t)$$

satisfies $u_i(y, t + T) = u_i(y, t)$ and $u_i(y, jT) = 0$ for $i = 1, \ldots, k$ for all $(y, t)$ and all integers $j$, and carries (5.1) into

$$(5.3) \qquad \dot{y} = \varepsilon g(y, t, \varepsilon) = \varepsilon g_1(y) + \cdots + \varepsilon^k g_k(y) + \varepsilon^{k+1} \hat{g}(y, t, \varepsilon)$$

with $g(y, t + T, \varepsilon) = g(y, t, \varepsilon)$. The truncated averaged system is denoted

$$(5.4) \qquad \dot{z} = \varepsilon h(z, \varepsilon) = \varepsilon g_1(z) + \cdots + \varepsilon^k g_k(z).$$

As before, the solution of (5.4) satisfying $z = a$ at $t = 0$ is denoted $z(t, a, \varepsilon)$, and similarly for $x$ and $y$.

In the last section, the order of averaging to be used was determined by the asymptotic accuracy desired for the approximate solutions. In this section, some results require hypotheses on $g_1$ only, and for these it suffices to use first order averaging ($k = 1$ in (5.1), (5.2), (5.3), and (5.4)). Other results impose conditions on $g_1, \ldots, g_k$ with $k$ unspecified. Therefore one will usually use the least value of $k$ for which the hypotheses of a theorem are satified (if it is possible to satisfy them at all for a given system (5.1)).

The basic hypothesis of this section is that there exists a simple zero $z_0$ of $g_1$, that is, a point at which $g_1(z_0) = 0$ and the matrix $A_0 = (\partial g_1/\partial z)(z_0)$ is non-singular. Since this hypothesis makes reference to $g_1$ only, it is meaningful for any order of averaging but never in itself requires more than first order. The basic conclusion is the existence of unique periodic solutions of (5.1) and (5.3) near $z_0$ for small $\varepsilon$. At the same time, we establish the existence of a rest point of (5.4) near $z_0$. In case $k = 1$, $z_0$ itself is such a rest point, but when $k > 1$, the rest point becomes a function of $\varepsilon$. It is important to be able to locate this rest point, because the stability properties of the rest point of (5.4) sometimes carry over to corresponding properties of the periodic solutions of (5.1) and (5.3). Even when first order averaging suffices to prove the existence of a periodic solution, higher order averaging may be needed to determine its stability (or more generally, its hyperbolicity).

5.1 THEOREM   If $z_0$ is a simple zero of $g_1$ then there exists an $\varepsilon_0 > 0$ and unique initial conditions $a(\varepsilon)$ and $b(\varepsilon)$, $0 \leqslant \varepsilon \leqslant \varepsilon_0$, such that $a(0) = b(0) = z_0$, $a(\varepsilon)$ is a rest point of (5.4), and $x(t, b(\varepsilon), \varepsilon)$ and $y(t, b(\varepsilon), \varepsilon)$ are periodic solutions of (5.1) and (5.3) with (not necessarily least) period $T$.

The proof is the implicit function theorem. A rest point of (5.4) for a given $\varepsilon$ is a point $a$ such that $h(a, \varepsilon) = 0$. Since $h(z_0, 0) = g_1(z_0) = 0$ and $h_a(z_0, 0) = (\partial g_1/\partial z)(z_0) = A_0$ is nonsingular, there is a unique $a(\varepsilon)$ for $\varepsilon$ near zero such that $a(0) = z_0$ and $h(a(\varepsilon), \varepsilon) = 0$, and this $a(\varepsilon)$ is of class $C^\infty$. The proof for the periodic solutions is slightly harder. A solution $y(t, b, \varepsilon)$ of (5.3) for a given $\varepsilon$ is periodic if $\phi(b, \varepsilon) = y(T, b, \varepsilon) - y(0, b, \varepsilon) = 0$, since in this case $y(t, b, \varepsilon)$ and $y(t + T, b, \varepsilon)$ are two solutions of (5.3) with the same initial conditions. (Notice that this fact depends on the periodicity of (5.3).) Now $\phi(b, \varepsilon) = \varepsilon \int_0^T g(y(t, b, \varepsilon), t, \varepsilon) \, dt$, and the implicit function theorem does not apply because $\phi$ is identically zero when $\varepsilon = 0$. But for $\varepsilon \neq 0$, $\phi$ vanishes if and only if $\psi(b,\varepsilon) = \int_0^T g(y(t,b,\varepsilon),t,\varepsilon)\,dt$ vanishes, and the implicit function theorem may be applied here. In fact, for any $b$, $\psi(b,0) = Tg_1(b)$ since $y(t,b,0) \equiv b$ and $g(b,t,0) \equiv g_1(b)$. So $\psi(z_0,0) = 0$ and $\psi_b(z_0,0) = TA_0$ is nonsingular, so there is a unique $b(\varepsilon)$ with $b(0) = b_0$ and $\psi(b(\varepsilon),\varepsilon) = 0$ for $\varepsilon$ near 0, and $b(\varepsilon)$ is of class $C^\infty$. Therefore $y(t,b(\varepsilon),\varepsilon)$ is periodic, and the same is true of $x(t,b(\varepsilon),\varepsilon)$ because (5.2) is periodic and stroboscopic. (If a nonstroboscopic averaging is used, putting $y(t,b(\varepsilon),\varepsilon)$ into (5.2) gives a periodic solution for $x$ having a different initial value $c(\varepsilon)$ but still with $c(0) = z_0$.)

The principal contrast between Theorem 5.1 and the results of Section 4 is that the latter are quantitative and valid for a limited time interval (although one which expands as $\varepsilon \to 0$), while this is qualitative and valid for all time. (Periodicity is an eternal quality.) A more subtle contrast is that Lemma 4.1 allows arbitrary specification of the interval of validity $0 \leqslant \varepsilon \leqslant \varepsilon_0$, whereas here the existence of such an interval is part of the conclusion and it is difficult to determine a suitable value for $\varepsilon_0$. To do so would require replacing the mere citation of the implicit function theorem with an actual proof of the implicit function theorem in the present setting, giving careful attention to the various quantities (such as Lipschitz constants) that enter into the estimates in the proof.

In the case $g_1$ has a zero $z_0$ which is not simple, higher order averaging can be used to determine whether periodic solutions near $z_0$ exist. We will not pursue this question, which belongs to bifurcation theory, except to give a brief example. Namely suppose that $n = 1$ (the problem is one-dimensional) and that (5.4) for $k = 2$ reads $\dot{z} = \varepsilon z^2 - \varepsilon^2 z$. Then $g_1(z) = z^2$ has a double root at $z_0 = 0$ which gives rise to two rest points $a_1(\varepsilon) = 0$ and $a_2(\varepsilon) = \varepsilon$, both reducing to $z_0$ at $\varepsilon = 0$. It seems reasonable in this case to expect that (5.1) and (5.3) have two periodic solutions near 0, but this expectation need not be true without further conditions. (For instance (5.3) could be $\dot{y} = \varepsilon y^2 - \varepsilon^2 y + \varepsilon^3$, which has no periodic solutions near 0.) Instead of addressing these issues we return to the hypothesis of Theorem 5.1, that $z_0$ is a simple zero of $g_1$, and study the hyperbolicity properties of the resulting rest point of (5.4) and periodic solution of (5.1) and (5.3).

If we had assumed real analyticity in $\varepsilon$, Theorem 5.1 could have been proved using power series in $\varepsilon$. In the $C^\infty$ case these series need not converge, so an iteration method which does converge was used instead. (The implicit function theorem is usually proved by iteration using the contraction mapping lemma.) But even in the $C^\infty$ case perturbation series have some advantages for computation. Knowing by Theorem 5.1 that they exist, such $C^\infty$ functions as $b(\varepsilon)$ and $y(t,b(\varepsilon),\varepsilon)$ can be expanded in asymptotic power series and the coefficients evaluated recursively. In this way we will show:

**5.2 LEMMA**  Let $z_0$, a simple zero of $g_1$, be given as in Theorem 5.1. Then:

(1) The functions $z(t, a, \varepsilon)$ and $y(t, a, \varepsilon)$, for constant $a$, are equal through order $\varepsilon^k$ when expanded in powers of $\varepsilon$.

Each of the following pairs of functions, when expanded in $\varepsilon$, are equal through order $\varepsilon^{k-1}$:

(2) $a(\varepsilon)$ and $b(\varepsilon)$.
(3) $a(\varepsilon)$ and $y(t, b(\varepsilon), \varepsilon)$.
(4) $h_z(a(\varepsilon), \varepsilon)$ and $g_y(y(t, b(\varepsilon), \varepsilon), t, \varepsilon)$.

In particular the series for $y(t, b(\varepsilon), \varepsilon)$ and $g_y(y(t, b(\varepsilon), \varepsilon), t, \varepsilon)$ are independent of $t$ through order $\varepsilon^{k-1}$.

To prove (1), let $z(t, a, \varepsilon) = z_0(t, a) + \varepsilon z_1(t, a) + \cdots$ and similarly for $y(t, a, \varepsilon)$. Substituting into (5.4) and (5.3) and expanding yields a sequence of differential equations for $y_i$ and $z_i$, subject to initial conditions $z_0(0, a) = y_0(0, a) = a$, $z_i(0, a) = y_i(0, a) = 0$ for $i \geq 1$. The differential equations for $y_i$ and $z_i$ coincide for $i \leq k$ since (5.3) and (5.4) agree to this order, and the initial conditions are the same, so the solutions are equal. (Notice that $z_0(t, a) = y_0(t, a) \equiv a$, but the higher orders usually depend on $t$.) To prove (2), recall that $a(\varepsilon)$ is the unique solution of $h(a(\varepsilon), \varepsilon) = 0$, $a(0) = z_0$. Thus the coefficients of $a(\varepsilon) = a_0 + \varepsilon a_1 + \cdots$ can be determined recursively from

$$0 = g_1(z_0) + \varepsilon \left[ \left( \frac{\partial g_1}{\partial z} \right)_{a_0} a_1 + g_2(a_0) \right] + \cdots;$$

taking $a_0 = z_0$ makes the first term vanish, the second can then be solved for $a_1$ since $(\partial g_1/\partial z)(z_0)$ is nonsingular, and the subsequent terms can be solved since each $a_i$ appears for the first time multiplied by $(\partial g_1/\partial z)(z_0)$. On the other hand, $b(\varepsilon)$ is the unique solution of $\psi(b(\varepsilon), \varepsilon) = 0$, $b(0) = z_0$, where $\psi(b, \varepsilon) = \int_0^T g(y(t, b, \varepsilon), t, \varepsilon)\, dt$. This can be expanded in powers of $\varepsilon$, and by (1) and (5.3), (5.4) the expansion will be unchanged through order $k - 1$ if $g$ is replaced by $h$ and $y(t, b, \varepsilon)$ is replaced by $z(t, b, \varepsilon)$. Let $\Psi(b, \varepsilon) = \int_0^T h(z(t, b, \varepsilon), \varepsilon)\, dt$. Since $z(t, a(\varepsilon), \varepsilon) \equiv a(\varepsilon)$ and $h(a(\varepsilon), \varepsilon) = 0$, we see that $\Psi(a(\varepsilon), \varepsilon) = 0$. Therefore $b = a(\varepsilon)$ is the unique solution of $\Psi(b, \varepsilon) = 0$; it is unique because, once again, $\Psi_b(z_0, 0) = (\partial g_1/\partial z)(z_0)$ is nonsingular. But the recursive determination of the coefficients of the series solution of $\Psi(b(\varepsilon), \varepsilon) = 0$ is the same as for $\psi(b(\varepsilon), \varepsilon) = 0$ through order $k - 1$. Therefore $a(\varepsilon) = b(\varepsilon)$ through order $k - 1$. Next, (3) and (4) follow immediately from (1), (2), (5.3) and (5.4).

The expressions in item (4) of Lemma 5.2 play an essential role in determining the qualitative behavior of solutions near the rest points and periodic solutions because they form the coefficient matrices of the linear variational equations of these solutions. We write them as

$$h_z(a(\varepsilon), \varepsilon) = A(\varepsilon) + \varepsilon^k B(\varepsilon)$$

(5.5)
$$g_y(y(t, b(\varepsilon), \varepsilon), t, \varepsilon) = A(\varepsilon) + \varepsilon^k C(t, \varepsilon)$$

$$A(\varepsilon) = A_0 + \varepsilon A_1 + \cdots + \varepsilon^{k-1} A_{k-1}$$

noticing that $A_0 = (\partial g_1/\partial z)(z_0)$ is the same as $A_0$ in Theorem 5.1, the non-singular matrix which keeps reappearing in our analysis, and that $C(t, \varepsilon)$ is periodic with period $T$. If the equations $\dot{z}(t, a, \varepsilon) = \varepsilon h(z(t, a, \varepsilon), \varepsilon)$ and $\dot{y}(t, b, \varepsilon) = \varepsilon g(y(t, b, \varepsilon), t, \varepsilon)$ are differentiated with respect to $a$ and $b$ respectively, and $a$ and $b$ replaced by $a(\varepsilon)$ and $b(\varepsilon)$, one finds that $\zeta = z_a(t, a(\varepsilon), \varepsilon)$ and $\eta = y_b(t, b(\varepsilon), \varepsilon)$ are the principal matrix solutions (i.e. the solutions equal to $I$ at $t = 0$) of the linear variational equations:

(5.6)
$$\dot{\zeta} = \varepsilon[A(\varepsilon) + \varepsilon^k B(\varepsilon)]\zeta$$

and

(5.7)
$$\dot{\eta} = \varepsilon[A(\varepsilon) + \varepsilon^k C(t, \varepsilon)]\eta.$$

A fundamental notion in qualitative theory of differential equations is hyperbolicity. A constant matrix will be called *hyperbolic* if its eigenvalues lie off the imaginary axis, in which case its *index* (of hyperbolicity) is the number of eigenvalues in the right half-plane. (The definition is frequently formulated for the exponential of the matrix, in which case the unit circle replaces the imaginary axis.) A rest point $a$ of an autonomous system $\dot{z} = H(z)$ is called a *hyperbolic rest point* if $H_z(a)$ is a hyperbolic matrix, and the index of $a$ is the index of $H_z(a)$. The matrix $H_z(a)$ is the coefficient matrix of the variational equation $\dot{\zeta} = H_z(a)\zeta$. The definition of hyperbolicity for a $T$-periodic orbit $y(t)$ of a $T$-periodic system $\dot{y} = G(y, t)$ is a little more involved than the definition for a rest point. The starting point is the variational equation $\dot{\eta} = G_y(y(t), t)\eta$ of the periodic solution. To this equation, which has period $T$, one applies Floquet theory: if the principal matrix solution is $U(t)$, there exists a complex matrix $\Gamma$ such that $e^{\Gamma T} = U(T)$, and a periodic matrix function $P(t)$ such that $U(t) = P(t)e^{\Gamma t}$. The periodic orbit $y(t)$ is called *hyperbolic* if $\Gamma$ is hyperbolic, with the index defined accordingly.

Hyperbolicity of the rest points $a(\varepsilon)$ of (5.4) and the corresponding periodic solutions $y(t, b(\varepsilon), \varepsilon)$ of (5.3) resulting from Theorem 5.1 depend, according to these definitions, upon the linear variational equations (5.6) and (5.7). We will show shortly that the Floquet theory of (5.7) yields $\Gamma(\varepsilon) = \varepsilon[A(\varepsilon) + \varepsilon^k D(\varepsilon)]$ for some matrix $D(\varepsilon)$. Thus hyperbolicity of $a(\varepsilon)$ depends upon the placement of the eigenvalues of $\varepsilon[A(\varepsilon) + \varepsilon^k B(\varepsilon)]$ relative to the imaginary axis, whereas hyperbolicity of $y(t, b(\varepsilon), \varepsilon)$ involves the matrix $\varepsilon[A(\varepsilon) + \varepsilon^k D(\varepsilon)]$. Our primary task will be to give conditions involving $A(\varepsilon)$ alone, which guarantee that both of these matrices are hyperbolic, with the same index, for sufficiently small $\varepsilon$. (The simplest sufficient condition—the standard one for first order averaging—is that $A_0$ be hyperbolic, but we will give weaker sufficient conditions involving higher order terms in $A(\varepsilon)$ when $A_0$ is not hyperbolic. It is not sufficient to require merely that $A(\varepsilon)$ be hyperbolic.)

Our first goal is to do the Floquet theory for (5.7). For generality and simplicity we will allow a constant term in the next two lemmas, replacing $\varepsilon A(\varepsilon)$ by $R(\varepsilon)$. Thus we consider an arbitrary linear system autonomous through order $k$:

$$(5.8) \qquad\qquad \dot{\eta} = [R(\varepsilon) + \varepsilon^{k+1} S(t, \varepsilon)]\eta.$$

**5.3 LEMMA** The principal matrix solution of (5.8) has the form $U(t, \varepsilon) = e^{R(\varepsilon)t}[I + \varepsilon^{k+1} V(t, \varepsilon)]$ for some matrix function $V(t, \varepsilon)$ with $V(0, \varepsilon) = 0$.

For analytic functions of $\varepsilon$ the lemma is obvious: expand $U(t, \varepsilon)$ in powers of $\varepsilon$ and determine the coefficients recursively from (5.8); $S$ cannot affect the solution until the term of order $k + 1$, so through order $k$ it coincides with the solution $e^{R(\varepsilon)t}$ when $S = 0$. For the $C^\infty$ case, substitute the desired form for $U$ into (5.8) and conclude that $V$ must satisfy $\dot{V} = e^{-R(\varepsilon)t} S(t, \varepsilon) e^{R(\varepsilon)t}[I + \varepsilon^{k+1} V]$ with $V(0, \varepsilon) = 0$. The anti-climactic punch line is that this differential equation (a nonperiodic, nonhomogeneous linear equation) has a smooth solution $V(t, \varepsilon)$ defined for all $t$, which is all that is required for the lemma.

**5.4 LEMMA** For sufficiently small $\varepsilon$, there exists a real periodic matrix $P(t, \varepsilon)$ of period $T$ and a real matrix $D(\varepsilon)$ such that the change of variables $\eta = P(t, \varepsilon)\xi$ reduces (5.8) to the autonomous system $\dot{\xi} = \Gamma(\varepsilon)\xi$, where $\Gamma(\varepsilon) = R(\varepsilon) + \varepsilon^{k+1} D(\varepsilon)$. In particular, the principal matrix solution of (5.8) has the form $U(t, \varepsilon) = P(t, \varepsilon) e^{\Gamma(\varepsilon)t}$.

The proof is a repeat of the proof of Floquet's theorem, using Lemma 5.3. $\Gamma(\varepsilon)$ should be taken to be $1/T$ times a logarithm of $U(T, \varepsilon) = e^{R(\varepsilon)T}[I + \varepsilon^{k+1} V(T, \varepsilon)]$. One choice of such a logarithm, smooth in $\varepsilon$, is $R(\varepsilon)T + \log[I + \varepsilon^{k+1} V(T, \varepsilon)]$, where the second term is evaluated from the familiar Taylor series for $\log(1 + x)$. The second term is therefore of order $k + 1$, so $\Gamma(\varepsilon) = R(\varepsilon) + \varepsilon^{k+1} D(\varepsilon)$. Then $P(t, \varepsilon)$ is defined to equal $U(t, \varepsilon) e^{-\Gamma(\varepsilon)t}$, and periodicity and the remaining properties are checked as usual. Notice that a complex logarithm was not required. (In standard Floquet theory one must either allow a complex logarithm or allow $P$ to have period $2T$ in some cases. We did not need this because $I + \varepsilon^{k+1} V$ is near the identity, and the exponential for real matrices is a diffeomorphism from a neighborhood of the zero matrix to a neighborhood of the identity, so is locally invertible. Our choice of logarithm gives this inverse.)

Applying Lemma 5.4 to 5.7 we see that $y(t, b(\varepsilon), \varepsilon)$ is hyperbolic if $\varepsilon[A(\varepsilon) + \varepsilon^k D(\varepsilon)]$ is hyperbolic. We have already seen that $a(\varepsilon)$ is a hyperbolic rest point if $\varepsilon[A(\varepsilon) + \varepsilon^k B(\varepsilon)]$ is hyperbolic. For $\varepsilon > 0$, the placement of the eigenvalues of both matrices relative to the imaginary axis is unchanged if the factor $\varepsilon$ is deleted. Both matrices will be hyperbolic if $A(\varepsilon)$ is $k$-hyperbolic in the following sense.

5.5 DEFINITION   $A(\varepsilon)$ is *k-hyperbolic* of index $i$ if for every matrix function $B(\varepsilon)$ there exists an $\varepsilon_0 > 0$ such that $A(\varepsilon) + \varepsilon^k B(\varepsilon)$ is hyperbolic of index $i$ for all $\varepsilon$ in the interval $0 < \varepsilon < \varepsilon_0$.

Under the rules governing this chapter, 'every $B(\varepsilon)$' means 'every smooth $B(\varepsilon)$,' but Lemma 5.6 and Theorem 5.7 remain valid under the interpretation 'every continuous $B(\varepsilon)$'. This definition says that $A(\varepsilon)$ is $k$-hyperbolic if it is hyperbolic and, in addition, its hyperbolicity cannot be destroyed by arbitrary perturbations of order $k$. An equivalent condition is that $e^{\varepsilon A(\varepsilon)}$ be hyperbolic in the sense of having no eigenvalues on the unit circle, and that this hyperbolicity be resistant to perturbations of order $k + 1$ (rather than $k$); this guarantees that the principal matrix solutions of (5.6) and (5.7) have no eigenvalues on the unit circle. We now turn to sufficient conditions for $k$-hyperbolicity.

Observe that if a portion $\tilde{A}(\varepsilon) = A_0 + \varepsilon A_1 + \cdots + \varepsilon^{\tilde{k}-1} A_{\tilde{k}-1}$ of $A(\varepsilon)$, with $\tilde{k} < k$, is $\tilde{k}$-hyperbolic, then $A(\varepsilon)$ is $k$-hyperbolic; the terms of order $> \tilde{k}$, whether they come from $A(\varepsilon)$ or from $\varepsilon^k B(\varepsilon)$, do not destroy the hyperbolicity. The following lemma covers the case $\tilde{k} = 1$, and implies the standard criterion (previously mentioned) for hyperbolicity in first order averaging.

5.6 LEMMA   If the constant matrix $A_0$ is hyperbolic then it is 1-hyperbolic, and hence $A(\varepsilon)$ is $k$-hyperbolic for all $k \geq 1$.

We must examine the eigenvalues of $A_0 + \varepsilon B(\varepsilon)$, which are roots of an $n$th degree polynomial in $\lambda$ having coefficients depending continuously on $\varepsilon$. For $\varepsilon = 0$ these roots lie off the imaginary axis. Place disjoint circles around each root, each circle being contained entirely in one half plane. By Rouché's theorem, for small $\varepsilon$ the roots remain inside the circles, with the sum of the multiplicities of roots in each circle equalling the multiplicities of the original root (for $\varepsilon = 0$) at the center of the circle. In particular, the number of roots on either side of the imaginary axis remains constant for small $\varepsilon$.

For $k > 1$, hyperbolicity of $A(\varepsilon)$ does not imply $k$-hyperbolicity. Two examples, of quite different character, are given by $A(\varepsilon) + \varepsilon^2 B$ with

$$(5.9) \qquad A(\varepsilon) = \begin{bmatrix} 2 & \varepsilon \\ \varepsilon & 0 \end{bmatrix}, \qquad B = \begin{bmatrix} 0 & 0 \\ 0 & 2a \end{bmatrix}$$

and

$$(5.10) \qquad A(\varepsilon) = \begin{bmatrix} \varepsilon & 0 \\ 1 & \varepsilon \end{bmatrix}, \qquad B = \begin{bmatrix} 0 & a^2 \\ 0 & 0 \end{bmatrix}.$$

In (5.9), $A(\varepsilon)$ has eigenavalues $1 \pm \sqrt{(1 + \varepsilon^2)}$; these have expansions $2 + \frac{1}{2}\varepsilon^2 + \cdots$ and $-\frac{1}{2}\varepsilon^2 + \cdots$. Thus $A(\varepsilon)$ is hyperbolic with index 1 for small $\varepsilon$. The eigenvalue in the right half-plane will remain under perturbation, but the eigenvalue in the left half-plane is of order $\varepsilon^2$ and is easily disrupted by $\varepsilon^2 B$. The eigenvalues of $A(\varepsilon) + \varepsilon^2 B$ are $1 + a\varepsilon^2 \pm \sqrt{[1 + (1 - 2a)\varepsilon^2 + a^2\varepsilon^4]}$. Choosing the negative square root and expanding in $\varepsilon$ gives $(-\frac{1}{2} + 2a)\varepsilon^2 + \cdots$, which lies in the right half-plane for $a > 1$. Thus $A(\varepsilon)$ is not 2-hyperbolic. In the second example (5.10), $A(\varepsilon)$ has double eigenvalue $\varepsilon$ and so is hyperbolic of index two for all positive $\varepsilon$. These eigenvalues might seem to be invulnerable to $\varepsilon^2 B$ since they are of order $\varepsilon$, but in fact the eigenvalues of $A(\varepsilon) + \varepsilon^2 B$ are $\varepsilon(1 \pm a)$: the perturbation, even though it is of order $\varepsilon^2$, affects the eigenvalues at order $\varepsilon$. If $a \geqslant 1$, the hyperbolicity of index 2 is disrupted.

These examples suggest that two conditions must be imposed upon $A(\varepsilon)$ for $k$-hyperbolicity, one to guarantee that the eigenvalues of $A(\varepsilon)$ are at a distance at least $0(\varepsilon^{k-1})$ from the imaginary axis, and one to guarantee that no $0(\varepsilon^k)$ perturbation of $A(\varepsilon)$ can affect the eigenvalues at a lower order. In addition, Lemma 5.6 suggests that eigenvalues whose distance from the imaginary axis is $0(1)$ pose no difficulty: only eigenvalues which are on the imaginary axis when $\varepsilon = 0$ are sensitive. These ideas will be seen reflected in the hypotheses of Theorem 5.7 below.

Let $S(\varepsilon)$ be a smooth matrix function which block diagonalizes $A(\varepsilon)$ into 'left', 'center', and 'right' blocks $L(\varepsilon)$, $C(\varepsilon)$, $R(\varepsilon)$ which for $\varepsilon = 0$ have their eigenvalues respectively in the left half-plane, on the imaginary axis, and in the right half-plane. Thus

$$(5.11) \qquad S(\varepsilon)^{-1} A(\varepsilon) S(\varepsilon) = \begin{bmatrix} L(\varepsilon) & 0 & 0 \\ 0 & C(\varepsilon) & 0 \\ 0 & 0 & R(\varepsilon) \end{bmatrix}.$$

(A simple way to show the existence and smoothness of $S(\varepsilon)$ is to draw curves $\gamma_l, \gamma_c, \gamma_r$ in the complex plane surrounding the eigenvalues of $A_0$ in the left, center, and right respectively. For small $\varepsilon$ the eigenvalues of $A(\varepsilon)$ remain within these curves and the projection operators

$$\frac{1}{2\pi i} \int_\gamma (\lambda I - A(\varepsilon))^{-1} \, d\lambda$$

define the invariant subspaces of $A(\varepsilon)$ which give the block diagonalization.) The eigenvalues of $C(\varepsilon)$ do not necessarily stay on the imaginary axis for $\varepsilon > 0$; it is these which are critical for $k$-hyperbolicity.

5.7 THEOREM   Let $C(\varepsilon)$ be the center block of $A(\varepsilon)$ and let its size be $m \times m$. Then $A(\varepsilon)$ is $k$-hyperbolic provided:

(a)   there is a polynomial $P(\varepsilon)$ of degree $k - 1$ with $m \times m$ matrix coefficients which is invertible for small $\varepsilon$ and diagonalizes $C(\varepsilon)$ through order $k - 1$:

$$P(\varepsilon)^{-1} C(\varepsilon) P(\varepsilon) = \begin{bmatrix} \tilde{\lambda}_1(\varepsilon) & & & \\ & \cdot & & \\ & & \cdot & \\ & & & \cdot & \\ & & & & \tilde{\lambda}_m(\varepsilon) \end{bmatrix} + \varepsilon^k E(\varepsilon)$$

   where $\tilde{\lambda}_i(\varepsilon)$ are polynomials of degree $k - 1$.

(b)   Each $\tilde{\lambda}_i$ has at least one coefficient which is not pure imaginary.

   In particular, (a) is implied by the much simpler condition

(a$'$)   $A_0$ has no multiple eigenvalues on the imaginary axis
   and (b) is equivalent (given (a)) to

(b$'$)   There exists $c > 0$ such that every eigenvalue $\lambda(\varepsilon)$ of $C(\varepsilon)$ satisfies $|\operatorname{Re} \lambda(\varepsilon)| \geqslant c\varepsilon^{k-1}$ for all small $\varepsilon$.

   In hypothesis (a) it is essential that $\tilde{\lambda}_i$ be polynomials of degree (less than or equal to) $k - 1$, because of the way we have formulated condition (b): thus $\tilde{\lambda}_i$ are truncations of the eigenvalue expansions. On the other hand it is not essential that $P(\varepsilon)$ be a polynomial, but it must be smooth, defined for $\varepsilon = 0$, and nonsingular for $\varepsilon = 0$ and for $\varepsilon$ small. (In counterexample (5.10), $A(\varepsilon)$, which is its own center block, can be diagonalized to any order by a matrix which either becomes unbounded or singular when $\varepsilon \to 0$.) Given such a smooth $P(\varepsilon)$, its Taylor polynomial of order $k - 1$ will satisfy hypothesis (a). It seems best to formulate (a) in terms of polynomials, since this facilitates searching for $P(\varepsilon)$ by solving sequentially for the coefficients $P_0, \ldots, P_{k-1}$, a process which can meet with obstructions at every step. (We will not pursue this except to point out that $P_0$ must diagonalize $C_0$; there is no hope to satisfy (a) unless $C_0$ is diagonalizable.)

   To prove Theorem 5.7, we will show first that (a$'$) implies (a). The eigenvalues of $A(\varepsilon)$ are roots of the $n$th degree polynomial $\det(A(\varepsilon) - \lambda I) = a_n(\varepsilon)\lambda^n + \cdots + a_0(\varepsilon) = Q(\varepsilon, \lambda)$. A simple eigenvalue $\lambda_0$ of $A(0)$ satisfies $Q(0, \lambda_0) = 0$ and $Q_\lambda(0, \lambda_0) \neq 0$. By the implicit function theorem, such a simple eigenvalue defines a smooth function $\lambda(\varepsilon)$ which is an eigenvalue for all small $\varepsilon$. If all imaginary eigenvalues of $A_0$ are simple, then the eigenvalues of the center block of $A(\varepsilon)$ are given by distinct functions $\lambda_j(\varepsilon)$. The matrix $C(\varepsilon)$, having distinct eigenvalues, is diagonalizable. The matrix $P(\varepsilon)$ whose columns are the eigenvectors of unit length can be shown to be smooth: each unit eigenvector $v(\varepsilon)$ is the solution of a system $[A(\varepsilon) - \lambda(\varepsilon)I]v = 0$, $v_1^2 + \cdots + v_n^2 = 1$; exactly one of the $n$ 'eigenvector' conditions on $v$ is redundant since $\lambda(\varepsilon)$ is a simple eigenvalue, but this condition is replaced by the 'unit length' condition, giving $n$ independent conditions on the $n$ unknowns; the implicit function theorem guarantees a smooth solution for small $\varepsilon$. Hence $P(\varepsilon)^{-1}C(\varepsilon)P(\varepsilon) = \Lambda(\varepsilon)$ is a smooth diagonalization to all orders, which implies (5.12) if the $\tilde{\lambda}_j(\varepsilon)$ are defined as the truncations of $\lambda_j(\varepsilon)$ at order $k - 1$; the $P(\varepsilon)$ may also be truncated. Thus (a') implies (a). Given that (a) holds, the equivalence of (b) and (b') is clear, for no term of order greater than $\varepsilon^{k-1}$ could impart the required rate of divergence from the imaginary axis.

It remains to show that (a) and (b) imply $k$-hyperbolicity. To this end we examine the eigenvalues of $A(\varepsilon) + \varepsilon^k B$, for arbitrary $B$. Applying the similarity (5.11) we may assume $A(\varepsilon)$ is already in block form, and applying the similarity given in (a), we may assume that the center block is diagonalized through order $k - 1$. The argument of Lemma 5.6 shows that the left and right eigenvalues of $A(\varepsilon)$ remain in their half-planes under perturbation by $B$. We will show that for each $\tilde{\lambda}_i(\varepsilon)$, $i = 1, \ldots, m$, there is an eigenvalue of $A(\varepsilon) + \varepsilon^k B$ of the form $\lambda_i(\varepsilon) = \tilde{\lambda}_i(\varepsilon) + \varepsilon^k \sigma_i(\varepsilon)$; that is, that $\varepsilon^k B$ affects the eigenvalues only at order $k$ and higher. This much depends only on (a). It then follows from (b) that each $\lambda_i(\varepsilon)$ moves off the imaginary axis into one of the half-planes, the same half-plane as $\tilde{\lambda}_i(\varepsilon)$, for small $\varepsilon$. (The half-plane is determined by the lowest order coefficient of $\tilde{\lambda}_i(\varepsilon)$ which is not pure imaginary.) This establishes $k$-hyperbolicity.

Let $\tilde{\lambda}(\varepsilon)$ denote any one of the $\tilde{\lambda}_i(\varepsilon)$. Since the $\tilde{\lambda}_i(\varepsilon)$ need not all be distinct, let $p$ be the number of $\tilde{\lambda}_i(\varepsilon)$ which are equal to $\tilde{\lambda}(\varepsilon)$. We must show that $A(\varepsilon) + \varepsilon^k B$ has $p$ eigenvalues of the form $\tilde{\lambda}(\varepsilon) + \varepsilon^k \sigma(\varepsilon)$. Consider the equation $f(\varepsilon, \lambda) = \det[A(\varepsilon) + \varepsilon^k B - \lambda I] = 0$ and make the change of variables $\lambda = \tilde{\lambda}(\varepsilon) + \varepsilon^k \sigma$ to obtain $g(\varepsilon, \sigma) = \det[A(\varepsilon) - \tilde{\lambda}(\varepsilon)I + \varepsilon^k(B - \sigma I)] = 0$. The matrix $A(\varepsilon) - \tilde{\lambda}(\varepsilon)I$ is block diagonal; its center block is diagonalized through order $k - 1$; and $p$ of these diagonal entries are zero through order $k - 1$. By rearranging rows and columns we may bring these zero entries into a $p \times p$ block in the upper left, so that $g(\varepsilon, \sigma) = 0$ takes the form

$$(5.12) \qquad \det\left[\begin{array}{c|c} \varepsilon^k(Q(\varepsilon) - \sigma I_p) & \varepsilon^k R(\varepsilon) \\ \hline \varepsilon^k S(\varepsilon) & T(\varepsilon) + \varepsilon^k(V(\varepsilon) - \sigma I_{n-p}) \end{array}\right] = 0.$$

Here the terms of order $\varepsilon^k$ (that is, $Q, R, S,$ and $V$) come from $\varepsilon^k B$ and also from $\varepsilon^k E$ (in hypothesis (a)), whereas $T(\varepsilon)$ is block diagonal and consists of the left and right blocks L and R of $A$, and an $(m - p) \times (m - p)$ block which is diagonal with entries $\tilde{\lambda}_i(\varepsilon) - \tilde{\lambda}(\varepsilon)$, where $\tilde{\lambda}_i(\varepsilon)$ are those $m - p$ of the $\tilde{\lambda}_i$ which are not equal to $\tilde{\lambda}(\varepsilon)$. Thus the eigenvalues of $T(\varepsilon)$ are all of order strictly less than $\varepsilon^k$, implying that det $T(\varepsilon)$ is of order strictly less than $\varepsilon^{(n-p)k}$. Dividing the first $p$ rows of (5.2) by $\varepsilon^k$, it is easy to see that the resulting equation is of the form $\det(Q(\varepsilon) - \sigma I_p)\det T(\varepsilon) + 0(\varepsilon^{(n-p)k})$. Since det $T(\varepsilon)$ is of order strictly less than $0(\varepsilon^{(n-p)k})$, this is equivalent to $\det(Q(\varepsilon) - \sigma I_p) + 0(\varepsilon) = 0$. This is a polynomial in $\sigma$ of degree $p$. It therefore has $p$ roots $\sigma_i(\varepsilon)$; then $\tilde{\lambda}_i(\varepsilon) + \varepsilon^k \sigma_i(\varepsilon)$ are eigenvectors of $A(\varepsilon) + \varepsilon^k B$, as claimed. (The nature of the dependence of $\sigma_i(\varepsilon)$ on $\varepsilon$ is explained by algebraic function theory: they are given, in general, by fractional power series and so are continuous, but not necessarily differentiable, at $\varepsilon = 0$.) This completes the proof of Theorem 5.7.

If a rest point or periodic solution of a differential equation is hyperbolic, there is a neighborhood in which solutions 'look' hyperbolic; that is, solutions near the stable manifold first approach the rest point or periodic orbit, and then veer off along the unstable manifold. (This statement can be formalized as Hartman's theorem.) If $A(\varepsilon)$ is $k$-hyperbolic then $a(\varepsilon)$ and $y(t, b(\varepsilon), \varepsilon)$ are hyperbolic for small $\varepsilon$, and so they have these 'hyperbolic neighborhoods'. But it is unclear how the size of these neighborhoods depends upon $\varepsilon$. We will show that if the hypotheses of Theorem (5.7) are satisfied, a condition we call *strong $k$-hyperbolicity*, there exist hyperbolic neighborhoods whose radius is of order $\varepsilon^{k-1}$. It is not clear whether this result holds under the assumption of (weak) $k$-hyperbolicity alone. In the special case $k = 1$, where 1-hyperbolicity is equivalent to hyperbolicity of $A_0$, strong 1-hyperbolicity is also equivalent. Also for $k = 1$, the neighborhoods are of size $\varepsilon^0$ and do not shrink. Therefore every 1-hyperbolic solution has a hyperbolic neighborhood of fixed size (Theorem 5.8 below). This theorem will be crucial in Section 6, and is the only use that will be made of hyperbolic neighborhoods. In order to avoid the details of Hartman's theorem our approach is a simple geometric one, which establishes some of the basic features of hyperbolic motion in the neighborhood without proving local topological conjugacy. But it is not difficult to combine some of these ideas with the proof of Hartman's theorem to obtain conjugacy on the shrinking neighborhoods.

The arguments to follow will be based on a pair of inequalities for a quadratic form defined by a hyperbolic matrix. We present these inequalities first for a fixed hyperbolic matrix $B$ not depending on a parameter $\varepsilon$. There exists a projection matrix $Q^+$ onto the 'expanding' subspace for $B$, that is, the direct sum of the generalized eigenspaces corresponding to eigenvalues of $B$ with positive real parts. There is a similar projection $Q^-$ onto the 'contracting' subspace. Since these subspaces are invariant, $Q^+$ and $Q^-$ commute with $B$, and also with $e^{Bt}$. We write $Q^+ u = u_+$, $Q^- u = u_-$, $u = u_+ + u_-$. Our aim is to establish that there is an inner product $\langle , \rangle$ and a constant $r$ such that

$$(5.13) \qquad \langle Bu_+, u_+ \rangle \geq r\langle u_+, u_+ \rangle \quad \text{and} \quad \langle Bu_-, u_- \rangle \leq -r\langle u_-, u_- \rangle$$

for $u_+ \neq 0$ and $u_- \neq 0$. This inner product will not always be the 'standard' inner product on $\mathbb{R}^n$. The meaning of (5.13) is that the vector field $Bu$ points 'away from the origin' in the expanding subspace and 'towards the origin' in the contracting subspace, in the following sense: the vector $Bu_+$ at the point $u_+$ makes an angle less than a right angle with the ray from the origin through $u_+$, when angles are defined using $\langle\,,\,\rangle$; the angle between $u_-$ and $Bu_-$ is greater than a right angle. It follows that the solutions $e^{Bt}u$ of $\dot{u} = Bu$ move away from the origin in the expanding subspace and towards the origin in the contracting subspace. (There are of course other ways to prove this, but for $A(\varepsilon)$ it is easy to prove a sharp form of (5.13) that will give us hyperbolic neighborhoods.)

To prove (5.13), begin with the real Jordan canonical form of $B$. In place of ones outside the diagonal blocks, any other number $\eta > 0$ may be used. Thus there exist an invertible matrix $T$ with the following properties:

(i)

$$\tilde{B} = T^{-1}BT = \begin{bmatrix} L & 0 \\ 0 & R \end{bmatrix}$$

where the eigenvalues of $L$ and $R$ are in the left and right half-planes respectively.

(ii) $\quad L = \begin{bmatrix} L_1 & & & & & \\ & L_2 & & & * & \\ & & \ddots & & & \\ & * & & & l_1 & \\ & & & & & l_2 \end{bmatrix}$

where $L_i$ is a two by two real matrix of the form

$$L_i = \begin{bmatrix} a_i & b_i \\ -b_i & a_i \end{bmatrix} \text{ with } a_i < 0,$$

and each $l_i$ is a real number $< 0$. The entries $*$ are either $0$ or $\eta$.

(iii) A similar statement holds for $R$ except that all diagonal entries are positive.

Of course the real Jordan form is more specific about the location of entries $\eta$, but we will not need this. The columns of $T$ form a basis, with the first few columns spanning the contracting subspace and the remainder, the expanding subspace. If $u$ is any column vector expressed in the standard basis, $\tilde{u} = T^{-1}u$ gives its components with respect to the $T$-basis. Let $(\,,\,)$ be the usual inner product $(x, y) = y^*x$, $*$ denoting transpose. Working entirely in the coordinates with respect to the $T$ basis, and using the standard inner product $(\,,\,)$ for these coordinates we consider the quadratic form $(\tilde{B}\tilde{u}, \tilde{u})$. The values of this quadratic form are not changed if $\tilde{B}$ is replaced by its symmetrization

$\frac{1}{2}(\tilde{B} + \tilde{B}^*)$. According to (ii) and (iii), this matrix is almost diagonal: the entries $\pm b_i$ are eliminated, and the only nonzero off-diagonal elements are $\pm \eta/2$. The diagonal entries are unchanged. Therefore if $\eta$ is taken sufficiently small, the eigenvalues of $(\tilde{B} + \tilde{B}^*)/2$ are close to the diagonal elements, and hence lie in the same half-planes. Since a quadratic form defined by a symmetric matrix is diagonalizable with its eigenvalues as coefficients, it is clear that $(\tilde{B}\tilde{u}, \tilde{u})$ is negative definite in the contracting space and positive definite in the expanding space. To obtain (5.13), it is only necessary to define the inner product $\langle u, v \rangle = (T^{-1}u, T^{-1}v) = (\tilde{u}, \tilde{v})$. This is the inner product in which the columns of $T$ are orthonormal. In terms of this inner product, $\langle Bu, u \rangle = (\tilde{B}\tilde{u}, \tilde{u})$. Notice that in this argument the matrix $T$, and hence the inner product $\langle\ ,\ \rangle$, depends upon the choice of $\eta$, which must be taken sufficiently small; how small depends upon the distance of the eigenvalues of $B$ from the imaginary axis. In general the eigenvalues of a non-symmetric matrix are not equal to the eigenvalues of the quadratic form it defines. In the case of $\tilde{B}$, the eigenvalues of the quadratic form $(\tilde{B}\tilde{u}, \tilde{u})$ are obtained from the eigenvalues of $\tilde{B}$ by dropping the imaginary parts and perturbing slightly (because of $\eta$). This is what makes the argument work.

Next we apply a similar argument to a strongly $k$-hyperbolic matrix $A(\varepsilon)$. The projections $Q^+(\varepsilon)$ and $Q^-(\varepsilon)$ depend continuously on $\varepsilon$, since they may be written as contour integrals around closed paths surrounding the eigenvalues. (The paths must vary with $\varepsilon$ if some of the eigenvalues approach the imaginary axis as $\varepsilon \to 0$.) Combining the similarity transformation (5.11) with the similarities putting $L(0)$ and $R(0)$ in real Jordan form and with the similarity in hypothesis (a) of Theorem 5.7, we see that there is a matrix $T(\varepsilon)$ such that

$$(5.14) \quad \tilde{A}(\varepsilon) = T(\varepsilon)^{-1}A(\varepsilon)T(\varepsilon) = \begin{bmatrix} L^{\#}(\varepsilon) & 0 & 0 & 0 \\ 0 & C_1^{\#}(\varepsilon) & 0 & 0 \\ 0 & 0 & C_2^{\#}(\varepsilon) & 0 \\ 0 & 0 & 0 & R^{\#}(\varepsilon) \end{bmatrix}$$

where $L^{\#}(0)$ and $R^{\#}(0)$ are in real Jordan form with $\eta$ outside the diagonal blocks, having eigenvalues in the left and right half-planes respectively; $L^{\#}(\varepsilon)$ and $R^{\#}(\varepsilon)$ differ from these by $0(\varepsilon)$; and $C_1^{\#}(\varepsilon)$ and $C_2^{\#}(\varepsilon)$ are diagonal to order $\varepsilon^{k-1}$, the diagonal elements of $C_1^{\#}(\varepsilon)$ having real parts $< -c\varepsilon^{k-1}$ and those of $C_2^{\#}(\varepsilon)$ having real parts $> c\varepsilon^{k-1}$. (These inequalities express hypothesis (b$'$) of Theorem 5.7.) It is important that the matrix $T(\varepsilon)$ is defined and nonsingular even for $\varepsilon = 0$. The direct sum of the invariant subspaces associated with the $L^{\#}$ and $C_1^{\#}$ blocks is the contracting subspace, the expanding subspace corresponding in the same way to $C_2^{\#}$ and $R^{\#}$. Of course these subspaces depend upon $\varepsilon$. Let $\langle\ ,\ \rangle$ be the inner product (also dependent upon $\varepsilon$) in which the columns of $T(\varepsilon)$ are orthonormal. We claim that

$$(5.15) \quad \langle A(\varepsilon)u_+, u_+ \rangle \geq \tfrac{1}{2} c\varepsilon^{k-1} \langle u_+, u_+ \rangle \quad \text{and}$$

$$\langle A(\varepsilon)u_-, u_- \rangle \leq -\tfrac{1}{2} c\varepsilon^{k-1} \langle u_-, u_- \rangle.$$

These estimates are easily proved in the coordinate system in which $A(\varepsilon)$ takes the form $\tilde{A}(\varepsilon)$ and $\langle \, , \rangle$ is $(\,,)$. There are four invariant subspaces to check. In the space associated with $L^+(\varepsilon)$, the argument is the same as before: taking $\eta$ small enough and restricting $\varepsilon$ to a small enough interval $0 < \varepsilon \leq \varepsilon_0$, the symmetrization of $L^{\#}(\varepsilon)$ is nearly diagonal with negative diagonal entries, hence is negative definite with an inequality $(L^{\#}(\varepsilon)x, x) \leq -r(x, x)$ with $r > 0$ independent of $\varepsilon$. Since $-r \leq -\tfrac{1}{2} c\varepsilon^{k-1}$ for small $\varepsilon$, this proves the second inequality of (5.15) on the invariant subspace associated with $L^{\#}$. Of course the necessity for the factor $-\tfrac{1}{2} c\varepsilon^{k-1}$ arises from the block $C_1^{\#}$. The symmetrization of this block has diagonal elements $< -c\varepsilon^{k-1}$ and off-diagonal elements $0(\varepsilon^k)$. It follows (by the same argument as in the proof of Theorem 5.17, using (5.12)) that the eigenvalues are equal to the diagonal elements plus $0(\varepsilon^k)$, and are therefore less than $-\tfrac{1}{2} c\varepsilon^{k-1}$ for small $\varepsilon$. This completes the proof of the second inequality in (5.14), and the first, involving the invariant subspaces for $C_2^{\#}$ and $R^{\#}$, is similar.

Having established (5.15), we now use it to obtain shrinking hyperbolic neighborhoods of strongly $k$-hyperbolic rest points of (5.4). It is convenient to work in the norm $|u| = |u|_\varepsilon = \langle u, u \rangle^{1/2}$, where $\langle \, , \rangle$ is the inner product occurring in (5.15). This norm is $\varepsilon$-dependent, but because the matrix $T(\varepsilon)$ in (5.14) is continuous even for $\varepsilon = 0$ (although $A(0)$ is not hyperbolic), there exists constants $\mu$ and $\nu$ independent of $\varepsilon$ (for small $\varepsilon$) such that $\mu\|u\| \leq |u| \leq \nu\|u\|$; here $\| \, \|$ is the Euclidean norm. Choose $0 < \sigma < \tfrac{1}{2}$ and define the *expanding cone* to be the set of $u$ such that $|u_+| \geq \sigma|u|$. Similarly, define the *contracting cone* by $|u_-| \geq \sigma|u|$. These cones overlap and every vector $u$ belongs to one or the other or both.

By Taylor's theorem and $h(a(\varepsilon), \varepsilon) = 0$, we can write $h(a(\varepsilon) + u, \varepsilon) = h_z(a(\varepsilon), \varepsilon)u + H(u, \varepsilon)$, where $H(u, \varepsilon)$ satisfies a bound $|H(u, \varepsilon)| \leq \beta|u|^2$ for $|u| \leq \gamma$, with $\beta, \gamma > 0$. Recall $h_z(a(\varepsilon), \varepsilon) = A(\varepsilon) + \varepsilon^k B(\varepsilon)$, $Q^+ Au = AQ^+ u = Au_+$, and $|\langle u, v \rangle| \leq |u||v|$ (Cauchy–Schwarz inequality). Then

$$\frac{1}{2} \frac{d}{dt} |Q^+ u|^2 = \langle Q^+(\varepsilon)h(a(\varepsilon) + u), Q^+ u \rangle = \langle Q^+ Au, u_+ \rangle$$

$$+ \varepsilon^k \langle Q^+ Bu, u_+ \rangle + \langle Q^+ H(u, \varepsilon), u_+ \rangle$$

$$\geq c\varepsilon^{k-1} |u_+|^2 - \varepsilon^k \delta |u| |u_+| - \beta |u|^2 |u_+|$$

for some $\delta > 0$ (the operator norm of $Q^+ B$ relative to $| \, |$). In the expanding cone $|u_+| > \sigma|u|$, this is $\geq (c\sigma\varepsilon^{k-1} - \varepsilon^k\delta - \beta|u|)|u||u_+|$. Now consider a shrinking neighborhood $|u| < \alpha\varepsilon^{k-1}$. In such a neighborhood the quantity in parentheses is positive for small $\varepsilon$ if $\alpha$ is sufficiently small that $c\sigma - \alpha\beta > 0$. Therefore $|Q^+u|$ increases along solutions of (5.4) in the intersection of the expanding cone with $|u| < \alpha\varepsilon^{k-1}$. Using $\mu\|u\| \leqslant |u| \leqslant \nu\|u\|$ we see that $\|Q^+u\|$ increases in the intersection of the expanding cone with $\|u\| < (\alpha/\nu)\varepsilon^{k-1}$. It is also clear why one cannot expect the hyperbolicity to manifest itself at greater distances from $a(\varepsilon)$. In a similar way, $\|u_-\|$ decreases in the intersection of the contracting cone with a shrinking neighborhood. Since the boundary of the expanding cone lies in the contracting cone, on this boundary $\|u_-\|$ decreases and $\|u_+\|$ increases. This shows that the solutions flow into the expanding cone (as time increases) on this boundary. Similarly solutions on the boundary of the contracting cone flow into that cone as time decreases. Every solution in the expanding cone remains in that cone until it leaves the shrinking neighborhood, due to the growth of $\|u_+\|$. Every solution in the contracting cone remains in that cone under the reverse flow until it leaves the shrinking neighborhood. So the only solution which remains in the shrinking neighborhood for all time is the rest point itself.

The proof of the existence of hyperbolic neighborhoods for $y(t, b(\varepsilon), \varepsilon)$ is similar. One can either examine $g(y(t, b(\varepsilon), \varepsilon) + u, t, \varepsilon)$ or $g(a(\varepsilon) + u, t, \varepsilon)$ for $u$ in a shrinking neighborhood of zero; the former can be written $[A(\varepsilon) + \varepsilon^k C(t, \varepsilon)]u + G(u, t, \varepsilon)$ where $G = 0(\|u\|^2)$, and the latter differs from this only to order $\varepsilon^{k+1}$ in view of Lemma 5.2. The second form is better because it yields neighborhoods $\|u - a(\varepsilon)\| < \alpha\varepsilon^{k-1}$ which do not depend on time. These neighborhoods are not centered on the periodic orbit, but on $a(\varepsilon)$.

The hyperbolic neighborhoods $\|z - a(\varepsilon)\| \leqslant \alpha\varepsilon^{k-1}$ and $\|y - a(\varepsilon)\| \leqslant \alpha\varepsilon^{k-1}$ not only shrink as $\varepsilon \to 0$, but their centers move. In the case $k = 1$, in which the size of the neighborhoods is constant, there exists a fixed neighborhood $U$ of $z_0 = a(0)$ which is contained in the hyperbolic neighborhoods for sufficiently small $\varepsilon$. Therefore what we have proved includes the following result, needed in Section 6:

5.8 THEOREM  If $(\partial g_1/\partial z)(z_0)$ is hyperbolic (and hence 1-hyperbolic), there exists a neighborhood $U$ of $z_0$ and an $\varepsilon_0 > 0$ such that for $0 < \varepsilon < \varepsilon_0$, every solution of (5.4) which contains a point of $U$ leaves $U$ either forward or backward in time, except for the constant solution $a(\varepsilon)$; and every solution of (5.3) which contains a point of $U$ leaves $U$ either forward or backward in time, except for the periodic solution $y(t, b(\varepsilon), \varepsilon)$.

## Notes

Theorem 5.1 has been proved many ways but is perhaps orginally due to Poincaré. Many generalizations and examples of this basic idea are given in

[11], not without some errors. The concept of hyperbolicity for rest points and periodic orbits is beautifully explained in [27].

The concept of $k$-hyperbolicity is due to Clark Robinson and myself and was first presented in [24]. The approach taken there was in terms of period maps rather than flows, and a technical result of Hirsch and Pugh played a major role. The method used here is more elementary, and also perhaps more satisfactory. The basic criterion for $k$-hyperbolicity here (Theorem 5.7) is more computable than that in [24]. It is based on our results in [23]. The example in section 5 of [24] contains an error, corrected in [16]. The concept of $k$-hyperbolicity has been generalized to exponential dichotomies in [12].

The discussion of hyperbolic neighborhoods given here is somewhat different than any that I have seen. For the linear system $\dot{u} = Bu$, the usual theorem (for instance in [11]) says that there exist constants $K$ and $\alpha$ such that $\| u_+(t) \| \leqslant K e^{\alpha t} \| u_+(0) \|$ for $t < 0$ for solutions in the $+$ space. This uses the Euclidean norm at the expense of needing the constant $K$. On the other hand the usual treatment of the same idea for maps instead of flows (see [26]) is based on changing the norm so that the expanding subspace is truly expanding, and the contracting subspace truly contracting, in the new norm. Inequality (5.13) is similar in that a new inner product, and hence a new norm, is used. The argument based on symmetrizing a real Jordan form was suggested to me by Richard Miller. The result on hyperbolic neighborhoods here is considerably weaker than that in [24]. This was done in order to avoid the technicalities of Hartman's theorem. The stronger version says that on the hyperbolic neighborhoods, which are of size $\varepsilon^{k-1}$, the flows of $x$, $y$, and $z$ are topologically conjugate by a homeomorphism which moves points only a small distance. These local conjugacies can be used to obtain some of the extended asymptotic estimates mentioned in the notes to Section 4; this is done in [29].


## 6  GLOBAL BEHAVIOR


In Section 5 it was shown that a rest point of an averaged system often gives rise to a periodic orbit of the original system, and that the stability type of the rest point often determines that of the periodic orbit. If the averaged system has several such rest points, the original system will have (for sufficiently small $\varepsilon$) several periodic orbits. The next question to ask is, does the structure of the orbits in a region around the rest points carry over to the original system? The most important feature of this structure is the interconnection pattern, consisting of the orbits which tend to one rest point as $t \to -\infty$ and to another as $t \to +\infty$. We will show that under suitable hypotheses this structure does carry over to the original system, provided that attention is confined to a finite set of rest points, in the case of first-order averaging. For second and higher order averaging this is not always the case without much stricter hypotheses.

(A general result for higher-order averaging is not available, but some examples will be treated in the next section.) The present section is restricted to first-order averaging of periodic systems, with period $T = 2\pi$ for convenience. (This can always be achieved by changing the unit of time.) The basic equations (5.1)–(5.4) take the simple form

$$(6.1) \qquad \dot{x} = \varepsilon f(x, t, \varepsilon); \qquad f(x, t + 2\pi, \varepsilon) = f(x, t, \varepsilon)$$

$$(6.2) \qquad x = y + \varepsilon u(y, t); \qquad u(y, t + 2\pi) = u(y, t)$$

$$(6.3) \qquad \dot{y} = \varepsilon g(y) + \varepsilon^2 \hat{g}(y, t, \varepsilon); \qquad g(y) = (1/2\pi) \int_0^{2\pi} f(y, t, 0)\, dt$$

$$(6.4) \qquad \dot{z} = \varepsilon g(z).$$

A special feature of first-order averaging is that the orbits of (6.4) are independent of $\varepsilon$; only the time scale along solutions depends upon $\varepsilon$. This may be seen by setting $\tau = \varepsilon t$ and $' = d/d\tau$, upon which (6.4) becomes

$$(6.5) \qquad\qquad\qquad z' = g(z),$$

completely independent of $\varepsilon$. We place the following assumptions on the solutions of (6.5):

(H1) The zeros $\alpha$ of $g$ are isolated and each matrix $g_z(\alpha)$ is hyperbolic.
(H2) Every solution $z(\tau)$ of (6.5) either approaches some zero $\alpha$ of $g$ as $\tau \to \infty$, or else $\| z(\tau) \| \to \infty$ as $\tau$ increases (either becoming infinite as $\tau \to$ some value beyond which $z(\tau)$ ceases to exist, or becoming infinite as $\tau \to \infty$). Similarly every $z(\tau)$ either approaches some $\alpha$ as $\tau \to -\infty$, or recedes to infinity as $\tau$ decreases.
(H3) Any intersections of stable and unstable manifolds of two rest points are transversal in the sense that the tangent vectors to the stable and unstable manifolds at each point of intersection span $\mathbb{R}^n$.

If the solutions of (6.5) satisfy (H1)–(H3) they are said to form a Morse–Smale flow on $\mathbb{R}^n$ (without periodic orbits). Such a flow (actually a partial flow, as solutions need not be defined for all $\tau$) does not have all the properties of a Morse–Smale flow as usually defined, because $\mathbb{R}^n$ is not a compact manifold. (In the course of the reasoning to follow, compactness will be achieved by restriction to a compact subset, at the expense of having to deal with a boundary.) The stable and unstable manifolds of $\alpha$ are denoted $W^s(\alpha)$, $W^u(\alpha)$.

Consider a finite collection $\alpha_1, \ldots, \alpha_q$ of zeros of $g$. A *connecting orbit* from $\alpha_i$ to $\alpha_j$ is a solution $z(\tau)$ of (6.5) which approaches $\alpha_i$ as $\tau \to -\infty$ and $\alpha_j$ as $t \to +\infty$. Connecting orbits exist if and only if $W^u(\alpha_i) \cap W^s(\alpha_j) \neq \emptyset$. The intersection of stable and unstable manifolds may consist of a single connecting orbit, or a whole family of them; but each connecting orbit, together with $\alpha_i$ and $\alpha_j$, is compact, which may not be true of the whole intersection of stable and unstable manifolds. If there is a connecting orbit from $\alpha_i$ to $\alpha_j$ and one from $\alpha_j$ to $\alpha_k$, then there is a connecting orbit from $\alpha_i$ to $\alpha_k$. (This is a fact which we take over from dynamical systems theory without proof. This section will contain a number of such facts, which we justify on the grounds that they are obvious in each application and their general proof would take us far afield. As usual, references are given in the notes at the end.) So the rest points $\alpha_1, \ldots, \alpha_q$ are partially ordered by the relation 'there is a connecting orbit from $\alpha_i$ to $\alpha_j$', denoted $\alpha_i > \alpha_j$, and we assume that the rest points are numbered compatibly with this partial order, so that $\alpha_i > \alpha_j$ implies $i > j$. (Note that the flow runs 'downhill' in terms of this ordering.) The strict inequality sign is used because no connecting orbit can run from a rest point to itself, or even from one rest point to another of the same index: if $\alpha_i > \alpha_j$ then $u_i = \dim W^u(\alpha_i) > u_j = \dim W^u(\alpha_j)$, since the fact that $W^u(\alpha_i)$ intersects $W^s(\alpha_j)$ transversally along at least a one-dimensional intersection (the connecting orbit) implies $u_i + (n - u_j) \geqslant n + 1$.

The partial order $>$ on a finite set of rest points is called its *diagram*, because it can be represented as a directed graph in which points labeled $\alpha_1$ to $\alpha_q$ are arranged with $\alpha_i$ above $\alpha_j$ if $i > j$, and an arrow drawn from $\alpha_i$ to $\alpha_j$ if $\alpha_i > \alpha_j$ and there is no $\alpha_k$ with $\alpha_i > \alpha_k > \alpha_j$.

A *containing domain $D$* for a diagram for (6.5) is an open subset of $\mathbb{R}^n$ with the following properties: $D$ contains $\alpha_1, \ldots, \alpha_q$ and no other rest points; if there exists a connecting orbit from $\alpha_i$ to $\alpha_j$ then there exists a connecting orbit contained entirely in $D$; and the boundary $\partial D$ is a smooth manifold. Figure 1 shows a simple containing domain in $\mathbb{R}^2$ which will be used as an example throughout the following discussion. Figure 2(a) is not a containing domain because the connecting orbit from $\alpha_2$ to $\alpha_1$ touches $\partial D$ and so is not contained in the open set $D$. The slightly smaller domain in Fig. 2(b) is not a containing domain either, but the slightly larger domain in Fig. 2(c) is. Figure 3 is a containing domain in spite of the connecting orbits tangent to $\partial D$, because there are other connecting orbits between the same rest points which are contained in $D$. Every diagram existing in an averaged system (6.5) has a containing domain since the set consisting of the points $\alpha_1, \ldots, \alpha_q$ and one connecting orbit between each pair with $\alpha_i > \alpha_j$ is compact and can be enclosed in an open set containing no other rest points.

Our aim is to establish results on the structure of solutions of (6.3), since such results will automatically carry over, via the transformation (6.2), to the original system (6.1). It is best to present (6.3) in the augmented, autonomous form

$$(6.6) \qquad \dot{y} = \varepsilon g(y) + \varepsilon^2 \hat{g}(y, \theta, \varepsilon) \qquad y \in \mathbb{R}^n$$
$$\dot{\theta} = 1 \qquad\qquad\qquad \theta \in S^1.$$

We know from Section 5 and hypothesis (H1) that for each $\alpha_i$ in the finite set $\alpha_1, \ldots, \alpha_q$ there exists an interval $0 < \varepsilon < \varepsilon_i$ in which (6.6) has a periodic orbit

$\beta_i = \beta_i(\varepsilon)$ which approaches $\{\alpha_i\} \times S^1$ as $\varepsilon \to 0$. By Theorem 5.8 each $\beta_i$ lies in a product neighborhood $U_i \times S^1$, independent of $\varepsilon$, such that every orbit remaining in the neighborhood as $t \to +\infty$ or as $t \to -\infty$ must approach $\beta_i$. Remember we do not know that (6.6) satisfies the analog of (H1)–(H3); this has been assumed only for (6.5). We will estabish the following result, which says that 'semi-globally', i.e. for any containing domain $D$, the behavior of (6.6) is the same as (6.5).

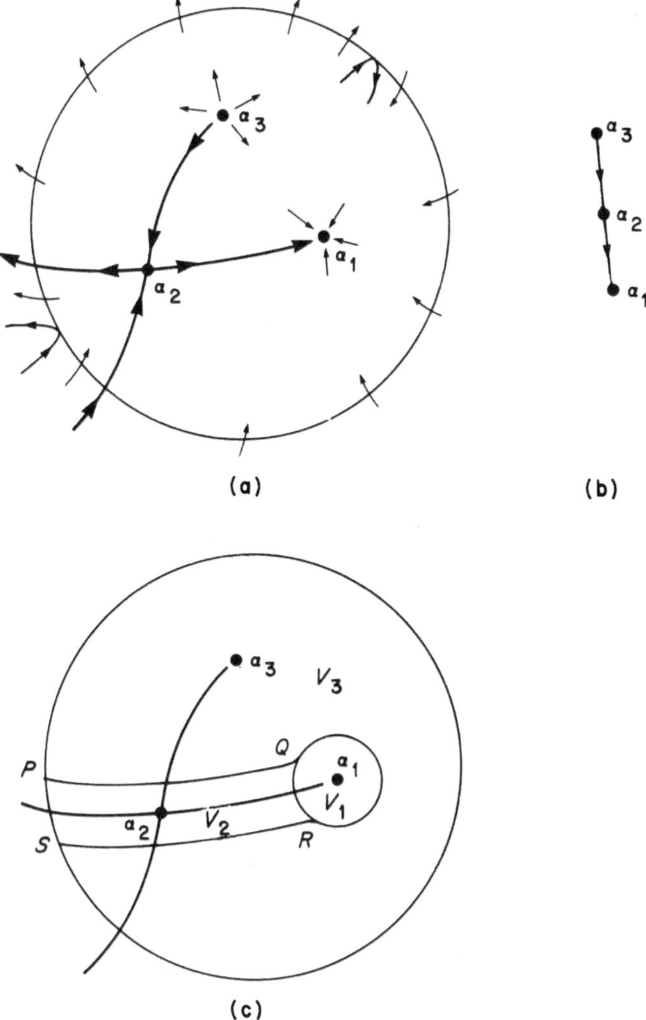

Fig. 1. (a) A containing domain; (b) The diagram of the containing domain; (c) A filtration for the same system

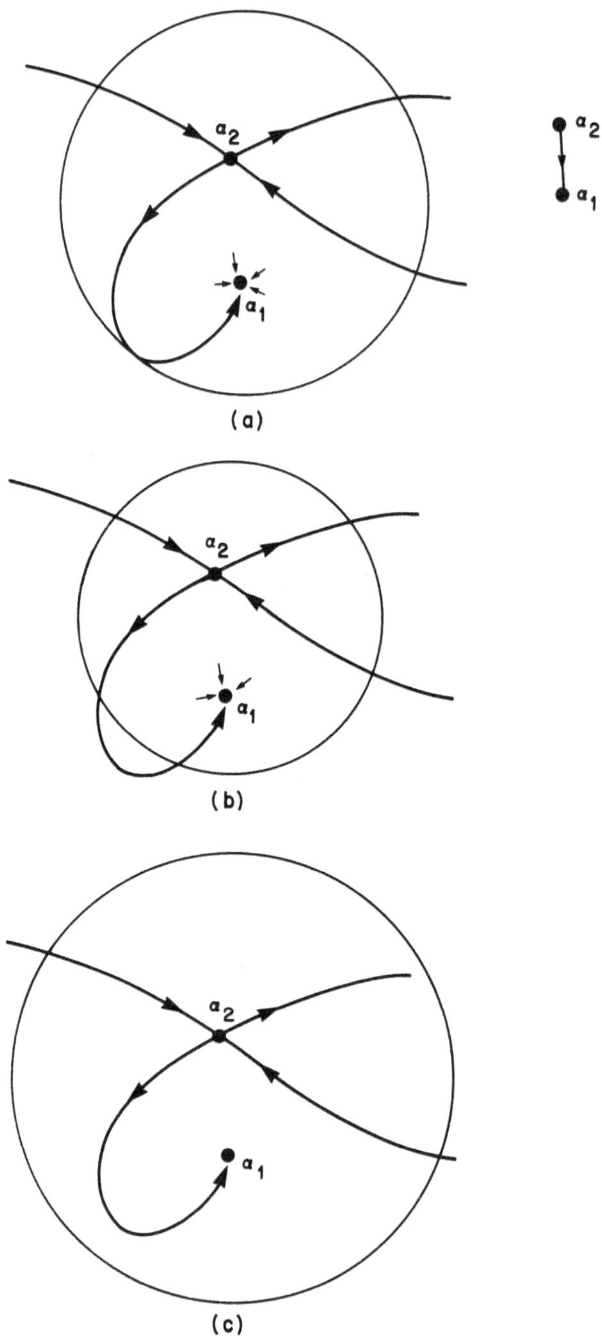

**Fig. 2.** (a) and (b) are not containing domains but (c) is

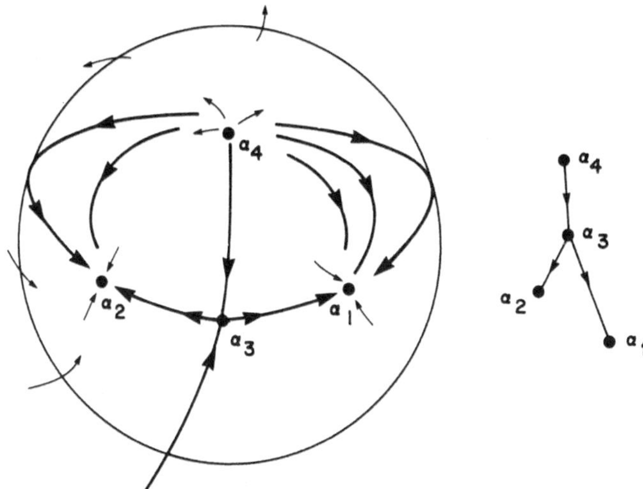

**Fig. 3.** A more complicated containing domain and its diagram. There are infinitely many connecting orbits from $\alpha_4$ to $\alpha_1$, one of which is tangent to $\delta D$; similarly for $\alpha_2$. These tangencies are acceptable because there exist other connecting orbits contained in $D$

**6.1 THEOREM** Let (6.5) satisfy (H1)–(H3), and let $D$ be a containing domain for the diagram of a finite set of rest points $\alpha_1, ..., \alpha_q$ of (6.5). Then there exists $\varepsilon_0 > 0$ such that for $0 < \varepsilon < \varepsilon_0$, the following statements are true:

(1) Every solution of (6.6) which passes through a point of $D \times S^1$ either approaches one of the orbits $\beta_1, ..., \beta_q$ as $t \to \infty$ or else leaves $D \times S^1$, and the same is true as $t \to -\infty$.
(2) The diagram of connecting orbits between the $\beta_i$ which are contained entirely in $D \times S^1$ is the same as the diagram of the $\alpha_i$.

An alternative formulation of this theorem, whose proof is the same, relaxes hypotheses (H1)–(H3) and the requirement that $D$ be a containing domain. Simply let $D$ be any bounded open set with smooth boundary in $\mathbb{R}^n$, containing finitely many rest points $\alpha_1, ..., \alpha_q$ of (6.5) all of which are hyperbolic, and such that if there is a connecting orbit from $\alpha_i$ to $\alpha_j$ in $\bar{D}$ then there is one in $D$. (This rules out Fig. 2(a) but allows 2(b), which was formely forbidden, as well as 2(c).) Replace the full diagram of $\alpha_1, ..., \alpha_q$ by the diagram of connecting orbits contained entirely in $D$. Replace (H2) by the requirement that each solution in $D$ either approach an $\alpha_i$ or leave $D$ as $t \to \pm \infty$, and replace (H3) by the requirement that the intersections of stable and unstable manifolds be transverse at least along the connecting orbits contained in $D$. Then the conclusions of Theorem 6.1 hold.

In the following discussion it is helpful to keep in mind that (6.6) can be regarded as a perturbation (at order $\varepsilon^2$) of

$$(6.7) \qquad\qquad \dot{z} = \varepsilon g(z) \qquad z \in \mathbb{R}^n$$
$$\dot{\theta} = 1 \qquad \theta \in S^1.$$

The flow of (6.7) is related trivially to the flow of (6.5): they differ in the time scale along orbits and in the presence of the $\theta$ variable. Most features of (6.5) carry over immediately to (6.7) by taking Cartesian products with $S^1$; for instance if $W$ is a stable or unstable manifold of a rest point $\alpha$ of (6.5), then $W^* = W \times S^1$ is a stable or unstable manifold of the periodic orbit $\{\alpha\} \times S^1$ of (6.7). We speak of comparing (6.5) with (6.6) because (6.5) is easier to picture, but it is always '(6.5) in the form (6.7)' that compares most directly with (6.6).

There exists a *relative filtration* for the flow of (6.5) in $D$. Before defining this notion, consider Fig. 1(c), which presents a relative filtration for Fig. 1. The sink $\alpha_1$ is contained in a disk $V_1$; solutions flow transversely across $\partial V_1$ into $V_1$. The saddle $\alpha_2$ is contained in a set $V_2$ (the 'rectangle' PQRS); the flow across $\partial(V_1 \cup V_2)$ is transverse into $V_1 \cup V_2$ except where $\partial(V_1 \cup V_2)$ coincides with $\partial D$. It is especially important to consider the 'corners' P, Q, R, S. At $Q$ and $R$ the corners are smoothed so that $\partial(V_1 \cup V_2)$ is a manifold at these points. At P and S the corners are not smoothed, and the flow is transverse to arc PQ at P although it happens not be directed into $V_1 \cup V_2$ at this point. (The corners at P and S could not be smoothed without losing transversality.) Along the arc PS where $\partial(V_1 \cup V_2)$ lies along $\partial D$, the flow need be neither transverse nor directed inward. Finally, $\alpha_3$ is contained in $V_3 = \overline{D - (V_1 \cup V_2)}$, and $V_1 \cup V_2 \cup V_3 = \overline{D}$. With this example in mind we define a *relative filtration* for a domain $D$ by the following requirements:

(1)  $\overline{D} = V_1 \cup \dots \cup V_q$, where the $V_i$ are closed sets with disjoint interior, and $\alpha_i \in \text{int } V_i$.
(2)  Let $F_i = V_1 \cup \dots \cup V_i$, so that $F_1 \subset F_2 \subset \dots \subset F_q = \overline{D}$. Then $\partial F_i = A_i \cup B_i$, where $A_i$ is a compact smooth manifold-with-boundary having its interior in $D$ and its boundary in $\partial D$, and $B_i$ is a subset of $\partial D$; $A_i$ and $B_i$ are joined along their common boundary.
(3)  The flow is transverse to $A_i$, and is directed *into* $F_i$ on $A_i - \partial A_i$.

Condition (3) says that the flow is 'downhill' in terms of the filtration as long as solutions stay in $D$. (This of course is an expression of the diagram structure reflected in the proper numbering of the $\alpha_i$.) The existence of a relative filtration is discussed briefly in the notes, but is usually obvious in the applications.

Given a relative filtration $F_1 \subset F_2 \subset \dots \subset F_q$ for (6.5) on $D$, let $F_i^* = F_i \times S^1$. Then clearly $F_1^* \subset \dots \subset F_q^*$ is a relative filtration for (6.7), with $\partial F_i^* = A_i^* \cup B_i^*$; henceforth $*$ denotes Cartesian product with $S^1$. (The definition of relative filtration is of course modified so that $V_i^*$ is to contain the periodic orbit $\alpha_i^*$ rather than the rest point $\alpha_i$.) Now it is easy to see that $F_i^*$ is also a relative filtration for (6.6) for small $\varepsilon$ (with $\alpha_i^*$ replaced by $\beta_i(\varepsilon)$). The key point is to show that the flow of (6.6) is transverse to $A_i^*$. The angle between the vector $(\varepsilon g(z), 1)$ and the normal $N$ to $A_i^*$ at any point $(z, \theta)$ is less than a right angle; we claim this holds also for $(\varepsilon g(y) + \varepsilon^2 \hat{g}(y, \theta, \varepsilon), 1)$ for small $\varepsilon$. Since $N$ has last component zero, an $\varepsilon$ may be factored out of the dot product and the result is then clear. Because $A_i^*$ is compact, such an estimate may be carried out uniformly, and there is an interval of $\varepsilon$ for which the flow of (6.6) is transverse to $A_i^*$. It follows at once that it is directed into $F_i^*$ along $A_i^* - \partial A_i^*$.

The next step is to show that every solution of (6.6) which remains in $V_i^*$ as $t \to \infty$ must approach $\beta_i$, and similarly as $t \to -\infty$. (The corresponding fact, for $\tau \to \pm\infty$, is obvious for (6.5) from (H2) and the filtration properties.) Recall that there is a neighborhood $U_i$ of $\alpha_i$ such that $U_i^*$ has the required property (everything remaining in $U_i^*$ approaches $\beta_i$). We may assume $U_i \subset V_i$.

It suffices to prove that any orbit which remains in $V_i^*$ for all time (both forward and backward) is contained in $U_i^*$. For suppose that this is true, and let $\gamma(t)$ be a solution which remains in $V_i^*$ as $t \to +\infty$. Then the $\omega$-limit set of $\gamma(t)$, which is an invariant set, consists of one or more orbits which remain in $V_i^*$ for all time, and so must be contained in $U_i^*$. It follows that $\gamma(t) \to \beta_i$. A similar argument holds if $\gamma(t)$ remains in $V_i$ as $t \to -\infty$.

So what needs to be proved is the following: every orbit of (6.6) passing through a point of $V_i^* - U_i^*$ leaves $V_i^*$ either forward or backward in time. Now every solution $z(\tau)$ of (6.5) passing through a point of $V_i - U_i$ leaves $V_i$ either forward or backward in finite $\tau$-time; in terms of $t$-time the solutions require time of order $1/\varepsilon$ to leave $V_i$. But this is exactly the kind of time interval on which solutions for $z$ remain $\varepsilon$-close to solutions for $y$ (Section 4), so solutions of (6.5) should pull corresponding solutions of (6.6) out of $V_i^*$. This argument must be formalized. For a fixed $i$, let $K = V_i - U_i$ and let $V$ be the $\delta$-halo of $V_i$ (the set of points at distance $\leqslant \delta$ from $V_i$) for some $\delta > 0$. For small enough $\delta$, there will exist $\mathbf{T} > 0$ such that every solution $z(\tau)$ of (6.5) lies outside $V$ for some $\tau$ in $-\mathbf{T} \leqslant \tau \leqslant \mathbf{T}$. Now apply Theorem 4.2(b) with this $K$, $V$, and $\mathbf{T}$, letting $\varepsilon_0$ be the current bound on $\varepsilon$ (the bound for which all statements made until now are true). Then $z$ and $y$ are closer than $c\varepsilon$ until one or the other leaves $V$ or until $|t|$ reaches $\mathbf{T}/\varepsilon$. Since $z$ leaves $V$ before $|t|$ reaches $\mathbf{T}/\varepsilon$, this is not the significant restriction. Either $y$ leaves $V$ before $z$, in which case we are done; or else, at the time $z$ leaves $V$, $y$ is within $c\varepsilon$ of $z$. For small enough $\varepsilon$, $c\varepsilon$ will be less than $\delta$, guaranteeing that $y$ left $V_i$ (and the solution of (6.7) has left $V_i^*$). So the current bound $\varepsilon_0$ should be reduced if necessary to make $c\varepsilon_0 < \delta$. Since the argument must be repeated for each $i$, this means finitely many reductions in $\varepsilon_0$.

At this point the situation is as follows. Any orbit of (6.6) passing through a point of $V_i^*$ has three possibilities as $t \to \infty$: it can approach $\beta_i$, it can descend to a lower level $V_j$ with $j < i$, or it can leave $D$. Similarly there are three possibilities as $t \to -\infty$: it can approach $\beta_i$, ascend to a higher level, or leave $D$. This implies conclusion (1) of Theorem 6.1.

Conclusion (2) is a consequence of the transversality hypothesis H3. Suppose there exists a connecting orbit in $D$ for 6.5 from $\alpha_i$ to $\alpha_j$. Then $W^u(\alpha_i) \cap W^s(\alpha_j) \neq \emptyset$. Let $K_i$ be the intersection of $D$ with a large compact disk in $W^u(\alpha_i)$, and $K_j$ the intersection of $D$ with a large compact disk in $W^s(\alpha_j)$. Then $K_i$ and $K_j$ are compact manifolds with boundary, and they can be chosen large enough that they intersect transversally at points other than boundary points. (The intersection will consist of one or infinitely many arcs of connecting orbits from $\alpha_i$ to $\alpha_j$, running from the boundary of $K_j$ to the boundary of $K_i$. An entire connecting orbit is not compact because it does not contain its endpoints.) $K_i^*$ and $K_j^*$ also intersect transversally. It can be shown that the stable and unstable manifolds of the $\beta_i$ are $C^1$-close to the stable and unstable manifolds of $\alpha_i^*$ (in (6.7)); this means that given the compact part $K_i^*$ of the unstable manifold of $\alpha_i^*$, there is a compact part of the unstable manifold of $\beta_i$ lying arbitrarily close (as $\varepsilon \to 0$) to $K_i^*$, and the tangent spaces of nearby points on the two manifolds are also close. It follows that $W^u(\beta_i)$ and $W^s(\beta_j)$ continue to intersect transversally in $D^*$, so that at least one connecting orbit exists from $\beta_i$ to $\beta_j$. (In cases in which the full set of connecting orbits from $\alpha_i$ to $\alpha_j$ includes some tangent to $\partial D$, those passing near $\partial D$ may be disrupted by the perturbation. This is because two compact manifolds with boundary intersecting transversally may be pulled apart along their boundaries. But as long as the intersection is not exclusively along the boundaries, some intersection will remain.)

Several arguments for the $C^1$ closeness of the stable and unstable manifolds of $\alpha_i^*$ and $\beta_i$ are possible; in fact any proof of the existence of the manifolds can probably be modified into a proof of their $C^1$ closeness. But it is important to understand that this result is not simply the continuous dependence of stable manifolds on parameters. It is a second-order result, since (6.6) is a second-order perturbation of a system having no stable manifolds at all when $\varepsilon = 0$. One proof is sketched in the notes. A key role is played by Corollary 4.5, which gives control over the $C^1$ distance between the flows of (6.6) and (6.7) on time intervals of order $1/\varepsilon$, which is sufficiently long to 'reach' any compact part of $W^u(\beta_i)$ from points near $\beta_i$.

The completes the proof of Theorem 6.1; it is possible to continue on to prove topological conjugacy (relative to $D$) of the averaged and original flows (see notes).

If a version of Theorem 6.1 is to be proved for higher order averaging, two difficulties must be dealt with. First, the truncated averaged system (5.4) depends upon $\varepsilon$ in a more serious way than does (6.4), in that the orbit positions, and not merely the time along orbits, vary with $\varepsilon$. So it is possible for the topological structure of the orbits not to be constant on any interval $0 < \varepsilon < \varepsilon_0$. An example of this phenomenon is given in Section 8. But because of the structural stability of Morse–Smale systems, this does not happen if (5.4) is Morse–Smale in an interval $0 < \varepsilon < \varepsilon_0$, and in this case there is probably a single filtration valid for (5.4) in such an interval. (The example in Section 8 suggests that the restriction 'Morse–Smale in an interval' is a fairly strong one, leaving out many interesting systems.) But even when there is such a filtration for (5.4), another problem arises which obstructs the proof that the filtration carries over to (5.3). Namely, if $\alpha_i$ is strongly $k$-hyperbolic, the neighborhood $U_i$ (in which every orbit approaches $\alpha_i$ or leaves $U_i$ as $t \to \pm \infty$) may shrink with $\varepsilon$, according to section 5. Therefore no bound may be placed on the $\tau$-time required for an orbit in $V_i - U_i(\varepsilon)$ to leave $V_i$ (or rather, any such bound must be $\varepsilon$-dependent). In terms of $t$, this 'escape time' is larger than $0(1/\varepsilon)$, and therefore longer than the time during which the asymptotic estimates of Section 4 are valid. It is not clear whether this problem can be overcome, or if the extension to $k$th order averaging is false. Section 8 contains an example in which the equivalence can be proved; all of the difficulties mentioned here are overcome by the use of a Lyapunov function which is available in this case.

**Notes**

The best starting place to learn the dynamical systems theory used in this section is [27]. The arguments in this section are based on the proof of structural stability of Morse–Smale systems, with modifications because $\mathbb{R}^n$ is not compact and because of the presence of the parameter $\varepsilon$. Reference [27] develops the background for the structural stability proof, and then carries out the proof in two dimensions. Unfortunately the restriction $n = 2$ is imposed just at the point of proving the existence of filtrations; the arguments in [27] will give the filtration in Fig. 4, but for $n$ dimensions something more is needed. In particular, it must be shown how to 'trim the edge' of a neighborhood of an unstable manifold to make it transverse to the flow. The necessary idea is contained in [33], Lemma 2.1. Another approach to filtrations is possible, using Lyapunov functions. A Lyapunov function for (6.5) is a function which decreases along orbits (except the rest points $\alpha_i$) and takes the value $i$ at $\alpha_i$. Given such a function, the surfaces on which the function takes half-integer values form filtration boundaries. But although the existence of Lyapunov functions in this sense is generally acknowledged, even in much more general (Axiom A) systems, a clear and complete exposition does not seem to exist. The best approach uses the notion of attractor/repeller pairs [7]. A complete exposition will appear in a forthcoming monograph on topological dynamics by Akin [1]. Incidentally, a filtration for Fig. 1 based on Lyapunov functions would look different because it is not possible to skip a level in flowing down the filtration; the boundary of $V_2$ would not join $V_1$ at $Q$ and $R$, but surround it.

It is not possible simply to reduce the proof of Theorem 6.1 to the structural stability theorem. Imagine a space in which points are dynamical systems. The structural stability theorem states that given a point representing a Morse–Smale system, nearby points represent systems with the same topological structure. Theorem 6.1 deals with two arcs in this space, parameterized by $\varepsilon$, beginning (for $\varepsilon = 0$) at a point representing the trivial flow (nothing moves). A neighborhood of this point contains every possible topological behavior. One of the arcs consists of Morse–Smale systems (except at $\varepsilon = 0$), and the structural stability theorem implies that this arc has a neighborhood (narrowing as $\varepsilon \to 0$) of equivalent systems. Theorem 6.1 says that the second arc is close enough to the first to lie inside this neighborhood.

The structural stability theorem for Morse–Smale systems gives more than diagram equivalence, it gives topological conjugacy. It is possible to continue on from Theorem 6.1 and prove that the flows of the averaged and exact systems are topologically conjugate; that is, there is a homeomorphism (depending upon $\varepsilon$) carrying orbits of one flow onto orbits of the other flow. It is necessary to use Robinson's notion of topological conjugacy for a manifold with boundary [30], which allows moving $\partial D$. The general lines of Robinson's proof may then be followed, with the difference that Corollary 4.8 must be used to give $C^1$ closeness on time intervals of order $1/\varepsilon$, in place of the estimates for finite time intervals which suffice in Robinson's case. One may think of topological conjugacy as saying that instead of comparing the exact and averaged solutions with the same initial conditions, as in Section 4, it is better to compare exact and averaged solutions whose initial points are related by the conjugating homeomorphism. These solutions will remain so related for all time, and their alpha and omega limit sets will correspond under the conjugacy. Unfortunately the conjugating homeomorphism is not computable, so it is not possible to use this idea in studying individual solutions. We have confined our proof to diagram equivalence, because this seems to be the most important part of the result for applications, and the step from diagram equivalence to topological conjugacy is a very big one. These results (Theorem 6.1 and topological conjugacy) seem to have existed in a nebulous form as folklore for some time, but the proof written out here is to my knowledge the most complete so far. A sketchy statement is given in [10], page 180, but the questions concerning $\partial D$ are not dealt with adequately (the correct definition of topological conjugacy for this situation is not given), and the brief sketch of a proof is somewhat misleading. In particular, the impression is given that the change of time scale from $t$ to $\tau$ (which carries (6.4) to (6.5)) can be applied to (6.3) in a straightforward way. However, the result is a system that does not have a fixed period, which is why the usual estimates in Morse–Smale theory must be replaced by the asymptotic averaging estimates. The authors were no doubt aware of this, but they provide no details.

It remains to address the issue of $C^1$ closeness of the stable and unstable manifolds of $\alpha_i^*$ and $\beta_i$ which was needed in the proof of Theorem 6.1. We sketch a proof relying on the $\lambda$-lemma given in [27]. Choose a disk $A$ of the same dimension as $W^u(\alpha_i)$, close to $\alpha_i$ and transverse to $W^s(\alpha_i)$. Flowing this disk forward for time $\tau = T$ by the flow of (6.5) brings it $C^1$-close to any compact part of $W^u(\alpha_i)$. Then flowing $A^*$ forward for time $t = T/\varepsilon$ by the flow of (6.7) brings it $C^1$ close to any compact part of $W^u(\alpha_i^*)$. Now $A^*$ is also transverse to $W^s(\beta_i)$ (this can be guaranteed by local arguments using hyperbolic estimates near $\alpha_i^*$, along the lines of Theorem 5.8 and Hartman's theorem) and so flowing $A^*$ forward for $T/\varepsilon$ by the flow of (6.6) brings it $C^1$-close to $W^u(\beta_i)$. But the flows of (6.6) and (6.7) are themselves $C^1$ close for time $T/\varepsilon$ (Corollary 4.5) and so compact parts of $W^u(\alpha_i^*)$ and $W^u(\beta_i)$ are $C^1$-close. Another argument can be constructed using the integral equations for stable and unstable manifolds developed in [11].

## 7 NEARLY INTEGRABLE SYSTEMS

With this section, the focus changes from nearly linear to nearly integrable systems, that is, systems of the form (1.2), repeated here as

$$
(7.1) \qquad
\begin{aligned}
\dot{r} &= \varepsilon f(r, \theta, \varepsilon) & r &\in \mathbb{R}^n \\
\dot{\theta} &= \Omega(r) + \varepsilon g(r, \theta, \varepsilon) & \theta &\in T^m.
\end{aligned}
$$

Many of the results developed for nearly linear systems will be brought to bear on this system, by way of localizations such as (1.3).

When $\varepsilon = 0$, $r$ is constant and every solution lies on an $m$-dimensional torus $r = r_*$. (The notation $r_*$ for a fixed value of $r$ is preferable to $r_0$ because $r_1, \ldots, r_n$ denote components.) The solutions on this torus take the form $\theta = \omega_* t + \theta_*$, with $\omega_* = \Omega(r_*)$, which are sometimes called 'winding lines' on the torus. The character of these orbits depends heavily on $\omega_*^\perp = \{\nu \in \mathbb{Z}^m : \nu\omega_* = 0\}$. As discussed in Section 2, $\omega_*^\perp$ is a group with a rank (or dimension) denoted by $s$, $0 \leqslant s \leqslant m$. The case $s = m$ only occurs if $\omega_* = 0$; then every point on the torus $r = r_*$ is a rest point (for $\varepsilon = 0$). The case $s = m - 1$ means that there are $m - 1$ independent linear relations with integer coefficients among the components of $\omega_*$, which implies that each $\omega_i$ is an integral multiple of a single real number $\lambda : \omega_i = p_i\lambda$. In this case every orbit $\theta = \omega_* t + \theta_*$ is periodic, with period $2\pi/\lambda$. One way to see this, and to find the generalization for other values of $s$, is to use the matrix $A$ given in Theorem 2.1, which satisfies $A\omega_* = \mathrm{col}(\lambda_1, \ldots, \lambda_{m-s}, 0, \ldots, 0)$ with $\lambda^\perp = 0$. Since $A$ is unimodular, each component of $\omega_*$ is an integral linear combination of $\lambda_1, \ldots, \lambda_{m-s}$. Any function defined on the torus becomes a quasiperiodic function of $t$ with these independent frequencies. The changes of variables $\phi = A\theta$, with the components of $\phi = (\phi_1, ..., \phi_m)$ renamed $\phi = (x, y) = (x_1, ..., x_{m-s}, y_1, ..., y_s)$, carries $\theta = \omega_* t + \theta_*$ into $x = \lambda t + x_*$, $y = y_*$ with $\lambda = (\lambda_1, ..., \lambda_{m-s})$. Therefore when the rank of $\omega_*^\perp$ is $s$, the orbits (for $\varepsilon = 0$) not only lie in the $m$-dimensional torus $r = r_*$, but in an $(m - s)$-dimensional subtorus, namely, $A^{-1}$ of the $x$ torus. By the same argument as in the proof of Theorem 2.2, the orbits are dense in this subtorus. When $s = m$ the 'subtori' are points; when $s = m - 1$ they are closed curves; when $s = 0$ the orbits are dense in the torus $r = r_*$.

For small $\varepsilon \neq 0$, we will construct approximate solutions to (7.1) and use them to draw qualitative conclusions about the actual solutions. Heuristically, these approximations are obtained as follows. As a first approximation to the solution with initial values $r_*, \theta_*$, use the solution with $\varepsilon = 0$, that is, $r = r_*$, $\theta = \omega_* t + \theta$. A second approximation to the motion of $r$ would then come from solving $\dot{r} = \varepsilon f(r_*, \omega_* t + \theta_*, 0)$ with initial values $r_*, \theta_*$. Instead we simplify this equation by averaging over $t$, obtaining an 'approximation of order one-and-one-half' by solving $\dot{r} = \varepsilon \bar{f}(r_*, \theta_*)$, where

$$(7.2) \qquad \bar{f}(r, \theta) = \lim_{T \to \infty} \frac{1}{T} \int_0^T f(r, \Omega(r)t + \theta, 0) \, dt.$$

This approximate solution is simply $r = r_* + \varepsilon \bar{f}(r_*, \theta_*)t$. An approximation for $\theta$ could be found similarly, but will not be needed since our focus will be on the drift of $r$; in the applications $r$ is the primary observable of interest, since the components of $\theta$ are rotating rapidly. The actual process of constructing and justifying the approximate solutions will, of course, rest on the averaging theory developed earlier (with some modifications). But continuing heuristically for the moment, it seems reasonable to say that if $\bar{f}(r_*, \theta_*) \neq 0$, the approximation $r = r_* + \varepsilon \bar{f}(r_*, \theta_*)t$ will reflect the behavior accurately for small $t$, and the solution beginning at $r_*, \theta_*$ will drift away from $r_*$. On the other hand if $\bar{f}(r_*, \theta_*) = 0$ we cannot conclude from the approximation $r = r_*$ that $r$ is actually constant; but we may suppose that $r$ drifts much more slowly near such an $r_*$, and that if periodic solutions or invariant tori are to be found for $\varepsilon \neq 0$ (survivals of the invariant tori that exist everywhere for $\varepsilon = 0$), they will be found near zeros of $\bar{f}$. In fact, let $K$ be a compact set which is the closure of an open set in $\mathbb{R}^n$ with $\bar{f} \neq 0$ on the boundary, and let $U \subset K$ be an open set containing the zeros of $\bar{f}$ (more precisely: containing all $r$ for which there exists $\theta$ with $\bar{f}(r, \theta) = 0$). Then on $K - U$ we can expect $\| \dot{r} \| > c\varepsilon$ for some $c > 0$; on this 'non-lingering set' $r$ will drift on a time scale of $\varepsilon t$. The set $U$ will contain any invariant tori and orbits drifting on a slower time scale, and may be called the 'lingering set', although not all orbits in this set will linger. Under fairly general circumstances (which, however, exclude purely Hamiltonian systems), $U$ will be a small set. Most of $K$ will consist of nonlingering points, near which the drift in $r$ is calculable from the first order average.

Inside $U$, neighborhoods of the points $r$ for which $\bar{f}(r,\theta)$ has zeros can be investigated by higher order averaging.

The full program outlined in the last paragraph can be carried out, but becomes rather technical. Considerable simplification is achieved by focusing on *pass-through points* rather than non-lingering points, that is, points $r_*$ such that not only is $\bar{f}(r_*,\theta) \neq 0$ for all $\theta$, but in fact $\bar{f}(r_*,\theta)$ has a *single component* $\bar{f}_j(r_*,\theta)$ which is $\neq 0$ for all $\theta$. In this case the approximation $r = r_* + \varepsilon\bar{f}(r_*,\theta_*)t$ implies that all solutions which touch $r = r_*$ pass through in a single direction, in the sense that the component $r_j$ increases (or decreases) along all such solutions. In contrast to this are the *non-pass-through points*, values of $r_*$ for which each component of $\bar{f}(r_*,\theta)$ vanishes for some $\theta$. One can enclose the non-pass-through points of $K$ in an open set $V$, show that $V$ is usually small, and prove rigorously that the points of $K - V$ have the expected pass-through behavior. It is this program which we shall carry out here; it falls short of the full program in that $V$ is not just an open set $U$ containing the zeros of $\bar{f}$, but must contain all non-pass-through points. After completing this work we will briefly sketch how to decrease $V$ by an inductive process to obtain a lingering set $U$.

We begin our careful work by studying $\bar{f}$ and the location of its zeros and, more generally, non-pass-through points. The integrand of (7.2) is quasi-periodic in $t$ for each fixed value of $r$, with frequencies which depend upon $r$ but not $\theta$. As a result the average $\bar{f}(r,\theta)$ is usually discontinuous in $r$. To see this clearly, begin with the Fourier expansion

$$(7.3) \qquad f(r,\theta,0) = \sum_{\nu} a_\nu(r)e^{i\nu\theta}.$$

Then according to Theorem 2.2 and (7.2),

$$(7.4) \qquad \bar{f}(r,\theta) = \sum_{\nu \in \Omega(r)^\perp} a_\nu(r)e^{i\nu\theta} = a_0(r) + \sum_{0 \neq \nu \in \Omega(r)^\perp} a_\nu(r)e^{i\nu\theta}$$

and this also equals the average of $f$ (with respect to a suitable measure) over the closure of the curve $\Omega(r)t + \theta$ in the torus determined by $r$. This closure is an embedded subtorus. The possible discontinuities in $\bar{f}$ result from changes in the dimension of the subtorus as the frequencies vary, or, equivalently, changes in the resonance group $\Omega(r)^\perp$ leading to inclusion or omission of different terms of the Fourier series (7.4). The term $a_0(r)$ is always present in (7.4) and may be thought of as the 'continuous part'.

The simplest case to analyze is that in which (7.3) contains finitely many harmonics, that is, $a_\nu(r) \equiv 0$ for all but finitely many $\nu$. In this case only the omission or retention of these $a_\nu$ is significant. For each $\nu$ such that $a_\nu(r)$ is not identically zero, the hypersurface $S(\nu) = \{r : \nu\Omega(r) = 0\}$ is called a *resonance surface* in $\mathbb{R}^n$. (This is only a hypersurface if $\nu\Omega(r)$ is nondegenerate, which we shall assume, although other cases can be considered.) Under the present hypothesis there are finitely many of these surfaces, and away from these, $\bar{f}(r,\theta)$ is continuous and in fact equal to $a_0(r)$. The resonance surfaces are surfaces of discontinuity of $\bar{f}$, on which $\bar{f}$ contains additional terms. Zeros of $\bar{f}$ occur at points off the resonance surfaces where $a_0(r) = 0$, and at points on resonance surfaces where the amplitude of the $\theta$-dependent terms which are present is sufficient to overcome the ever-present term $a_0(r)$. Similarly, non-pass-through points other than those where $a_0(r) = 0$ can occur only on resonance surfaces: any point off the resonance surfaces satisfies $\bar{f}(r,\theta) = a_0(r)$, which is independent of $\theta$, and if this is $\neq 0$ it must have a component $\bar{f}_j$ which is $\neq 0$ for all $\theta$, making $r$ a pass-through point.

When there are infinitely many harmonics, $\mathbb{R}^n$ is densely criss-crossed by resonance surfaces, and it would seem that zeros of $\bar{f}$, or more generally, non-pass-through points, could occur almost anywhere. This is indeed the case if $a_0(r) \equiv 0$. But because of the decay of Fourier coefficients, we can show that away from zeros of $a_0(r)$, only finitely many resonance surfaces are significant.

7.1 THEOREM Let $K_0 \subset \mathbb{R}^n$ be a compact region in which $a_0(r) \neq 0$. There exist at most finitely many resonance surfaces $S_1, \ldots, S_l$ such that $(S_1 \cup \ldots \cup S_l) \cap K_0$ contains all the non-pass-through points in $K_0$.

For this proof let $|||r||| = \max\{|r_1|, \ldots, |r_n|\}$ be the maximum norm on $\mathbb{R}^n$, and as usual let $\|\nu\| = |\nu_1| + \cdots + |\nu_m|$ for integer row vectors. Choose $M > 0$ so that $|||a_0(r)||| \geqslant 2M$ for $r \in K_0$. Then choose $N$ so that $\Sigma_{\|\nu\| > N} |||a_\nu(r)||| \leqslant M$ for $r \in K_0$. Divide the terms of (7.4) into three groups: (i) the ever-present $a_0(r)$; (ii) terms with $\|\nu\| > N$; (iii) terms with $0 < \|\nu\| \leqslant N$. If $\Omega(r)$ is such that $\bar{f}$ contains only terms from the first two groups, then $|||\bar{f}(r,\theta)||| \geqslant |||a_0(r)||| - \Sigma_{\|\nu\| > N} |||a_\nu(r)||| \geqslant 2M - M = M$. We claim that this can be strengthened: there exists a $j$ such that $|\bar{f}_j(r,\theta)| \geqslant M$ for all $\theta$. In fact $|||a_0(r)||| > 2M$ means that *there exists* a component $a_{0j}(r)$ which is bigger than $2M$; $\Sigma_{\|\nu\| > N} |||a_\nu(r)||| \leqslant M$ means that *every* component of the sum is $\leqslant M$. Therefore $|\bar{f}_j(r,\theta)| \geqslant |a_{0j}(r)| - \Sigma_{\|\nu\| > N} |a_{\nu j}(r)| \geqslant M$. We have proved that every point for which $\bar{f}$ involves only terms from groups (i) and (ii) is a pass-through point. Thus non-pass-through points can arise only where $\bar{f}$ contains terms from group (iii), that is, on resonance surfaces $S(\nu)$ with $\|\nu\| \leqslant N$. But there are finitely many of these, and they comprise the $S_1, \ldots, S_l$ of the theorem.

There is an important special case in which the hypothesis of Theorem 7.1 cannot be satisfied because $a_0(r) \equiv 0$. This is the case of Hamiltonian systems, for which $m = n$ and $f_i(r,\theta) = \partial H(r,\theta)/\partial\theta_i$, $i = 1, \ldots, n$, for some function $H : R^n \times T^n \to \mathbb{R}$. Differentiation of $H$ with respect to $\theta_i$ eliminates the constant term, whence $a_0(r) \equiv 0$. It follows that every component $\bar{f}_i(r,\theta)$ has zeros for every $r$, and every point is a non-pass-through point. The Hamiltonian case is covered by the theorems of Poincaré–Birkhoff and Kolmogoroff–Arnol'd–Moser, which guarantee the existence of many periodic solutions and invariant tori. In contrast to this, when Lemma 7.1 applies most points are pass-through points where, as we will now show, invariant tori cannot exist.

The method of averaging (Sections 3 and 4) can be brought to bear on (7.1) by means of the localization (1.3) at a given $r_*$. Since (1.3) is quasiperiodic with (in general) infinitely many harmonics, it is the method of Theorem 3.4 which should be used. Thus a near-identity transformation of $(R, \Theta)$ to $(\rho, \psi)$ will carry the first equation of (1.3) into $\dot{\rho} = \mu\{\bar{f}(r_* + \mu\rho, \psi) + h_N(\psi)\} + 0(\mu^2)$, where $h_N$ consists of high harmonics of $f(\|\nu\| > N)$. If $r_*$ is a pass-through state, there is a component $\bar{f}_j(r_*, \psi)$ which is $\neq 0$ for all $\psi$; by taking $N$ large enough and $\mu$ small enough, we guarantee $\dot{\rho}_j \neq 0$ in a neighborhood $\|\rho\| \leqslant c_1$. Thus the torus $\rho = 0$ has a neighborhood which is passed through by all solutions, with $\rho_i$ increasing (or decreasing) along every solution in the neighborhood. This neighborhood corresponds under the near-identity transformation to a neighborhood $\|R\| \leqslant c_2$, and hence under the localization to a neighborhood $\|r - r_*\| \leqslant c_2\sqrt{\varepsilon}$ in (7.1). Therefore this 'shrinking neighborhood' of the pass-through state $r_*$ is passed through in the $j$ direction by all solutions; in particular, it contains no invariant tori. We would like to extend this result to a *fixed* neighborhood of $r_*$. To do so requires a different form of localization from (1.3).

Beginning, then, with (7.1), we set $r = r_* + R$ instead of $r = r_* + \sqrt{(\varepsilon)}R$, and obtain, after expanding in $\varepsilon$ and $R$,

$$\dot{R} = \varepsilon f(r_*, \theta, 0) + 0(\varepsilon\|R\|) + 0(\varepsilon^2)$$
$$\dot{\theta} = \omega_* + 0(\|R\|) + 0(\varepsilon).$$
(7.5)

(We have retained in detail only those terms needed for Theorem 7.3 below; the form of localization and averaging that we are doing can be carried out in much greater detail, including splitting $\theta$ into 'slow' and 'fast' angles by means of the matrix $A$ associated to $\omega_*$.) System (7.5) differs from (1.3) in that amplitudes are not dilated, angles are not spun, and errors are controlled by $\varepsilon$ and $\|R\|$ separately rather than by linking both to $\mu = \sqrt{\varepsilon}$.


**7.2 LEMMA**   For any integer $N > 0$ there exists a near-identity transformation $R \leftrightarrow \rho$ of the form $R = \rho + \varepsilon u(\theta)$ carrying (7.5) into

$$\dot{\rho} = \varepsilon\{\bar{f}(r_*, \theta) + h_N(\theta)\} + 0(\varepsilon\|\rho\|) + 0(\varepsilon^2)$$
$$\dot{\theta} = \omega_* + 0(\|\rho\|) + 0(\varepsilon)$$
(7.6)

where $h_N(\theta)$ consists of certain terms in the Fourier series of $f(r_*, \theta, 0)$ having $\|\nu\| > N$. If the rank of $\omega_*^\perp$ is $m - 1$ then $h_N$ may be omitted.

The proof parallels closely the reasoning in Section 3. It is easily checked that $R = \rho + \varepsilon u(\theta)$ carries (7.5) into $\dot{\rho} = \varepsilon\{f(r_*, \theta, 0) - u'(\theta)\omega_*\} + 0(\varepsilon\|\rho\|) + 0(\varepsilon^2)$,

$\theta = \omega_* + 0(\|\rho\|) + 0(\varepsilon)$. (Here $u'(\theta)$ is a matrix and $\omega_*$ a column vector.) To achieve (7.6) it is only necessary to choose $u(\theta)$ so that $u'(\theta)\omega_* = f(r_*, \theta, 0) - \bar{f}(r_*, \theta) - h_N(\theta)$. Calling the right-hand side $g(\theta)$, taking the $k$th components, and expanding the right-hand side in a Fourier series, the equation for $u$ reads

$$(7.7) \qquad \sum_j \omega_{*j} \frac{\partial u_k(\theta)}{\partial \theta_j} = g_k(\theta) = \sum_n c_{k,\nu} e^{i\nu\theta}.$$

(This partial differential equation should be compared with the ordinary differential equation $u_t = h$ whose solution is (3.11).) If the solution to (7.7) is $u_k(\theta) = \sum_\nu b_{k,\nu} e^{i\nu\theta}$, then $i\nu\omega_* b_{k,\nu} = c_{k,\nu}$ for $\nu \neq 0$; the constant term of $u_k$ disappears from the left-hand side of (7.7) so the equation can only be solved if $c_{k,0} = 0$, $k = 1, ..., n$. It is possible to solve for $b_{k,\nu}$ provided $\nu\omega_* \neq 0$ whenever $c_{k,\nu} \neq 0$, and the resulting series for $u_k$ will certainly converge (in a trivial way) if it is finite. These conditions are satisfied if $g(\theta) = f(r_*, \theta, 0) - \bar{f}(r_*, \theta) - h_N(\theta)$; subtracting $\bar{f}$ eliminates the constant term and all terms for which $\nu\omega_* = 0$, and subtracting $h_N$ eliminates all remaining high harmonics. (Note that $h_N$ should not consist of all high harmonic terms in $f$, but only those remaining after $\bar{f}$ is subtracted.) In case the rank of $\omega_*^\perp$ is $m - 1$, transformation $A\theta = (x, y) = (x_1, y_1, ..., y_{m-1})$ carries (7.7) into the differential equation $\lambda(\partial/\partial x)u(A^{-1}(x, y)) = g(A^{-1}(x, y))$, where $A\omega = (\lambda, 0)$, $\lambda \in \mathbb{R}$; this may be viewed as an ordinary differential equation with single independent variable $x$ and parameters $y$, which can be solved by Theorem 2.3 case 1 without the need of $h_N$.

We are now in a position to prove the existence of regions of fixed size that are passed through 'on the time scale $\varepsilon t$'.

7.3 THEOREM Let $r_*$ be a pass-through point of (7.1). Then there exist constants $\delta > 0$, $\varepsilon_0 > 0$, $c > 0$ such that no orbit of (7.1) for $0 < \varepsilon < \varepsilon_0$ remains continuously in the neighborhood $\|r - r_*\| < \delta$ for a time longer than $c/\varepsilon$.

Since $r_*$ is a pass-through point, and since $T^m$ is compact, there is an index $j$ and $M > 0$ for which $|f_j(r_*, \theta)| > 2M$ for all $\theta$. It follows from (7.6) that there exist $\varepsilon_1 > 0$, $\delta_1 > 0$ such that $|\dot{\rho}_j| > \varepsilon M$ for $\|\rho\| < \delta_1$ and $0 \leqslant \varepsilon < \varepsilon_1$; for $\dot{\rho}_j = \varepsilon[\bar{f}_j(r_*, \theta) + h_{Nj}(\theta) + 0(\|\rho\|) + 0(\varepsilon)]$, and $N$ can be chosen large enough that $|h_{Nj}(\theta)| < M/3$, after which (since they depend upon $N$) the terms $0(\|\rho\|)$ and $0(\varepsilon)$ can each be made less than $M/3$. Therefore if a solution of (7.7) for $0 < \varepsilon < \varepsilon_1$ remains in $\|\rho\| < \delta_1$ for time $t$, $\rho_j$ travels a distance greater than $\varepsilon Mt$. If the solution were to remain in $\|\rho\| < \delta_1$ for time $t = 2\delta_1/M\varepsilon$, $\rho_j$ would have traveled a distance of greater than $2\delta_1$, the diameter of the region, which is impossible. Therefore no solution of (7.6) for $0 < \varepsilon < \varepsilon_1$ remains continuously in $\|\rho\| < \delta_1$ for a time longer than $c/\varepsilon$, with $c = 2\delta_1/M$. It only remains to use the near-identity transformation given by Lemma 7.2 to transfer this result to system (7.1). Thinking of

$\rho = r - r_* - \varepsilon u(\theta)$ as a function of $r$, $\theta$, and $\varepsilon$, the solid torus $\| r - r_* \| < \delta$ will be contained in the 'wavy solid torus' $\| \rho(r, \theta, \varepsilon) \| < \delta_1$ provided $\delta = \delta_1/2$ and $0 < \varepsilon < \varepsilon_0 = \min\{\varepsilon_1, \delta_1/2k\}$ where $k$ is the maximum value of $\| u(\theta) \|$. Since nothing remains in the 'wavy solid torus' for longer than $c/\varepsilon$, the same is true for $\| r - r_* \| < \delta$.

Theorems 7.1 and 7.3 together provide the following picture of the drift of $r$ in any compact set $K \subset \mathbb{R}^n$. First locate the subset of $K$ on which $a_0(r) = 0$. Since this is $n$ equations in $n$ unknowns, the zero set will be a finite number of points unless there are degeneracies. Let $V_0$ be a (relatively) open subset of $K$ containing the zeros of $a_0$; in the generic case this will consist of finitely many small open balls. The remainder of $K$ forms a set $K_0 = K - V_0$ which satisfies the hypotheses of Theorem 7.1. Therefore there are (at most) finitely many resonance surfaces $S_1, \ldots, S_l$ which contain all of the non-pass-through points in $K_0$. Let $V_1$ be a (relatively) open subset of $K_0$ containing $(S_1 \cup \ldots \cup S_l) \cap K_0$; this will be (again in the generic case, assuming the resonance surfaces are actually nondegenerate hypersurfaces) a set of thickened sheets cutting through $K_0$. Let $K_1 = K_0 - V_1$. We have decomposed $K$ into three disjoint sets, $K = V_0 \cup V_1 \cup K_1$. $K_1$ is compact and consists of pass-through points. It can be covered by finitely many of the sets given by Theorem 7.3. Therefore there is a lower bound for the rate of drift of $r$ on $K_1$ (on the time scale $\varepsilon t$), and $K_1$ can be called a 'nonlingering set'. The set $V = V_0 \cup V_1$ is a 'small' set containing any 'steady state' values of $r$ which may exist as well as any 'lingering' points where $r$ drifts on a slower time scale than $\varepsilon t$. (By 'steady states' we mean invariant tori lying close to invariant tori $r = r_*$ of the unperturbed system; these will not actually have constant $r$, but will have $r$ liberating in a neighborhood of $r_*$ which is small with $\varepsilon$.) The advantage gained from this analysis is that throughout $K_1$ it is unnecessary to examine the local averaged equations in detail. This would be impossible anyway since $K_1$ is criss-crossed by infinitely many resonance surfaces on each of which the average takes a different form. These resonances are all *inactive* in the sense that they do not alter the qualitative description of the motion, although they do have small quantitative influence on the drift rate.

One warning is in order. The set $K_1$ contains no steady state solutions close to those for $\varepsilon = 0$. This does not imply that $K_1$ contains no invariant sets or recurrent behavior. Near any point $r_* \in K_1$, $r$ must drift; and all $r$ near $r_*$ must drift in the same general direction (since one component must either be increasing for all such $r$, or else decreasing). But in the large, $r$ could drift back around to an $r_*$ which it has visited before, possibly even forming a periodic solution. Of course any such recurrent behavior must take place on a very long time scale (period approaching $\infty$ as $\varepsilon \to 0$). In the example of Section 8, the $r$-space is one-dimensional and so no such large-scale circulation is possible; the same holds in $n$ dimensions if it is possible to cover $K_1$ by sets from Theorem 7.3 using the same index $j$ for all sets, so that a single component of $r$ increases (or decreases) throughout $K_1$. (More generally, the ideas of

Theorem 7.1 can be used to prove that if $V(r)$ is a Lyapunov function for $\dot{r} = a_0(r)$ in a compact set, then it is a Lyapunov function for (7.1) except near finitely many resonance surfaces. This gives large regions that are swept through in the same general direction.)

In the introductory remarks to this section, it was indicated that one can further reduce the set $V$ where lingering can occur, to a smaller set $U$ which is a neighborhood of the zero set of $\bar{f}$ rather than a neighborhood of the non-pass-through set. Without details, we will sketch the ideas involved. If $r_*$ is a point for which $\bar{f}(r_*, \theta)$ is never zero but there is no single component which is never zero, then different solutions near $r_*$ will flow past $r_*$ in quite different directions. By using the matrix $A$ associated with $\omega_*$ and decomposing $A\theta = (x, y)$ into 'fast' angles $x$ (those which rotate when $\varepsilon = 0$) and 'slow' angles $y$, $\bar{f}(r_*, \theta) = \bar{f}(r_*, A^{-1}(x, y))$) becomes a function of $y$ alone. Each set $y = y_*$ is an embedded torus of lower dimension in the torus $r = r_*$, on which $\bar{f}$ has a component which is nonzero. By the ideas of Theorem 7.3 this torus has a neighborhood that is passed through. So if it is possible (in a finite number of steps) to enclose the zero set of $\bar{f}$ in an open set $U \times T^m$, $U \subset K$, then the remainder of $K \times T^m$ can be covered by finitely many such neighborhoods of tori of various dimensions, implying a lower bound to the rate of flow of $r$ (on time scale $\varepsilon t$). The construction of $U$ is inductive. Given $K$, recall that we first constructed $V_0$ as a neighborhood of the zero set of $a_0$. In $K - V_0$ there were finitely many significant resonance surfaces $S_1, ..., S_l$. Instead of simply adding a neighborhood $V_1$ of these to $V_0$ to obtain $V$, we analyze each one more carefully. In each $S_j \times T^m$, $\bar{f}$ can be decomposed into a continuous part consisting of those terms of (7.4) which are present for all $r \in S_j$, and a discontinuous part reflecting the intersection of $S_j$ with other resonance surfaces. By the ideas of Theorem 7.1, $\bar{f}$ can vanish on $S_j \times T^m$ only (i) near zeros of the continuous part or (ii) on finitely many intersections of $S_j$ with other resonance surfaces (possibly including some not in the original list $S_1, ..., S_l$.) The open sets (i) are added to $V_0$ as the first contribution to $U$, and the intersections (ii) are subjected to further analysis. On each intersection $\bar{f}$ again decomposes into a continuous and a discontinuous part. Neighborhoods of the zeros of this new continuous part form the next contribution to $U$. The process terminates with consideration of intersections of $m - 1$ resonance surfaces, the maximum number of commensurabilities possible among $m$ frequencies. A crucial point is that all together only finitely many resonance surfaces need to be considered, so that zeros of finitely many different functions (the various continuous parts of $\bar{f}$) need to be located. The set $U$ is a neighborhood of the set of these zeros.

Motion in the set $U$ cannot be accounted for as a simple drift in $r$. In the neighborhood of any point $r_*$ for which $\bar{f}(r_*, \theta)$ has zeros, first order averaging is not sufficient to describe the motion, and higher order averaging is not generally possible (unless the localized system satisfies one of the conditions of Theorem 3.3). If $\omega_*$ satisfies $m - 1$ commensurabilities, however, the localized system in the form (1.3) is periodic in time, and the theory of Sections 5 and 6 is applicable. Instead of developing this in a general setting, we turn in the next section to an illustrative example.

**Notes**

This section is based entirely on [18], with several simplifications. In particular, [18] contains the details of the inductive process leading to the smaller set $U$ containing the zeros of $\bar{f}$; the simpler but larger set $V$ containing the non-pass-through points was not used there. Because of this [18] contains greater detail about the averaged form of the angle equations, splitting $\theta$ into fast and slow angles near resonances using the matrix $A$.

## 8 SPIN/ORBIT RESONANCE AND DUFFING'S EQUATION

The example to be studied in this section illustrates many points of the general theory set forth in the previous sections. At the same time it has several special features which make it possible to go beyond what seems to be provable in general. In particular, there is a Lyapunov function which overcomes the barrier to proving diagram equivalence (and hence topological conjugacy) with second-order averaging.

In greatest generality, the problem to be studied is

$$(8.1) \qquad \ddot{\theta} = \varepsilon f(\dot{\theta}, \theta, t)$$

where $\theta$ is a (scalar) angle and $f$ is periodic in $\theta$ and $t$. As we proceed, the scope of the analysis will be narrowed to

$$(8.2) \qquad \ddot{\theta} = \varepsilon[\, P(\theta, t) + T(\dot{\theta}, t)]$$

in which no terms involve both $\theta$ and $\dot{\theta}$, and finally to specific forms for the relevant averages of $P$ and $T$. These specifics result from a model which has been proposed for the spin/orbit resonance of the planet Mercury. But the problem is of much more general interest, so before outlining the Mercury problem we will show how (8.1) could arise from the more familiar Duffing equation.

Duffing's equation in its general form is

$$\ddot{x} + C\dot{x} + (Ax + Bx^3) = F \cos \omega t$$

with $A > 0$; here $x$ is a scalar variable. It is most common to study a nearly linear form in which $C$, $B$, and $F$ are small and $\omega$ is close to $\sqrt{A}$; introducing a small parameter $\varepsilon$ to control $C$, $B$, $F$, and $\omega$, this becomes a standard example in the method of averaging. The problem we have in mind is slightly different in that $B$ is not assumed small. Thus the unperturbed oscillator $\ddot{x} + Ax + Bx^3 = 0$ is already nonlinear, although it is completely integrable (using elliptic functions). If $B > 0$ the phase plane for $\varepsilon = 0$ is filled with closed curves of constant energy $E = \dot{x}^2/2 + Ax^2/2 + Bx^4/4$ surrounding the origin, on which solutions circulate with a frequency which varies from curve to curve. It is possible to introduce distorted polar coordinates (action/angle variables) $r, \theta$ in terms of which the unperturbed motion is $\dot{r} = 0$ and $\dot{\theta} = \Omega(r)$, or $\ddot{\theta} = 0$. The perturbed motion then becomes (8.1) for a suitable $f$.

For the spin/orbit problem, Mercury is assumed to orbit the sun in a Keplerian ellipse, and to spin on an axis perpendicular to the orbit. Since the orbit is taken as fixed, the only unknown is the angle $\theta$ between a line fixed in the planet and a line fixed in space (i.e. relative to the fixed stars). If the planet were rigid and symmetrical about its spin axis, the sun would exert no torque and the spin rate would be constant ($\ddot{\theta} = 0$). If the planet is slightly asymmetrical, but still rigid, the torque will depend periodically on time (due to the 'yearly' variation in distance from the sun as Mercury completes its orbit) and $\theta$ (which determines the orientation of the asymmetry), but not on $\dot{\theta}$ (since the system is conservative). This leads to $\ddot{\theta} = \varepsilon P(\theta, t)$. If in addition the planet is not quite rigid, the sun can raise a so-called 'tidal bulge', an elongation of the planet along the line from the center of the planet to the sun. This creates an additional asymmetry. If the internal response of Mercury to the sun's attraction is instantaneous, the tidal bulge will always point directly toward the sun, and the sun will be unable to exert any additional torque on the planet due to this asymmetry. But if the planet's response is delayed, the tidal bulge will be carried by the spin ahead of the sun−planet line and a tidal torque will appear. If the tidal torque mechanism is assumed to operate independently of the permanent asymmetry the tidal term is independent of $\theta$, giving (8.2), in which $P$ stands for 'permanent asymmetry' and $T$ for 'tidal torque'. If the two effects are coupled, we have (8.1).

It is important to note that the tidal torque can change direction: although delay tends to cause the tidal bulge to move ahead of the sun−planet line, this line is itself turning as the planet moves in its orbit. This motion can sweep the tidal bulge behind the sun−planet line. When the bulge is ahead, the tidal torque is directed oppositely to the spin and tends to decrease the spin rate. When it lies behind, the spin rate increases. In terms of energy, the tidal mechanism itself is always dissipative (energy is lost to heat as the planet creaks and groans internally), and yet the spin kinetic energy sometimes increases (when the angle is behind). Since the energy ledger does not balance, the extra energy that is sometimes put into the spin must come at the expense of orbital energy, thus violating the assumption that the orbit is a fixed Keplerian ellipse. But this inherent self-contradiction in the model, which comes from failing to treat the spin and orbit together as a coupled system, is not fatal: a small transfer of energy between orbit and spin can have a significant effect on the spin and almost none on the orbit, so the model can be reasonably accurate over a reasonably long time period. Other inaccuracies in the model (ignoring the slight axis tilt, or perturbation by other planets) may be of comparable or greater magnitude. So without further anguish over physical reality, we turn our attention to equation (8.1).

For simplicity, let time be scaled in (8.1) so that the period of $f$ in $t$ is $2\pi$. Then (8.1) is equivalent to the system

$$
\begin{aligned}
\dot{r} &= \varepsilon f(r, \theta_1, \theta_2) \\
\dot{\theta}_1 &= r \\
\dot{\theta}_2 &= 1
\end{aligned}
$$

(8.3)

which is of the form (7.1) with $n = 1$, $m = 2$, $\Omega(r) = (r, 1)$ and $g = 0$. (Actually only solutions satisfying $\theta_2 = 0$ at $t = 0$ correspond to solutions of (8.1).) Our first task is to apply the analysis of Section 7 to this system to locate regions of pass-through points and to find the values of $r$ that must be studied in more detail.

The Fourier series (7.3), applied to (8.3), takes the form

$$
f(r, \theta_1, \theta_2) = \sum_{\nu} a_{\nu}(r) e^{i(\nu_1\theta_1 + \nu_2\theta_2)}
$$

(8.4)

where each $a_{\nu}(r)$ is a scalar-valued function of a single scalar variable. To compute $\bar{f}$, observe that

$$
\Omega(r)^{\perp} = (r, 1)^{\perp} = \{\nu : \nu_1 r + \nu_2 = 0\} = \{(0, 0)\} \cup \{\nu : -\nu_2/\nu_1 = r\}.
$$

There are two cases: if $r$ is irrational, $\Omega(r)^{\perp} = \{(0, 0)\}$, whereas if $r = p/q$ in lowest terms, $\Omega(r)^{\perp} = \{(kq, -kp) : k \in \mathbb{Z}\}$. Thus

(8.5)
$$
\bar{f}(r, \theta_1, \theta_2) =
\begin{cases}
a_0(r) = \dfrac{1}{4\pi^2} \displaystyle\int_0^{2\pi} \int_0^{2\pi} f(r, \theta_1, \theta_2)\, d\theta_1\, d\theta_2 \\
\qquad\qquad\qquad\text{if } r \text{ is irrational} \\
\displaystyle\sum_{k=-\infty}^{\infty} a_{kq,\,-kp}(p/q) e^{ik(q\theta_1 - p\theta_2)} \\
\qquad\qquad\qquad\text{if } r = p/q.
\end{cases}
$$

Let a compact interval $K = [A, B]$ of $r$ be given. Assume that the zeros of $a_0$ in $K$ are isolated, and surround each one with a small interval. According to Theorem 7.1, any non-pass-through points in the remaining set $K_0$ will be confined to a finite number of resonance 'surfaces'. In this instance a resonance 'surface' $S(\nu_1, \nu_2) = \{r : \nu_1 r + \nu_2 = 0\}$ is simply a rational point $r = p/q$, and Theorem 7.1 says that there is an upper bound $|q| \leqslant N$ to the denominators for which $p/q$ can be a non-pass-through point, due to the fact that for large $q$, the terms in (8.5) other than $a_0(r)$ are small.

In the case of the Mercury problem, the various forms that have been proposed for $f$ have the following features: $a_0(r)$ has a single zero $r = r_0$, with $a_0(r) < 0$ for $r > r_0$ and $a_0(r) > 0$ for $r < r_0$, and the only nonzero Fourier coefficients $a_\nu$ have $\nu_1 = \pm 2$. Thus the effect of the $a_0$ term is to cause $r$ to drift toward the 'ground state' at $r = r_0$, while resonances occurring at half integer values of $r$ ($p/q, q = 1$ or $2$) can disrupt this influence. Since there are only finitely many harmonics in $f$, the case is the simple one discussed as motivation prior to Theorem 7.1. But in fact it is unrealistic to assume that the other harmonics are absent entirely; they are certainly present, but small. Theorem 7.1 provides the justification for ignoring these terms: if they are sufficiently small, the cutoff for significant resonances will occur at $N = 2$. All points will be pass-through points (flowing towards the ground state) except possibly at half integer resonances and in a small interval around the ground state itself, where $a_0$ is small enough that very small higher harmonics may still have an effect. All solutions entering this interval around the ground state remain trapped there forever. (The proof of Theorem 7.1 shows that there is a relation between the size of this interval around the ground state, the amplitude of the higher harmonics in $f$, and the cutoff $N$. In order to have $N = 2$, the interval must be chosen large enough that outside it, $a_0$ dominates the harmonics of order $\geqslant 3$. If these harmonics are truly very small, the interval can be chosen small. The interval itself can then be regarded as the attracting ground state of the system, and the possibly complicated behavior of solutions within it can be ignored. In this respect the system may be regarded as qualitatively the same as the system obtained by deleting the higher harmonics altogether—as must be done, since they are entirely unknown.)

Having examined the over-all pattern of the motion, the next step is to study the neighborhood of a resonance $r = p/q$, in order to determine what happens at the finitely many 'active' resonances where $\bar{f}$ has zeros. At such a resonance we will localize and average to second order, finding (for the Mercury problem under nondegeneracy hypotheses) an even number of periodic orbits, half of which are 1-hyperbolic of index 1 (saddles) and half 2-hyperbolic of index 0 (sinks). It is of course the 2-hyperbolicity which necessitates second order averaging. Since second order averaging is required, Theorem 6.1 about qualitative equivalence of the exact and averaged systems does not apply. Nevertheless it is possible to find a Lyapunov function which fills in part of what is missing: we can prove that there are no other periodic solutions near resonance besides those which are found through averaging, and that every orbit either tends towards one of these or passes through the resonance toward the ground state. It is not always true (as it is for first order averaging) that the diagram of connections between periodic solutions is the same for the averaged and exact systems, as we will show by example. But with additional hypotheses, even this can be guaranteed. We will begin the analysis with the general case of (8.1) or (8.3), introducing additional hypotheses as needed and ending with phase portraits for specific functional forms of $f$.

In order to obtain results valid in a fixed interval about $r_* = p/q$ we will average (8.3) without dilation, as in Section 7. Write (8.3) as

$$\begin{aligned} \dot{r} &= \varepsilon f(r, \theta, s) & r \in R \\ \dot{\theta} &= r & \theta \in S^1 \\ \dot{s} &= 1 & s \in S^1 \end{aligned}$$

(8.6)

and the averaging transformation $(r, \theta, s) \leftrightarrow (\eta, \phi, s)$ as

$$\begin{aligned} r &= p/q + \eta + \varepsilon a(\phi, s) + \varepsilon \eta b(\phi, s) \\ \theta &= \phi + \varepsilon A(\phi, s) + \varepsilon \eta B(\phi, s) \\ s &= s. \end{aligned}$$

(8.7)

The transformed equation, like the transformation (8.7), is to be expanded in powers of $\varepsilon$ and $\eta$ (the latter being a measure of distance from resonance), with terms of order $\varepsilon$ and $\varepsilon\eta$ calculated explicitly. The deleted terms will be of order at least $\varepsilon^2$ or $\varepsilon\eta^2$. Calculations in the spirit of those in Sections 3 and 7 (left this time to the reader) show that if $a$, $b$, $A$ and $B$ can be chosen to satisfy

$$\frac{p}{q} a_\phi + a_s = f\left(\frac{p}{q}, \phi, s\right) - \bar{f}\left(\frac{p}{q}, \phi, s\right)$$

(8.8)

$$\frac{p}{q} b_\phi + b_s = f_r\left(\frac{p}{q}, \phi, s\right) - \bar{f}_r\left(\frac{p}{q}, \phi, s\right) - a_\phi$$

$$\frac{p}{q} A_\phi + A_s = a$$

$$\frac{p}{q} B_\phi + B_s = b - A_\phi$$

then the transformed equations will be

$$\begin{aligned} \dot{\eta} &= \varepsilon \bar{f}(p/q, \phi, s) + \varepsilon \eta \bar{f}_r(p/q, \phi, s) + 0(\varepsilon^2) + 0(\varepsilon\eta^2) \\ \dot{\phi} &= p/q + \eta + 0(\varepsilon^2) + 0(\varepsilon\eta^2) \\ \dot{s} &= 1. \end{aligned}$$

(8.9)

The first equation of (8.8) can be solved by double Fourier series in $\phi$ and $s$ since the right-hand side is free of resonant terms. The Fourier coefficients of $a$ are thereby determined uniquely except for the resonant terms, which are arbitrary (except that the resulting series must converge). The next two equations in (8.8) dictate that the resonant terms in $a$ must be chosen to be zero in order to be able to solve for $b$ and $A$. Similarly, choosing the resonant terms in $b$ and $A$ to be zero enables solving the final equation for $B$. The convergence of the Fourier series for $a$, $b$, $A$, and $B$ is guaranteed by simple estimates.

Although (8.9) is the most powerful form of the averaged equations, the 'dilated and spun' version is more suited to the first part of our investigation. If we set $\eta = \mu\rho$ and $\phi = sp/q + \psi$ then (8.9) yields

$$\dot{\rho} = \mu \bar{f}(p/q, sp/q + \psi, s) + \mu^2 \rho \bar{f}_r(p/q, sp/q + \psi, s) + 0(\mu^3)$$
$$\dot{\psi} = \mu \rho + 0(\mu^4)$$
$$\dot{s} = 1.$$

The estimates $0(\mu^3)$ and $0(\mu^4)$ are valid uniformly in any interval $|\rho| < c$. The change of variables $\phi = sp/q + \psi$ causes any function $2\pi$-periodic in $\phi$ and $s$ to become $2\pi$-periodic in $\psi$ and $2\pi q$-periodic in $s$. An examination of (8.5) shows that $\bar{f}(p/q, sp/q + \psi, s)$ is actually independent of $s$, and similarly for $\bar{f}_r$. In fact (8.5) suggests the introduction of functions $h$ and $k$ for each $p$ and $q$, defined by

(8.10)
$$h(q\theta_1 - p\theta_2) = -\bar{f}(p/q, \theta_1, \theta_2)$$
$$k(q\theta_1 - p\theta_2) = -\bar{f}_r(p/q, \theta_1, \theta_2).$$

In terms of these functions,

(8.11)
$$\dot{\rho} = -\mu h(q\psi) - \mu^2 \rho k(q\psi) + 0(\mu^3)$$
$$\dot{\psi} = \mu \rho + 0(\mu^4)$$
$$\dot{s} = 1.$$

Note that $h$ and $k$ are $2\pi$-periodic in their argument, so the truncation of (8.11) has period $2\pi/q$ in $\psi$; the big-O terms however need only have period $2\pi$ in $\psi$, and $2\pi q$ in $s$. If the last equation in (8.11) is solved to give $s = s_0 + t$ and $s$ is then eliminated from the big-O terms, the resulting system is

(8.11′)
$$\dot{\rho} = -\mu h(q\psi) - \mu^2 \rho k(q\psi) + 0(\mu^3)$$
$$\dot{\psi} = \mu \rho + 0(\mu^4),$$

which looks like the first two lines of (8.11) except that now the big-O terms are $2\pi q$-periodic in $t$ (not $s$) and contain $s_0$ as a parameter. (The original system (8.1) actually corresponds to (8.11′) with $s_0 = 0$.) The truncation of (8.11′) (deleting the big-O terms) can be put in the form

(8.12)
$$\psi'' + \mu k(q\psi)\psi' + h(q\psi) = 0$$

where $' = d/d\tau$ with $\tau = \mu t$. For some $f$, (8.12) takes on a recognizable form (such as a damped pendulum or synchronized motor equation) leading to quick recognition of the phase portrait of the truncated averaged system. (The reader may wish to look ahead to the paragraphs following equation (8.23) for an example.) Of course (8.11) (and (8.12)) have the disadvantage that any arguments based upon them require confining attention to $|\rho| < c$, that is, $|\eta| \leqslant \sqrt{(\varepsilon)}c$, a shrinking interval (as $\varepsilon \to 0$) around the resonance. That is the reason we sometimes prefer (8.9), which for the record we repeat here using $h$ and $k$:

$$\dot{\eta} = -\varepsilon h(q\phi - ps) - \varepsilon\eta k(q\phi - ps) + 0(\varepsilon^2) + 0(\varepsilon\eta^2)$$

(8.13) $$\dot{\phi} = p/q + \eta + 0(\varepsilon^2) + 0(\varepsilon\eta^2)$$

$$\dot{s} = 1.$$

We will not introduce new variables (corresponding to $z$ in Sections 3–6) for the truncated versions of (8.11) and (8.13).

The form (8.11′) is the correct form to use in searching for hyperbolic periodic solutions by the methods of Section 5. (In fact an alternate way to obtain (8.11′) is to set $r = p/q + \mu R$, $\theta = tp/q + \Theta$, and $s = t + s_0$ in (8.6), drop the last equation, and average to second order in $\mu$. This reveals (8.11′) to be exactly the sort of averaged system studied in Sections 3, 4, and 5, and explains why we refer even to (8.13) as a 'second order' averaged system.) In view of Theorem 5.1, the first step is to look for simple zeros $(\rho_0, \psi_0)$ of the first-order terms in (8.11′) (the points called $z_0$ in Section 5). The equations are $h(q\psi) = 0$ and $\rho = 0$, and the Jacobian determinant equals $qh'(q\psi)$. So $(0, \psi_0)$ is a simple zero from which a periodic solution of (8.11′) arises, provided $h(q\psi_0) = 0$ and $h'(q\psi_0) \neq 0$. Along with each periodic solution of (8.11′), there is also, of course, a rest point of the truncation of (8.11′). Assume that $h$ has only simple zeros. Since $h$ has period $2\pi$, every zero of $h(\theta)$ at which $h'(\theta) > 0$ is matched by a zero at which $h'(\theta) < 0$. The total number of zeros of $h(\theta)$ is therefore an even number, $2J$. On the other hand each zero of $h(\theta)$, $0 \leqslant \theta < 2\pi$, occurs $q$ times for $h(q\psi)$, $0 \leqslant \psi < 2\pi$. So there are $2qJ$ simple zeros $(0, \psi_i^*)$, $i = 1, ..., 2qJ$ each giving rise to a $2\pi q$-periodic solution of (8.11′). Let these periodic solutions, for the case of (8.11′) with $s_0 = 0$, be denoted $\rho_i(t, \mu)$, $\psi_i(t, \mu)$. According to part (3) of Lemma 5.2, these differ from the rest points of the truncation of (8.11′) by order $\mu^2$. But the rest points are independent of $\mu$ and are exactly $(0, \psi_i^*)$. Therefore we have

(8.14)
$$|\rho_i(t, \mu)| = 0(\mu^2)$$
$$|\psi_i(t, \mu) - \psi_i^*| = 0(\mu^2)$$

uniformly for all $t$. (Actually the estimate for $\rho_i$ can be sharpened to $0(\mu^3)$ by calculating the perturbation series as in the proof of Lemma 5.2, but the estimate for $\psi_i$ cannot.)

When these solutions are carried over to equations (8.13), their interpretation changes somewhat. This is most easily seen by looking first at the periodic solutions of the truncated version of (8.13). These solutions are exactly $(\eta, \phi, s) = (0, (t + s_0)p/q + \psi_i^* \bmod 2\pi, t + s_0 \bmod 2\pi)$, where $s_0$ is the integration constant which arises in solving $\dot{s} = 1$. These solutions have period $2\pi q$. The solution starting at $(0, \psi_i^*, 0)$ at time $t = 0$ arrives at $(0, \psi_i^* + 2\pi np/q, 0) = (0, \psi_{i+2npJ}^*, 0)$ at time $t = 2\pi n$, $n = 0, ..., q - 1$, and so differs from the solution starting at $(0, \psi_{i+2npJ}^*, 0)$ at $t = 0$ only by a time shift (a choice of $s_0$). Thus there are actually only $2J$ distinct periodic *orbits* and not $2qJ$. (An *orbit* is the underlying set of a *solution*, sometimes called its locus or trace.) Each of these *orbits* carries infinitely many periodic *solutions* differing in $s_0$; of these there are exactly $q$ for which $s = 0$ mod $2\pi$ when $t = 0$ (and then $\phi = \psi^*_{i+2nJ}$ for some $n$). The situation is simplified by viewing (8.13) on a different space (another copy of $\mathbb{R} \times T^2$) in which $s$ is reduced modulo $2\pi q$ rather than $2\pi$. In this space the periodic solutions (we are still looking at (8.13) truncated) are $(\eta, \phi, s) = (0, (t + s_0)p/q + \psi^*_i$ mod $2\pi$, $t + s_0$ mod $2\pi q)$, and there are $2qJ$ distinct periodic *orbits* (each again carrying infinitely many solutions). From now on we will look at (8.13), and also (8.6), in this way. For the original equation (8.1), which is non-autonomous, the distinction between orbits and solutions does not arise since time-shifting in a solution is not allowed. Solutions of (8.1) correspond to solutions of (8.6) having $s = 0$ at $t = 0$, and there are $2qJ$ of these. So $2qJ$ is the correct count for the application.

Turning to the full system (8.13) (including the big-O terms), there will exist $2qJ$ distinct periodic orbits (when $s$ is taken mod $2\pi q$). Each orbit carries one solution for which $s = 0$ at $t = 0$, and these solutions we denote by $(\eta, \phi, s) = (\eta_i(t, \varepsilon), \phi_i(t, \varepsilon)$ mod $2\pi$, $t$ mod $2\pi q)$, $i = 1, ..., 2qJ$. According to (8.14) and the change of variable formulas $\eta = \rho\mu$ and $\phi = sp/q + \psi$, these solutions satisfy

$$|\eta_i(t, \varepsilon)| = 0(\varepsilon^{3/2})$$

(8.15)

$$|\phi_i(t, \varepsilon) - tp/q - \psi^*_i| = 0(\varepsilon)$$

uniformly for all $t$. (The former can be sharpened to $0(\varepsilon^2)$ if (8.14) is sharpened.)

Having established the existence of these orbits, the next question is their hyperbolicity. At this point we restrict our analysis to the case of (8.2), that is, $f(\dot\theta, \theta, t) = P(\theta, t) + T(\dot\theta, t)$. The effect of this hypothesis is that $k$ (defined in (8.10)) is now a constant, and we further assume $k > 0$, in agreement with the situation for Mercury.

Hyperbolicity depends on the matrix $A_i(\mu)$ equal to $1/\mu$ times the Jacobian of the terms of order $\mu$ and $\mu^2$ in (8.11), evaluated at $(0, \psi^*_i)$.

(8.16)
$$A_i(\mu) = \begin{bmatrix} -\mu k & -qh'(q\psi^*_i) \\ 1 & 0 \end{bmatrix}.$$

The eigenvalues of $A_i(\mu)$ are $\{- \mu k \pm \sqrt{[\mu^2 k^2 - 4qh'(q\psi^*_i)]}\}/2$. Recall that $h'(q\psi^*_i)$ alternates in sign for $i = 1, ..., 2qJ$. When $h'(q\psi^*_i) < 0$ the eigenvalues are real and of opposite sign, even for $\mu = 0$. Then Lemma 5.6 says that $A_i(\mu)$ is 1-hyperbolic of index 1. But when $h'(q\psi^*_i) > 0$ the eigenvalues are pure imaginary when $\mu = 0$, moving into the left half-plane for small $\mu > 0$, with Re $\lambda(\mu) < -c\mu$ for some $c$. Then conditions $a'$ and $b'$ of Theorem 5.7 imply that $A_i(\mu)$ is 2-hyperbolic of index 0. Thus the periodic solutions are alternately 1-hyperbolic saddles and 2-hyperbolic sinks. The meaning of this is as follows. Consider first the system obtained from (8.11) by deleting all terms of order higher than $\mu$. This system is equivalent to $\psi'' + h(q\psi) = 0$, a conservative system whose rest points are alternately saddles and centers (neutrally stable). Next consider (8.11) truncated at order $\mu^2$. This is equivalent to (8.12), a damped system (since $k > 0$) in which the centers have become sinks; the saddles, which are 1-hyperbolic, survive the perturbation and remain saddles. The 2-hyperbolicity of the sinks implies that when the full system (8.11) is considered, the sinks as well as the saddles survive the $0(\mu^3)$ perturbation. They become periodic solutions rather than rest points, but their hyperbolicity type does not change.

From this point on we discard (8.11) and work only with (8.13) (with $k > 0$ constant and $s$ computed modulo $2\pi q$). This system has $2qJ$ periodic orbits satisfying (8.15). It is helpful to make an $\varepsilon$-dependent change of variables on $\mathbb{R} \times T^2$ such that in the new coordinates the periodic solutions do not depend on $\varepsilon$, and in fact coincide with those of the truncation of (8.13). The change of coordinates will take place only in a neighborhood of the periodic orbits. Along the $i$th orbit, $\phi - sp/q$ is close to $\psi_i^*$. Choose a smooth function $\beta_i(\psi)$ such that $0 \leqslant \beta_i(\psi) \leqslant 1$ for all $\psi, \beta_i$ has period $2\pi$ in $\psi$, $\beta_i(\psi) = 1$ for $\psi$ in a small interval around $\psi_i^*$, and $\beta_i(\psi) = 0$ outside of a slightly larger interval around $\psi_i^*$. In particular, $\beta_i(\psi_j^*)$ should equal zero for $j \neq i$. Then introduce the change of variables

$$\tilde{\eta} = \eta - \sum_{i=1}^{2qJ} \beta_i(\phi - sp/q)\eta_i(s, \varepsilon)$$

(8.17) $$\tilde{\phi} = \phi - \sum_{i=1}^{2qJ} \beta_i(\phi - sp/q)\{\phi_i(s, \varepsilon) - sp/q - \psi_i^*\}$$

$$\tilde{s} = s.$$

In view of (8.15) this is a near-identity transformation, and so by Theorem 3.1 (*mutatis mutandis*) defines a valid (i.e. smoothly invertible) change of coordinates in the compact subset $|\eta| \leqslant c$ of $\mathbb{R}^2 \times T^2$ for sufficiently small $\varepsilon$. For sufficiently small $\varepsilon$, this transformation carries the periodic solutions $\eta = \eta_j(t, \varepsilon)$, $\phi = \phi_j(t, \varepsilon)$, $s = t$ to $\tilde{\eta} = 0$, $\tilde{\phi} = tp/q + \psi_j^*$, since for small enough $\varepsilon$, $\phi_j$ lies (according to (8.15)) in the region where $\beta_j = 1$ and $\beta_i = 0$ for $i \neq j$. Upon calculating the transformed version of (8.13) (using the idea expressed in (3.5)), it is found that the new equations are the same as the old, with regard to the terms written out in (8.13); the terms expressed in big-O notation are changed and some terms of slightly lower order are added. In doing the calculation, use (8.15) and the following consequences of (8.13) and (8.15): $|\dot{\eta}_i| = 0(\varepsilon^2)$ and $|\dot{\phi}_i - p/q| = 0(\varepsilon^{3/2})$. The result (omitting $\sim$ on the new variables) is

$$\dot{\eta} = -\varepsilon h(q\phi - ps) - \varepsilon\eta k + 0(\varepsilon^2) + 0(\varepsilon\eta^2) + 0(\varepsilon^{3/2}\eta)$$

(8.18) $$\dot{\phi} = p/q + \eta + 0(\varepsilon^{3/2}) + 0(\varepsilon\eta)$$

$$\dot{s} = 1.$$

This is the system with which we will henceforth be concerned. It has the same qualitative properties as (8.13), to which it is related by (8.17), and hence the same qualitative properties as (8.1). Its periodic orbits (taking $s = 0$ at $t = 0$) are given exactly by

$$(8.19) \qquad (\eta, \phi, s) = (0, tp/q + \psi_i^* \bmod 2\pi, t \bmod 2\pi q)$$

and therefore all perturbation terms in (8.18) expressed in big-O notation vanish along the solutions (8.19). System (8.18) is perhaps unusual in that the term of order $\varepsilon\eta$ is written explicitly in the $\dot{\eta}$ equation and not in the $\dot{\phi}$ equation, but this is exactly the degree of accuracy required in the subsequent analysis.

The next step is to introduce a Lyapunov function for (8.18). Remembering that $k > 0$, choose $\lambda > 0$ sufficiently small that $qk - \lambda qh'(q\psi) > a > 0$ for all $\psi$. Define

$$H(\theta) = \int_0^\theta h(u)\, \mathrm{d}u$$

(8.20)

$$V(\eta, \phi, s, \varepsilon) = \tfrac{1}{2} q\eta^2 + \varepsilon H(q\phi - ps) + \varepsilon\lambda\eta h(q\phi - ps).$$

The function $H(\theta)$ is not periodic in $\theta$ (unless the average value of $h$, which equals $a_0(p/q)$, is zero, an unusual situation, since it says that the resonant value $p/q$ is simultaneously a ground state). Therefore $V$ is not well-defined (or is multiple-valued) on $\mathbb{R} \times T^2$, and should instead be considered on $\mathbb{R}^2 \times T^1$: that is, $\phi$ should be regarded as a real variable not reduced modulo $2\pi$. So when we claim (as we now do) that $V$ decreases along orbits of (8.18) in the region $|\eta| \leqslant c$ (in which the coordinate change leading to (8.18) is valid), the meaning is this: if a unique value of $\phi$ is chosen (from its mod $2\pi$ equivalence class) at one point of any solution, then $\phi$ evolves according to (8.18), taking a unique value (not mod $2\pi$) at each time $t$. When this $\phi(t)$, along with $\eta(t)$ and $s(t)$, is substituted into (8.20), the resulting function of time is decreasing, for sufficiently small $\varepsilon$, as long as $|\eta| \leqslant c$.

To prove this result, differentiate $V$ using (8.18) and setting $q\psi = q\phi - ps$ to obtain (since $qk - \lambda qh' > a$)

$$(8.21) \quad \dot{V}/\varepsilon \leqslant - a\eta^2 - \varepsilon\lambda h(q\psi)^2 + 0(\varepsilon^{3/2}) + 0(\varepsilon\eta) + 0(\varepsilon^{1/2}\eta^2) + 0(\eta^3).$$

Divide $|\eta| \leqslant c$ into two parts (depending upon $\varepsilon$): the 'outer' region $\sigma\varepsilon^{1/2} \leqslant |\eta| \leqslant c$ and the 'inner' region $|\eta| \leqslant \sigma\varepsilon^{1/2}$, where the constant $\sigma$ is to be determined (much later). (More precisely: divide the set of pairs $(\eta, \varepsilon)$ satisfying $0 \leqslant \varepsilon \leqslant \varepsilon_0$, $|\eta| \leqslant c$ into two sets defined by these inequalities. Here $\varepsilon_0$ and $c$ are the 'current' values of these bounds, which will be further reduced in the following arguments. The division into inner and outer regions is only meaningful if $\varepsilon_0 \leqslant c^2/\sigma^2$, so if necessary we reduce $\varepsilon_0$ now to make this true, and again later if $c$ is reduced. The considerations involved in choosing $\sigma$ are totally unrelated to these, and $\sigma$ may be regarded here as already fixed.) In the outer region we show that $\dot{V} < 0$ by estimating it in terms of $\eta$. Since the term $-\varepsilon\lambda h^2$ in (8.21) is negative we may drop it, and using $\sigma\varepsilon^{1/2} \leqslant |\eta|$ we obtain $\dot{V}/\varepsilon \leqslant -a\eta^2 + 0(|\eta|^3)$ in the outer region; this is negative if $c$ is sufficiently small.

The inner region is more difficult because of the presence there of the periodic orbits. Because $|\eta| \leqslant \sigma\varepsilon^{1/2}$ in the inner region, (8.12) takes the simpler form

$$(8.22) \qquad\qquad \dot{V}/\varepsilon \leqslant -a\eta - \varepsilon\lambda h^2 + 0(\varepsilon^{3/2}).$$

Let $\delta > 0$ be a constant, which will be chosen shortly. The inner region breaks down into subregions as follows: points where $|\eta| \geqslant \delta$ (there are no such points for very small $\varepsilon$); points where $|\eta| < \delta$ but $|\psi - \psi_i^*| \geqslant \delta$ for all $i$; and points where $|\eta| < \delta$ and $|\psi - \psi_i^*| < \delta$ for some $i$. In the first subregion, the first term of (8.22) dominates for small enough $\varepsilon$, and $\dot{V}$ is negative. In the second subregion, the first term is ignored (it can only help to make $\dot{V}$ negative, but it may not suffice to do so), and the second term dominates the third since $h$ is bounded away from zero when $\psi$ is bounded away from the $\psi_i^*$. Again, $\dot{V}$ is negative. There remain the subregions near the periodic orbits, where both $|\eta|$ and $|\psi - \psi_i^*|$ are small. Here it is a question of how rapidly the first two terms of (8.22) decrease near the periodic orbits compared with the big-O term. The big-O term results from the dot product of $\nabla V$ with the big-O terms in (8.18). By Theorem 2.4, each big-O term in (8.18) can be written as a product of the quantity in the big-O symbol times a function of $\eta$, $\psi$, and $\varepsilon$; for instance $0(\varepsilon\eta) = \varepsilon\eta F(\eta, \psi, \varepsilon)$. Furthermore the functions $F$ vanish on the periodic orbits, as remarked after equation (8.19): $F(0, \psi_i^*, \varepsilon) = 0$. It is an easy check that $\nabla V$ also vanishes on the periodic orbits. Therefore the big-O terms in (8.21) can be written as the quantities in the big-O symbols times functions that vanish to second order on the periodic orbits. Using Taylor's theorem, these functions can be bounded in absolute value by a constant times $\| (\eta, \psi - \psi_i^*) \|^2$ in the region $|\eta| \leqslant \delta$, $|\psi - \psi_i^*| \leqslant \delta$, for small enough $\delta$. Using $|\eta| \leqslant \sigma\varepsilon^{1/2}$ to reduce (8.21) to (8.22), we obtain $0(\varepsilon^{3/2}) \leqslant \varepsilon^{3/2}\beta(\eta^2 + (\psi - \psi_i^*)^2)$. (Theorem 2.4 cannot be applied directly to (8.22) to factor $\varepsilon^{3/2}$ from $0(\varepsilon^{3/2})$, because this term is not of order $\varepsilon^{3/2}$ for all $\eta$, but only for $|\eta| \leqslant \sigma\varepsilon^{1/2}$. It is necessary to go back to (8.21) for this argument.) Similarly, $h^2$ vanishes to second order on the periodic orbits, and since $h'(\psi_i^*) \neq 0$, we can bound $h^2$ below by a positive multiple of $(\psi - \psi_i^*)^2$ in $|\psi - \psi_i^*| \leqslant \delta$ for $\delta$ sufficiently small. (This completes the determination of $\delta$, and hence of the subregions of the inner region that we are examining.) So in the part of the inner region defined by $|\eta| \leqslant \min\{\delta, \sigma\varepsilon^{1/2}\}$, $|\psi - \psi_i^*| \leqslant \delta$, (8.22) can be written

$$\dot{V}/\varepsilon \leqslant -a\eta^2 - \varepsilon b(\psi - \psi_i^*)^2 + \varepsilon^{3/2}\beta(\eta^2 + (\psi - \psi_i^*)^2)$$
$$= (-a + \varepsilon^2\beta)\eta^2 + (-\varepsilon b + \varepsilon^2\beta)(\psi - \psi_i^*)^2$$

which is negative for sufficiently small $\varepsilon$. This completes the proof that $\dot{V}$ is negative except on the periodic solutions.

The arguments to follow, making use of the Lyapunov function just constructed, will be easier to follow with a specific example in mind. It is not necessary to give the original form of the equations (8.2) for our example; instead we suppose $P$ and $T$ are such that

$$(8.23) \qquad h(q\psi) = A + B \cos q\psi + 2C \cos 2q\psi$$

with $0 < A < B$ and $C \geqslant 0$. Assume also that $k > 0$ as before. Then the Lyapunov function is

$$(8.24) \qquad \begin{aligned} V = \tfrac{1}{2}q\eta^2 &+ \varepsilon\{ Aq\psi + B \sin q\psi + C \sin 2q\psi\} \\ &+ \varepsilon\eta\lambda\{ A + B \cos q\psi + 2C \cos 2q\psi\}. \end{aligned}$$

Recall that $\psi$ must not be reduced modulo $2\pi$ since $V$ is not periodic. Therefore we regard (8.18) (replacing $\phi$ by $\psi + tp/q$ and $s$ by $t$) as a nonautonomous system in the strip $|\eta| \leqslant c$ in the $\psi, \eta$ plane.

First consider the case $C = 0$ in (8.23) and (8.24). (With $q = 2$ this is the usual model for Mercury.) The approximate form (8.12) of these equations becomes

$$(8.25) \qquad \psi'' + \mu k\psi' + B \cos q\psi + A = 0.$$

The solution curves of this equation are readily drawn. Figure 4 shows the solutions for $\mu = 0$, a case which does not actually arise since the dilation leading to the form (8.12) is not meaningful for $\mu = 0$. When $\mu = 0$ system (8.25) is conservative and the solution curves can be obtained from an integral. For small $\mu > 0$ it is intuitively clear that the damping changes Fig. 4 into Fig. 5. The lines $|\psi'| = \sigma$ are drawn in Fig. 5 such that the orbits approaching the saddles cross $\psi' = \sigma$ only once, and $\psi' = -\sigma$ not at all. We will show later that the full system (8.18) has a phase portrait similar to this (assuming of course that (8.23) holds, with $C = 0$). Namely, see Fig. 6. Here the curves do not represent solutions (recall (8.18) is nonautonomous in the strip, or is a flow in the strip cross $S^1$). Instead they are invariant curves of the period map; if the solution is viewed stroboscopically, the points appear in successive positions along these curves. The band $|\psi'| \leqslant \sigma$ in Fig. 5 has become the shrinking band $|\eta| \leqslant \sigma\varepsilon^{1/2}$ in Fig. 6.

When $C = 0$, (8.23) has $2q$ zeros ($J = 1$) which are the rest points in Figs 5 and 6. If $C$ is increased, the number of zeros remains the same (and the phase portrait remains similar) until $C$ reaches $(B - A)/2$. At this point the zeros of $h$ are not simple, and the argument for the existence of periodic solutions fails. For $C$ slightly beyond this point there are $4q$ simple zeros ($J = 2$) and the phase portrait of (8.12) is given in Fig. 7. (We say 'slightly beyond' because for some $A$ and $B$ there is another, larger, critical value of $C$.) As $C$ crosses the critical value, then, the sink in Fig. 5 bifurcates into a saddle and two sinks. There is

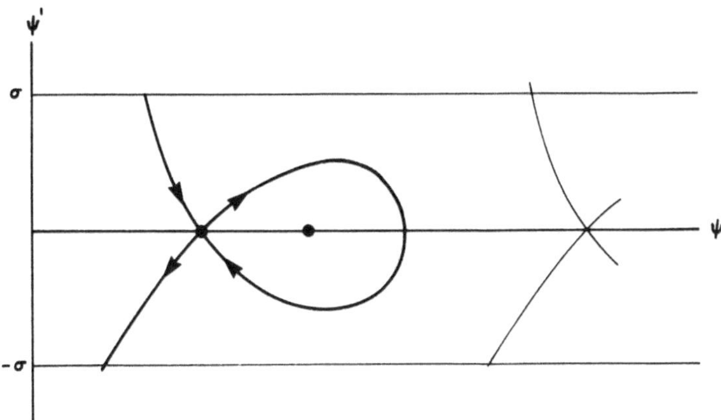

**Fig. 4.** A typical phase portrait in dilated coordinates for $\mu = 0$. The lines $\pm\sigma$ cross the unbounded components of the stable and unstable manifolds only once, and the bounded components not at all

a dramatic difference between Fig. 5 and Fig. 7 which explains why the arguments from the Lyapunov function (to be given below) establish much more in the former case than in the latter. Namely, in Fig. 5 the diagram (in the sense of Section 6) is quite clear: there is a connecting orbit from each saddle to its nearest sink. In Fig. 6 it is unclear which of the three rest points the unstable manifold of the 'major' saddle approaches. (The Lyapunov function is of no help in deciding this: the one sink in Fig. 5 was a minimum of $V$ and this bifurcates into two equal minima and a saddle, all of which are at a lower level than points near the major saddle.) In all probability the diagram

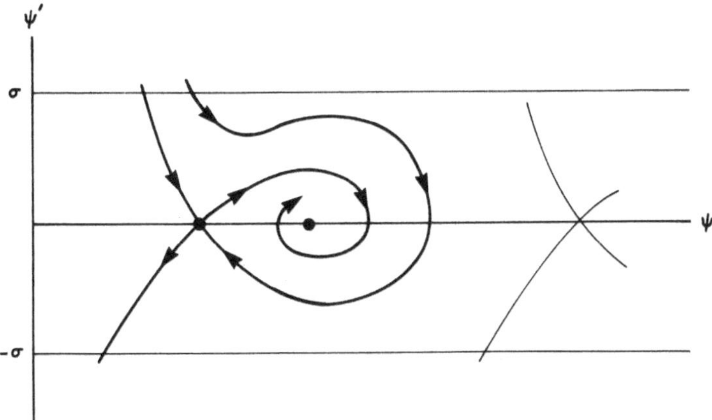

**Fig. 5.** A phase portrait in dilated coordinates for $\mu > 0$

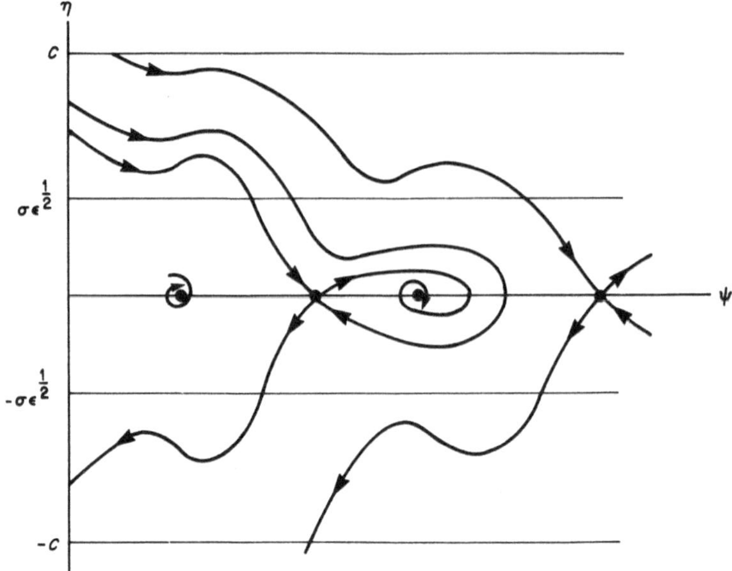

**Fig. 6.** The same phase portrait as Fig. 5, but in non-dilated coordinate, showing upper, inner, and lower regions

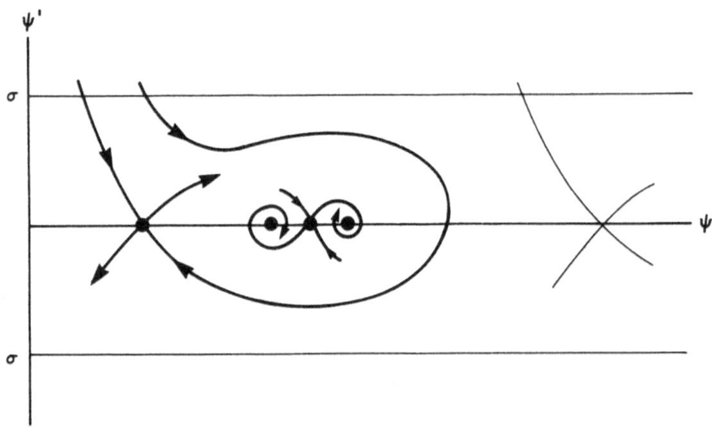

**Fig. 7.** A phase portrait in dilated coordinates with major and minor saddles. The captured portion of the unstable manifold of the major saddle might approach either of the two sinks or the minor saddle

depends upon $\varepsilon$ (for fixed $C$), in the sense that as $\varepsilon \to 0$ there are countably many changes in the destination of the connecting orbit. If this is true then there is no filtration for this system that is independent of $\varepsilon$, and a major ingredient in the argument of Section 6 breaks down. (Whenever the orbit in question connects to the 'minor' saddle, the system is not even even Morse–Smale.)

We are going to prove (in the generality of (8.18), not restricted to (8.23)) that every solution which remains in $|\eta| \leqslant c$ approaches one of the periodic solutions. During the argument, it will be shown that there exist 'major saddles' whose stable and unstable manifolds divide the strip $|\eta| \leqslant \sigma\varepsilon^{1/2}$ into cells; Fig. 8 shows what we mean by a cell in the case of Fig. 6. When applied to (8.23) with $0 \leqslant C < (B - A)/2$, our argument will imply that the connecting orbit is as shown in Fig. 6, so in this case qualitative equivalence between averaged and exact systems will be proved (in the same sense as in Section 6, namely, diagram equivalence.) When applied to (8.23) with $C > (B - A)/2$ (but less than the next critical value, when it exists) it follows only that the solutions have the characteristics of Fig. 7. The usual model for Mercury has $C = 0$. The principal concern in the spin/orbit resonance problem is that the line $\eta = c$ is divided into 'capture' and 'escape' intervals bounded by the stable manifolds of the major saddles. Solutions crossing $\eta = c$ in a capture interval tend toward a resonant periodic orbit, while solutions crossing an escape interval either 'bounce out' (and re-cross $\eta = c$ later) or 'pass through' (crossing $\eta = -c$). These features remain true in Fig. 7. This analysis shows that the

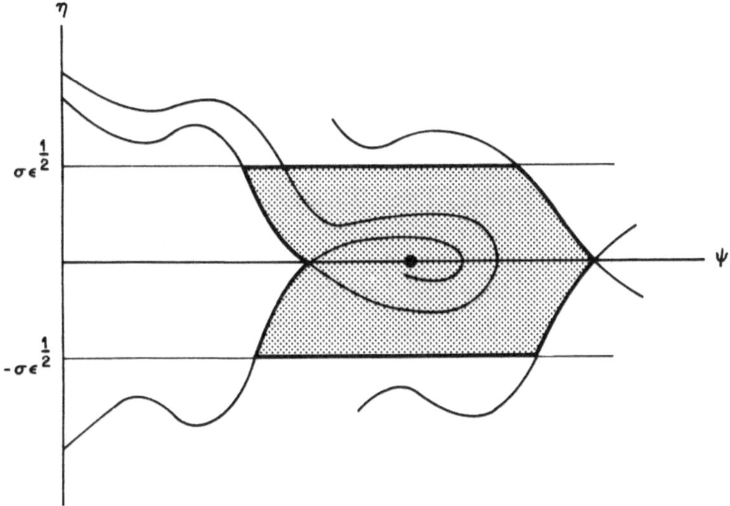


Fig. 8. The dark lines form the boundary of a cell in the inner region. (The example depicted is the same as Fig. 6.)

most important features of the motion are not changed by inclusion of higher harmonics. (This question—the effect of terms such as the $C$ term on the phase portrait—is not to be confused with the basic question we are addressing, which is the qualitative justification of averaging.)

In outline, the argument we are about to give is as follows. A solution of (8.18) enters the upper part of the outer region, $\sigma\varepsilon^{1/2} \leqslant \eta \leqslant c$, across the line $\eta = c$. It may bounce out; if it remains, we will show that it eventually crosses $\eta = \sigma\varepsilon^{1/2}$ and enters the inner region. It may bounce back and forth between the inner and outer regions, but it eventually remains below $\eta = \sigma\varepsilon^{1/2}$. In the inner region, solutions remain in a 'cell' until they either approach a periodic solution in that cell or cross $\eta = -\sigma\varepsilon^{1/2}$, again possibly with some bouncing. Finally a solution cannot remain in the lower part of the outer region, but must eventually cross $\eta = -c$. The Lyapunov function figures in each part of the argument, but none the less a different argument is required in each of the three regions (which we call upper, inner, and lower). The argument will be presented under the hypothesis that $H(\theta + 2\pi) - H(\theta)$, which is a constant (equal to $2\pi$ times the average value of $h$), is positive; that is, $H(q\psi)$ increases by a fixed amount when $\psi$ increases by $2\pi/q$. If it is negative, solutions drift upward (on the average) rather than downward, and the arguments for upper and lower regions are switched. (If it is zero, $p/q$ is both a resonance and a ground state, and we ignore this case here.)

The first step is to choose $\sigma$ and define the cells in the inner region. This is best done using the dilated form of (8.18), which coincides with (8.11). Now (8.11) can be truncated by omitting the big-O terms, which gives a system equivalent to (8.12); or it can be truncated more severely, retaining only the terms of order $\mu$. In this case the result is the same as (8.12) with the $\mu$ term deleted (although we have not set $\mu = 0$). This system is conservative, with integral $I(\rho, \psi) = \frac{1}{2}q\rho^2 + H(q\psi)$. A typical graph of $H(q\psi)$ and its associated phase portrait is shown in Fig. 9. Recall that $H$ is the integral of $h$, and that we are assuming $h$ has a positive mean, or that $H(\theta + 2\pi) > H(\theta)$. Therefore $H$ is the sum of an increasing linear function and a periodic function, and certain of its local maxima have the property that they are absolute maxima on the half-infinite interval to the left. These we call *major maxima*. Now the saddles occur at $\rho = 0$ and $\psi =$ any local maximum of $H$. Saddles occurring at major maxima we call major saddles; the others, if any, are minor saddles. Only the major saddles are involved in the choice of $\sigma$ and the definition of cells. The integral $I(\rho, \psi)$ is constant on the stable and unstable manifolds of a major saddle, and it follows from this, and from the definition of a major maximum, that the stable and unstable manifolds of a major saddle are unbounded to the left of the saddle and bounded to the right. (Furthermore, to the right the stable and unstable manifolds coincide and form a closed loop, called a *homoclinic loop*, surrounding at least one center and any other centers and minor saddles that exist up to the next major saddle, together with their

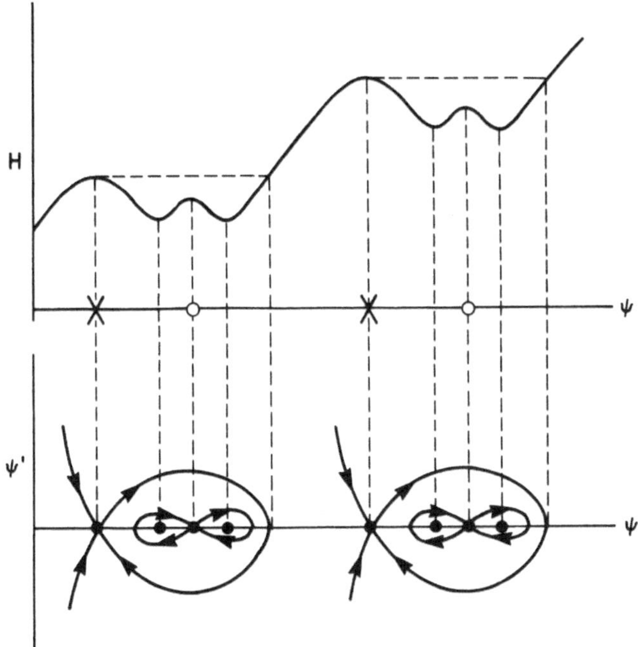

**Fig. 9.** Graph of a typical $H$. Maxima are saddles, minima are sinks. Major saddles (designated x) are those for which H attains an absolute maximum on the half-infinite interval to the left. Other maxima are minor saddles (designated O)

stable and unstable manifolds, which are bounded and which themselves form homoclinic loops.) Because of the boundedness to the right and the periodicity of the phase portrait, there exist lines $\rho = \pm\sigma$ such that $\rho = +\sigma$ cuts each stable manifold of a major saddle once (on the unbounded side) and $\rho = -\sigma$ cuts each unstable manifold once. In addition the strip $|\rho| \leqslant \sigma$ contains the homoclinic loops of the major saddles (see Fig. 9). Restoring the order $\mu^2$ term in (8.11) destroys the conservative character of the phase portrait. Intuitively, Fig. 9 becomes Fig. 7, but we do not wish to rely on this. However we do know that the order $\mu^2$ term does not greatly move compact segments of the stable and unstable manifolds. (This follows from the proof of the stable manifold theorem.) Therefore the portion of the stable manifold entering a major saddle from the left still crosses $\rho = \sigma$, and the portion of the unstable manifold leaving a major saddle to the left still crosses $\rho = -\sigma$. There arcs define the side boundaries of cells in the strip $|\rho| \leqslant c$. One modification remains before we have exactly the cells we need: the big-O terms in (8.11) must be restored. As these terms are $s$-dependent, the picture must be crossed with $S^1$ (in the manner familiar from Section 6), so that the arcs which form the side boundaries of the cells become (slightly $s$-dependent) surfaces in the space $\mathbb{R}^2 \times S^1$ of variables $(\rho, \psi, s)$, where $\psi$ is not taken as an angle.

There exist solutions which enter the strip $|\rho| \leqslant \sigma$ across the line $\rho = \sigma$ and soon afterwards 'bounce out' of the strip across the same line. It is easy to see from the conservative truncation that there is a lower bound to the value of $\rho$ along such arcs. Because compact segments of solutions move only slightly when the higher order terms are restored, there exists a line $\rho = \sigma'$, with $0 < \sigma' < \sigma$, such that any any solution of (8.18) which crosses $\rho = \sigma$ and $\rho = \sigma'$ remains below $\rho = \sigma$ thereafter. A similar line $\rho = -\sigma''$, $-\sigma < -\sigma'' < 0$, can be drawn such that solutions that bounce in and out of $|\rho| \leqslant \sigma$ at the bottom edge are confined to $-\sigma \leqslant \rho \leqslant -\sigma''$. (We speak of these as 'lines' although the figure must be crossed with $S^1$.)

Now we return to the undilated system (8.18). For $\eta > c$, we cannot expect the explicitly written terms in (8.18) to dominate, so nothing can be deduced about the motion. On the line $\eta = c$, (8.18) does not imply that $\dot{\eta}$ is always negative, so although we expect $\eta$ to be decreasing on the average, solutions can leave the strip by crossing $\eta = c$ upwards. After such a crossing they are lost to further study via (8.18). Therefore we confine ourselves to solutions remaining below $\eta = c$ on an interval $t \geqslant t_0$. Our aim is to show that each such solution either approaches one of the periodic orbits as $t \to \infty$, or else crosses $\eta = -c$. This will be done in the following steps:

(1) Solutions in the upper region must enter the inner region. They can bounce back and forth between the upper and inner regions finitely many times, but must eventually leave the upper region without returning.
(2) Solutions in the inner region must either approach a periodic orbit or enter the lower region.
(3) Solutions which enter the lower region may bounce in and out of the inner region finitely many times, but must eventually cross the lower region and exist across its bottom edge, $\eta = -c$.

To establish (1), we consider solutions in the 'extended' upper region $\varepsilon^{1/2}\sigma' \leqslant \eta \leqslant c$ (note $\sigma'$, not $\sigma$). This includes the upper region and the upper part of the inner region (to which 'bouncing' solutions are confined). First we show that in this extended upper region, $\psi$ is strictly increasing along solutions. In fact, (8.18) together with $\varepsilon^{1/2}\sigma' < \eta$ imply $\dot{\psi} = \eta + 0(\eta^3)$, which is positive and bounded away from zero if $c$ is chosen small enough. (Remember that we have left open the possibility of reducing $c$, and that this may entail reducing $\varepsilon_0$. This is not the last time that $c$ will be reduced.) Therefore $\psi$ increases without bound if the solution remains in the extended upper region. On the other hand, $V$ is decreasing. The term $\varepsilon H(q\psi)$ becomes larger than any bound (although not monotonically) if $\psi$ increases, while the other terms in $V$ are bounded. Therefore $\psi$ cannot increase indefinitely, and the solution must cross

$\eta = \varepsilon^{1/2}\sigma'$. By the construction of $\sigma'$, the solution remains below $\eta = \varepsilon^{1/2}\sigma$ for all future time. This proves (1).

After entering the inner region for the last time, $V$ continues to decrease. The motion is now confined to a cell, the only escape being through the bottom edge into the lower region. If the solution remains in the cell, $V$ decreases asymptotically to some value $V_0$. Because the cell is compact, the solution has an $\omega$-limit set, on which $\dot{V} = 0$. But $\dot{V} = 0$ only on the periodic orbits, so the solution must approach one of these. This establishes (2).

If the solution enters the lower region $-c \leqslant \eta \leqslant -\varepsilon^{1/2}\sigma$, it remains beneath $-\varepsilon^{1/2}\sigma''$ thereafter. So to prove (3), it suffices to show by contradiction that no solutions can remain in the extended lower region $-c \leqslant \eta \leqslant -\varepsilon^{1/2}\sigma''$ on a semi-infinite time interval $t \geqslant t_0$. To do so it is convenient to use the Lyapunov function $L = \frac{1}{2}q\eta^2 + \varepsilon H(q\psi)$ in place of $V$; $L$ differs from $V$ in the omission of the term $\varepsilon\lambda\eta h(\psi)$, which (as we saw following equation (8.21)) was not needed in proving that $V$ decreases in the lower region, and is also not needed in the extended lower region. The derivative of $L$ along orbits of (8.18) in the extended lower region is given by the following simpler analog of (8.22): $\dot{L}/\varepsilon = -qk\eta^2 + O(|\eta|^3)$. Since this is an equality, not an inequality as in (8.22), it implies both upper and lower bounds for $\dot{L}$, namely,

$$(8.26) \qquad -2qk\eta^2 \leqslant \dot{L}/\varepsilon \leqslant -\tfrac{1}{2}qk\eta^2$$

in the extended lower region, for small enough $c$. Now assume that a solution $(\eta(t), \psi(t))$ passing through $(\eta_0, \psi_0)$ in the extended lower region at time $t_0$ remains in this region for all $t \geqslant t_0$. The same argument showing that $\psi$ increases without bound in the extended upper region shows that it decreases without bound in the extended lower region. Let $t_i, i = 1, 2, \ldots$, denote the times at which $\psi(t_i) = \psi_i = \psi_0 - 2\pi i/q$. Let $\eta_i = \eta(t_i)$, $L_i = L(\eta_i, t_i)$, and $H_i = H(q\psi_i)$. Then from the definition of $L$ we have

$$(8.27) \qquad (L_i - L_0) - \varepsilon(H_i - H_0) = \tfrac{1}{2}q(\eta_i^2 - \eta_0^2).$$

By our hypothesis (to be contradicted), the right-hand side is bounded as $i \to \infty$. Both $L_i$ and $H_i$ approach $-\infty$ as $i \to \infty$, since $\psi \to -\infty$, so to show that the left-hand side is unbounded requires showing that $L_i$ and $H_i$ decrease at different rates. Now $H_i - H_0 = -i\alpha$ where $\alpha = H(\theta + 2\pi) - H(\theta) > 0$. On the other hand

$$L_i - L_0 = \int_{t_0}^{t_i} \dot{L}\, dt = \int_{\psi_0}^{\psi_i} (\dot{L}/\dot{\psi})\, d\psi = -\int_{\psi_i}^{\psi_0} (\dot{L}/\dot{\psi})\, d\psi.$$

(The latter form is preferred because $\psi_i < \psi_0$.) Since $\dot{L}$ and $\dot{\psi}$ are both negative, $\dot{L}/\dot{\psi} = |\dot{L}/\dot{\psi}|$. From (8.26), $|\dot{L}| \leqslant 2\varepsilon qk\eta^2$. From (8.18), we find that $|\dot{\psi}| \geqslant \frac{1}{2}|\eta|$ in the extended lower region for small enough $c$. Therefore $\dot{L}/\dot{\psi} \leqslant 4\varepsilon qk|\eta| \leqslant 4\varepsilon qkc$, and it follows that $0 > L_i - L_0 \geqslant 4\varepsilon qkc(\psi_i - \psi_0) = -8\varepsilon kc\pi i$. Therefore the left-hand side of (8.27) is greater than $i\varepsilon(\alpha - 8kc\pi)$.

Choosing $c$ small enough that $\alpha - 8kc\pi > 0$ (this is the last restriction on $c$), the left-hand side of (8.27) approaches $+\infty$ as $i \to \infty$. This contradiction establishes (3).

Next we consider the stable and unstable manifolds of the major saddles. We already know that the part of the stable manifold to the left of the saddle crosses $\eta = \varepsilon^{1/2}\sigma$, and the part of the unstable manifold to the left of the saddle crosses $\eta = -\varepsilon^{1/2}\sigma$. Consider the parts lying to the right. For the conservative truncation, these form a homoclinic loop. The energy function for the conservative truncation is $E = \frac{1}{2}q\eta^2 + \varepsilon H(q\psi)$, and this is constant on the homoclinic loop, less inside it, and greater outside it. In particular, $E$ is greater everywhere on $\eta = \pm \varepsilon^{1/2}\sigma$ than it is at the major saddle. For the full system, $V$ as defined in (8.20) decreases along orbits for any $\lambda > 0$ sufficiently small that $qk - \lambda qh'(q\psi)$ is strictly positive. Since $\varepsilon\eta h(q\psi)$ is bounded for $0 \leqslant \varepsilon \leqslant \varepsilon_0$ and $|\eta| \leqslant c$, the term $\varepsilon\lambda\eta h(q\psi)$ in $V$ can be made uniformly arbitrarily small by taking $\lambda$ small. Since this is the only term distinguishing $V$ from $E$, $\lambda$ can be chosen small enough that $V$ is greater everywhere on $\eta = \pm \varepsilon^{1/2}\sigma$ than at the major saddle. Since $V$ decreases along orbits, this proves that the unstable manifold of the major saddles (which is a 2-dimensional manifold in $\mathbb{R}^2 \times S^1$ filled with solution curves) is contained entirely in the inner region. By (2), each orbit on the manifold must approach a periodic orbit. If there are no minor saddles (which implies a unique sink in each cell), this completely determines the diagram with $|\eta| \leqslant c$. This is the case for (8.23) with $0 \leqslant C \leqslant (B - A)/2$, and establishes Fig. 6 (crossed with $S^1$) in·this situation. When minor saddles exist (Fig. 7), the Lyapunov function is insufficient to determine the diagram, as previously remarked.

## Notes

The models for spin/orbit resonance of Mercury are discussed in detail in [8]. The present section is a reworking of material presented in [20], [22], and [31]. In the first of these papers the first order averaging was carried out and it was shown that there are only finitely many active resonances. Also the periodic solutions were constructed. In the second paper, the second order averaged system was analyzed and its phase portraits studied using dilated coordinates. An early form of the Lyapunov function considered here was introduced. At this time the notion of $k$-hyperbolicity was not available, and an *ad hoc* analysis of stability was given. In the final paper the exact system was proved to have the same behavior, in some respects, as the second order averaged system, and the examples were given which suggest that the full topological equivalence is not always true. Some of the arguments in that paper have been simplified here. In particular, we used one Lyapunov function on $|\eta| \leqslant c$ with two proofs that it decreases (one proof in the inner region and another in the outer), where [31] used different Lyapunov functions in the two regions. (The argument using two Lyapunov functions was unable to deal adequately with the question of orbits bouncing between the upper and inner regions.) On the other hand [31] contains results beyond those presented here, in particular, an analysis of the phase portraits near ground states, both when the ground states are resonant and when they are nonresonant, and a discussion of topological conjugacy rather than just diagram equivalence.

Ideally the spin/orbit resonance problem should be formulated with two small parameters $\alpha$ and $\beta$, in the form

$$(8.28) \qquad \ddot{\theta} = \beta P(\theta, t) + \alpha T(\dot{\theta}, t).$$

For the case of Mercury it is estimated that $\alpha \cong 10^{-8}$ and $\beta \cong 10^{-4}$ (for suitable functions $P$ and $T$ which are roughly of the same order of magnitude). Since $\alpha \cong \beta^2$, most studies have replaced $\alpha$ by $\beta^2$ and used $\beta$ as the small parameter. More generally, we can consider the following two perturbation problems:

$$(8.29) \qquad \ddot{\theta} = \varepsilon\beta_0 P(\theta, t) + \varepsilon\alpha_0 T(\dot{\theta}, t)$$

$$(8.30) \qquad \ddot{\theta} = \varepsilon\beta_0 P(\theta, t) + \varepsilon^2\alpha_0 T(\dot{\theta}, t)$$

with constants $\alpha_0$ and $\beta_0$ (not necessarily equal in the two equations). The actual problem (8.28) is in this way embedded in two families of problems, each family being parameterized by $\varepsilon$. For instance if $\alpha = 10^{-8}$ and $\beta = 10^{-4}$, then in (8.29) we could take $\beta_0 = 1$, $\alpha_0 = 10^{-4}$, and $\varepsilon = 10^{-4}$, while in (8.30) we could take $\beta_0 = \alpha_0 = 1$, $\varepsilon = 10^{-4}$. In the $(\alpha, \beta)$ plane, (8.29) then corresponds to a straight line from $(0, 0)$ to $(10^{-8}, 10^{-4})$, whereas (8.30) corresponds to a parabolic arc between these two points. Each problem may be analyzed by perturbation methods, with results valid for some interval of $\varepsilon$, that is, for some segment of the straight line or parabola from $(0, 0)$ to $(10^{-8}, 10^{-4})$. It is not clear a priori whether one or the other (or both or neither) will give results valid at the point $(10^{-8}, 10^{-4})$, nor is there any convincing reason why one or the other is more likely to do so. On behalf of (8.29) it can be argued that a straight line path is aimed directly at the desired point while (8.30) starts off in a different direction. On the side of (8.30) is the long-standing feeling in perturbation theory that everything except the perturbation parameter should be of 'order of magnitude 1', which would rule out $\alpha_0 = 10^{-4}$ in (8.29). In fact this is a prejudice unsupported by the mathematical justification of perturbation methods. It results from a confusion between 'order of magnitude 1' and 'asymptotic order $0(\varepsilon^0)$'. A quantity is of 'order of magnitude 1' if it is 'very roughly equal to 1', say between $1/10$ and $10$. A quantity is of strict asymptotic order $0(1)$ if it remains bounded, and bounded away from zero, as $\varepsilon \to 0$. Therefore $\alpha_0 = 10^{-4}$ is not of order of magnitude 1, but is of strict asymptotic order $0(1)$ merely because it is a nonzero constant. Therefore all of the theorems justifying the method of averaging apply without hesitation to (8.29) even for $\alpha_0 = 10^{-4}$. (As we have remarked, this does not guarantee validity of the results out to $\varepsilon = 10^{-4}$, but merely on some unspecified interval $0 < \varepsilon < \varepsilon_0$.) The equation (8.2) discussed in this section is (8.29), with $\beta_0$ and $\alpha_0$ absorbed into $P$ and $T$.

Under the circumstances, the best course of action is to pursue the mathematical consequences of both (8.29) and (8.30). Our results for (8.29) have been presented here. In regard to (8.30), the following can be said. First, if $P$ is allowed to contain all harmonics in $\theta$ then the arguments of Section 7, showing that only finitely many resonances are active, fail. This is because as $\varepsilon$ is decreased, the term $\varepsilon^2\alpha_0 T$ which destroys active resonances decreases more rapidly than the term $\varepsilon\beta_0 P$ which creates them. Therefore (8.30) will admit infinitely many resonant periodic solutions, each existing for some interval of $\varepsilon$; however for any specific value of $\varepsilon$, only finitely many of these solutions will exist. Therefore the qualitative results for (8.30) are not constant for any interval of $\varepsilon$, making impossible the sort of rigorous qualitative analysis based on structural stability arguments which we have carried out for (8.29). In this respect (8.29) is mathematically simpler than (8.30).

On the other hand, if $P$ is assumed to contain only finitely many harmonics, as has been done in [8] for instance, the qualitative behavior of (8.30) seems to be similar to that of (8.29). In addition, according to the work of Burns [5,6], additional results for (8.30) are possible in this case: Burns argues that there exists a single approximation, analogous to (8.12), valid in an interval of $\dot{\theta}$ containing both the principal resonance in the Mercury problem and the ground state. The approximation takes the form of a 'synchronous motor equation' having 'periodic solutions of the first kind' corresponding to the principal active resonance ($p/q = 3/2$) and a 'periodic solution of the second kind' corresponding to the ground state. (Our equation (8.12) may also have periodic solutions of the second kind, but not within the interval in which (8.12) has been shown to be relevant to (8.29).) In this respect (8.30) may be mathematically simpler than (8.29), although the possibility of a simultaneous aprpoximation of a resonance and a ground state in (8.29), when these are close enough, deserves further study.

In conclusion, each of the equations (8.29) and (8.30) seems to have certain mathematical advantages. If both are valid out to the actual values of $\alpha$ and $\beta$, these results may be combined. In this case (8.29) justifies truncating the harmonics in $\theta$, after which (8.30) permits approximating the ground state and principal resonance simultaneously. But this situation has not been completely clarified.

## REFERENCES

[1] E. Akin, untitled manuscript in progress, private communication.
[2] V. I. Arnol'd, *Mathematical Methods of Classical Mechanics.* Springer-Verlag,

NY, 1978.

[3] N. N. Bogoliubov and Y. A. Mitropolskii, *Asymptotic Methods in the Theory of Nonlinear Oscillations*. Gordon and Breach, NY, 1961.

[4] H. Bohr, *Almost Periodic Functions*. Chelsea, NY, 1947.

[5] T. J. Burns, On the rotation of Mercury. *Celestial Mechanics*, **19** (1979), 297–313.

[6] T. J. Burns, On a dissipative model of the spin-orbit resonance of Mercury, unpublished manuscript.

[7] C. Conley, *Isolated Invariant Sets and the Morse Index*. A.M.S., Providence, 1978.

[8] C. C. Counselman and I. I. Shapiro, Spin-orbit resonance of Mercury. *Symposia Mathematica*, **3** (1970), 121–69.

[9] A. M. Fink, *Almost Periodic Differential Equations*. Springer-Verlag, Berlin, 1974.

[10] J. Guckenheimer and P. Holmes, *Nonlinear Oscillations, Dynamical Systems and Bifurcations of Vector Fields*. Springer-Verlag, NY, 1983.

[11] J. K. Hale, *Ordinary Differential Equations*. Wiley, NY, 1969.

[12] J. K. Hale and L. C. Pavlu, Dynamic behavior from asymptotic expansions. *Quart. Appl. Math.*, **41** (1983–84), 161–8.

[13] M. Hall, Jr., *The Theory of Groups*. Macmillan, NY, 1959.

[14] S. Kobayashi and K. Nomizu, *Foundations of Differential Geometry*, Vol. I. Wiley, NY, 1963.

[15] N. Krylov and N.N. Bogoliubov, *Introduction to Nonlinear Mechanics*. Annals of Mathematics Studies, No. 11. Princeton Univ. Press, Princeton, NJ, 1947.

[16] J. T. Montogomery, Existence and stability of periodic motion under higher order averaging. *J. Diff. Eq.*, **64** (1986), 67–78.

[17] J. Moser, *Stable and Random Motions in Dynamical Systems*. Annals of Mathematics Studies, No. 77. Princeton Univ. Press, Princeton, NJ, 1973.

[18] J. Murdock, Nearly Hamiltonian systems in nonlinear mechanics: averaging and energy methods. *Indiana U. Math. J.*, **25** (1976), 499–523.

[19] J. Murdock, Nested attractors near nonlinear centers. *J. Diff. Eq.*, **25** (1977), 115–29.

[20] J. Murdock, Resonance capture in certain nearly Hamiltonian systems. *J. Diff. Eq.*, **17** (1975), 361–74.

[21] J. Murdock, Some asymptotic estimates for higher order averaging and a comparison with iterated averaging. *SIAM J. Math. Anal.*, **14** (1983), 421–4.

[22] J. Murdock, Some mathematical aspects of spin-orbit resonance. *Celestial Mechanics*, **18** (1978), 237–53.

[23] J. Murdock and C. Robinson, A note on the asymptotic expansion of eigenvalues. *SIAM J. Math. Anal.*, **11** (1980), 458–9.

[24] J. Murdock and C. Robinson, Qualitative dynamics from asymptotic expansions: local theory. *J. Diff. Eq.*, **36** (1980), 425–41.

[25] A. Nayfeh, *Perturbation Methods*. Wiley, NY, 1973.

[26] Z. Nitecki, *Differentiable Dynamics*. MIT Press, Mass., 1971.

[27] J. Palis Jr. and W. deMelo, *Geometric Theory of Dynamical Systems: An Introduction*. Springer-Verlag, NY, 1982.

[28] L. M. Perko, Higher order averaging and related methods for perturbed periodic and quasiperiodic systems. *SIAM J. Appl. Math.*, **17** (1968), 698–724.

[29] C. Robinson, Stability of periodic solutions from asymptotic expansions, in *Classical Mechanics and Dynamical Systems (Medford, Mass., 1079) Lecture Notes in Pure and Appl. Math.*, 70, Dekker, NY, 1981, 173–85.

[30] C. Robinson, Structural stability on manifolds with boundary. *J. Diff. Eq.*, **37** (1980), 1–11.

[31] C. Robinson and J. Murdock, Some mathematical aspects of spin–orbit resonance. II. *Celestial Mechanics*, **24** (1981), 83–107.

[32] J. A. Sanders and F. Verhulst, *Averaging Methods in Nonlinear Dynamical Systems*. Springer-Verlag, NY, 1985.

[33] S. Smale, On gradient dynamical systems. *Ann. Math.*, **74** (1961), 199–206.

[34] S. Sternberg, *Celestial Mechanics*, Part I. Benjamin, NY, 1969.

[35] J. J. Stoker, *Nonlinear Vibrations in Mechanical and Electrical Systems*. Wiley, NY, 1950.

*Dynamics Reported, Volume 1*
Edited by U. Kirchgraber and H. O. Walther
© 1988 John Wiley & Sons and B. G. Teubner

# 4

# An Algorithmic Approach for Solving Singularly Perturbed Initial Value Problems

**K. Nipp**
*ETH-Zentrum, Zurich*

## CONTENTS



## 1  INTRODUCTION

There is a vast literature on singular perturbations both from the point of view of applications as well as of results concerning the theoretical foundations. For a general survey the reader is referred to the books by Cole [3], Eckhaus [5], [6], Kaplun [15], O'Malley [27], Van Dyke [32], Wasow [35] and to the articles by Fraenkel [10], Hoppensteadt [14], Kevorkian [16], Lagerstrom and Casten [17] and Vasil'eva [33], [34]. A good deal of work in singular perturbations is devoted to boundary value problems. In this paper, however, we will restrict ourselves to initial value problems (IVP's) although we believe that the ideas derived may be useful for boundary value problems as well.

Many of the techniques used in singular perturbations have their roots in applied sciences, particularly in fluid dynamics, and were derived by engineers.

So, a large group of authors in the field treat a special problem using formal methods and *ad hoc* arguments and without proving the validity of the formal results in a rigorous mathematical sense (compare e.g. [3], [15], [16], [32]). Moreover, there is no systematic approach for solving a wider class of problems not even on the formal level (a first attempt in this direction has been made in [17]). The number of publications containing rigorous results on singularly perturbed ODE's are comparatively small (see e.g. [9], [29], [33], [34], [35], [14]); and again a general view seems to be missing.

In this paper an attempt is made towards a systematic approach for solving a quite general type of singularly perturbed IVP's which combines the formal and the rigorous aspects. In the formal part (Sections 2 and 3) an algorithm is derived which allows to determine a formal approximation to the solution of a singularly perturbed IVP. In the rigorous part (Section 5) we provide typical error estimates needed for proofing that a formal approximation indeed is an approximation in a rigorous mathematical sense. In Section 4 the formal algorithm is applied to the Belousov–Zhabotinskii chemical reaction. A formal approximation is constructed to the relaxation oscillation of the Field–Noyes model, and it is shown that the approximate period obtained is more accurate than a corresponding result due to Stanshine and Howard [28].

Sections 2, 3 and 4 are a revised version of the author's thesis [22] (a brief note was already published in [21]). The concepts in Section 2 were stimulated by the article by Lagerstrom and Casten [17]. For simplicity, we restrict ourselves to autonomous systems; our methods carry over to the non-autonomous case without essential further modifications.

In the preliminary part below we introduce the notions to be used and give a detailed description of the problem we are going to solve.

We consider the system

$$\frac{\mathrm{d}x}{\mathrm{d}t} = f(x, y, \varepsilon)$$

(1.1)

$$\varepsilon \frac{\mathrm{d}y}{\mathrm{d}t} = g(x, y, \varepsilon)$$

together with the initial conditions

(1.2) $$x(0, \varepsilon) = x^0(\varepsilon), \quad y(0, \varepsilon) = y^0(\varepsilon)$$

where $x$ and $y$ are $m$- and $n$-vectors, respectively, and $\varepsilon$ is a small non-negative parameter, i.e. $\varepsilon \in [0, \varepsilon_0]$, $\varepsilon_0 < 1$. We assume that $f$ and $g$ are sufficiently smooth with respect to all variables in the domain considered as are $x^0$, $y^0$ for $\varepsilon \in [0, \varepsilon_0]$. Moreover, let the solution $z(t, \varepsilon) = (x(t, \varepsilon), y(t, \varepsilon))$ of the IVP (1.1), (1.2) exist for $t \in J := [0, T]$. (Precise statements on existence of such a solution are given in Theorem 1.1 below and in Section 5.) As usual in the presence of a small parameter instead of solving the IVP (1.1), (1.2) directly, which in general is not possible, we try to approximate the solution $z(t, \varepsilon)$ on $J$ for $\varepsilon$ small. More precisely, we try to find a vector function $Z(t, \varepsilon)$ such that

$$(1.3) \qquad \lim_{\varepsilon \to 0} |z\,(t, \varepsilon) - Z(t, \varepsilon)| = 0 \qquad \text{uniformly for } t \in J.$$

$Z(t, \varepsilon)$ is then called an *approximation* (to order unity) *to $z(t, \varepsilon)$ on the interval J*. This approach of course only makes sense if the function $Z(t, \varepsilon)$ is in a way easily determined.

Usually in a perturbation problem an approximation is found by solving the so-called *reduced problem* obtained from the original problem by putting $\varepsilon = 0$. If we put $\varepsilon = 0$ in Eq.(1.1) we get the reduced system

$$(1.4) \qquad \begin{aligned} \frac{\mathrm{d}x}{\mathrm{d}t} &= f(x, y, 0) \\[2ex] 0 &= g(x, y, 0) \end{aligned}$$

which consists of $m$ differential equations and $n$ algebraic equations. If we try to solve the algebraic equations for $y$ and if we insert this solution into the $x$-system we obtain an $m$-dimensional autonomous system of differential equations whose solutions from an $m$-dimensional manifold $S$ in $\mathbb{R}^{m+n}$. Hence, in general, only the initial data on $x$

$$(1.5) \qquad x(0) = x^0(0)$$

can be satisfied. And it is obvious that in this case the solution of the reduced problem (1.4), (1.5) in general is not an approximation to the solution $z(t, \varepsilon)$ of the full problem on the whole interval $J$. This is the reason why (1.1), (1.2) is called *singularly perturbed* (contrary to a regularly perturbed problem).

However, since the solutions of Eq.(1.1) often approach the manifold $S$ in a very short time it is likely that the solution of the reduced problem (1.4), (1.5) approximates $z(t, \varepsilon)$ on some subinterval $J_1$ of $J$, which does not contain the origin.

Let us become more precise. Assume there exists a smooth function $p(x)$ defined for $x$ in some domain $G \subset \mathbb{R}^m$, such that $g(x, p(x), 0) = 0$ for $x \in G$. The set $\{(x, y) \mid y = p(x), x \in G\}$ is then called a *reduced manifold* (or slow manifold) of Eq.(1.1). We suppose that the reduced manifold is *asymptotically stable*, i.e. there is a positive constant $b$ such that each eigenvalue $\lambda$ of the Jacobian $g_y(x, p(x), 0)$ satisfies $\mathrm{Re}(\lambda) \leqslant -b$ for all $x \in G$.

Moreover, assume that the corresponding solution $(X(t), Y(t))$, $Y(t) = p(X(t))$, of the reduced problem (1.4), (1.5), with $x^0(0) \in G$ exists for $t \in [0, t_1]$. Then the following theorem holds ($|\,.\,|$ denoting some norm in $\mathbb{R}^m$ and $\mathbb{R}^n$, respectively).

THEOREM 1.1   If $|y^0(\varepsilon) - Y(0)|$ is sufficiently small, then for $\varepsilon$ small enough the solution $(x(t, \varepsilon), y(t, \varepsilon))$ of (1.1), (1.2) exists and is unique for $t \in [0, t_1]$ and satisfies

$$|x(t, \varepsilon) - X(t)| = O(\varepsilon) \qquad \text{uniformly for } t \in [0, t_1]$$

$$|y(t, \varepsilon) - Y(t)| = O(\varepsilon) \qquad \text{uniformly for } t \in [\delta, t_1], \delta \in (0, t_1). \qquad \square$$

This is a classical result due to A. N. Tikhonov (cf. [29], [30], [35]). It is given here, however, in a somewhat sharper form than in the original paper; an even sharper result is contained in Theorem 5.4. Theorem 1.1 answers the question of existence of the solution of the IVP (1.1), (1.2) in the case where the reduced manifold is asymptotically stable and the initial point $(x^0(\varepsilon), y^0(\varepsilon))$ lies in its domain of attraction. We will see in Section 5 *how* the solution approaches the reduced manifold. Theorem 1.1 states that the solution $(X(t), Y(t))$ of the reduced problem is an approximation to the solution of the full problem on every interval $[\delta, t_1]$, $\delta > 0$. $[\delta, t_1]$ is called a *domain of validity* of the approximation $(X(t), Y(t))$, or $(X(t), Y(t))$ is said to be *valid* on every interval $[\delta, t_1]$, $\delta > 0$ (as an approximation to $z(t, \varepsilon)$). However, this is not the maximal domain of validity. It will be shown in Section 5, Theorem 5.4, that $(X(t), Y(t))$ is actually valid on an $\varepsilon$-dependent interval $[\delta^*(\varepsilon), t_1]$ with $\delta^*(\varepsilon) \to 0^+$, as $\varepsilon \to 0$. For the proof of Theorem 1.1 the reader is again referred to Section 5.

Theorem 1.1 gives some insight into our problem, yet it does not provide an approximation to the solution $z(t, \varepsilon)$ of (1.1), (1.2) on the whole interval $J = [0, T]$.

Introducing a new independent variable in Eq.(1.1) by means of the *scaling transformation* $t = \varepsilon \tau$ we obtain the system

(1.6)

$$\frac{d\tilde{x}}{d\tau} = \varepsilon f(\tilde{x}, \tilde{y}, \varepsilon)$$

$$\frac{d\tilde{y}}{d\tau} = g(\tilde{x}, \tilde{y}, \varepsilon)$$

with the reduced system

(1.7)

$$\frac{d\tilde{x}}{d\tau} = 0$$

$$\frac{d\tilde{y}}{d\tau} = g(\tilde{x}, \tilde{y}, 0)$$

Eq.(1.7) is of full order and the initial conditions (1.2) with $\varepsilon = 0$ can be satisfied. It is well known (compare Theorem 5.1) that the solution $(\tilde{X}(\tau), \tilde{Y}(\tau))$ of this reduced problem is an approximation to the solution

$\bar{z}(\tau, \varepsilon)$ of the full problem (1.6), (1.2) on every finite $\tau$-interval $[0, \rho], \rho > 0$ (error estimate $O(\varepsilon)$). However, without strong additional assumptions on the right-hand sides of Eq.(1.1) $(\bar{X}(\tau), \bar{Y}(\tau))$ will not be an approximation on a long $\tau$-interval $[0, L/\varepsilon]$. Hence, the domain of validity (in $t$) of this approximation (to $z(t, \varepsilon) = \bar{z}(t/\varepsilon, \varepsilon))$ will in general be a small $\varepsilon$-dependent interval $[0, \rho^*(\varepsilon)]$ with $\rho^*(\varepsilon) \to 0^+$, as $\varepsilon \to 0$. (The precise statement is given in Section 5, Theorem 5.2). If an approximation is only valid in such a shrinking (with $\varepsilon$) region this situation is usually called a (boundary) *layer*.

So far we have found two approximations to the solution $z(t, \varepsilon)$ of (1.1), (1.2), one valid on a $t$-interval $[0, \rho^*(\varepsilon)]$ the other one valid for $t \in [\delta^*(\varepsilon), t_1]$. Now, of course, the following question arises: Are these two intervals disjoint, or do the two approximations have an *overlap domain*, i.e. $\delta^*(\varepsilon) < \rho^*(\varepsilon), \varepsilon \in (0, \varepsilon^*)$, for some $\varepsilon^* \leqslant \varepsilon_0$? In the case of an overlap domain the two approximations together would form an approximation to $z(t, \varepsilon)$ on the interval $[0, t_1]$. However, there would still remain one more question: Do the two approximations cover the whole interval $J$? As we shall see lateron the stability of the reduced manifold may break down and $t_1$ may be smaller than $T$. In this case we would have to look for further approximations to $z(t, \varepsilon)$ valid on other subintervals of $J$.

Let us summarize the *characteristic properties* derived above *of a singular perturbation problem*:

—There is no single reduced system whose solution approximates the solution of the full system on the whole $t$-interval of interest.
—Different approximations are needed on different subintervals, and they usually have different time scales.
—Some approximations are only valid on small (with $\varepsilon$) $t$-intervals: layers.

We are now ready to state the problem we are going to attack in the sequel: *Given a singularly perturbed IVP of the form (1.1), (1.2) find a chain of approximations (i.e. with overlapping domains of validity) to its solution on the t-interval of interest in a systematic way.*

Let us conclude this preliminary part by introducing some useful notions and by making some *comments*.

—In the sequel the following type of functions of $\varepsilon$ will play an important role. A function $\eta:(0, \varepsilon_0) \to \mathbb{R}^+ := \{s \mid s > 0\}$ having a limit as $\varepsilon \to 0$ (which may be infinity) is called an *order function*.
One can define an equivalence relation between two order functions $\eta(\varepsilon)$ and $\zeta(\varepsilon)$: $\eta(\varepsilon)$ and $\zeta(\varepsilon)$ are said to be of the *same order* as $\varepsilon \to 0$ (*o*-equivalent) if $\lim_{\varepsilon \to 0} \eta/\zeta$ exists and is different from zero and infinity. If two order functions $\eta(\varepsilon)$ and $\zeta(\varepsilon)$ are in the same equivalence class this will be denoted by

$$\eta(\varepsilon) \approx \zeta(\varepsilon) \quad \text{as} \quad \varepsilon \to 0.$$

We also introduce the notations

$$\eta(\varepsilon) \ll \zeta(\varepsilon) \quad \text{as} \quad \varepsilon \to 0,$$

if $\eta = o(\zeta)$ as $\varepsilon \to 0$, i.e. $\lim_{\varepsilon \to 0} \eta/\zeta = 0$ (then, $\eta$ is said to be of *higher order* than $\zeta$, or $\zeta$ to be of *lower order* than $\eta$, as $\varepsilon \to 0$); and

$$\eta(\varepsilon) \ll \approx \zeta(\varepsilon) \quad \text{as} \quad \varepsilon \to 0,$$

if $\eta \ll \zeta$ or $\eta \approx \zeta$ as $\varepsilon \to 0$.

    Finally
$$\eta(\varepsilon) < \zeta(\varepsilon) \quad \text{as} \quad \varepsilon \to 0$$

of course means that there is $\varepsilon_1 \leqslant \varepsilon_0$ such that $\eta(\varepsilon) < \zeta(\varepsilon)$ for all $\varepsilon \in (0, \varepsilon_1)$.

—The reader might have noticed that the error estimate of Theorem 1.1 (as well as the one for the second approximation, i.e. the solution of Eq. (1.7)) is sharper than what we require in the definition (1.3) of an approximation. Theorem 1.1 implies that also for all order functions $\zeta(\varepsilon)$, with $\varepsilon \ll \zeta \ll 1$ as $\varepsilon \to 0$,

$$\lim_{\varepsilon \to 0} \frac{|z(t, \varepsilon) - Z(t))|}{\zeta(\varepsilon)} = 0 \quad \text{uniformly for } t \in [\delta, t_1],$$

i.e. $Z(t) := (X(t), Y(t))$ is actually a better approximation on every interval $[\delta, t_1]$ than required in (1.3) (an approximation to an order higher than 1 but lower than $\varepsilon$). However, this is not the maximal domain of validity of this approximation, and, as can be seen in Section 5, if one tries to sharpen the estimate for the domain of validity of an approximation one obtains a weaker error estimate on the extended interval. (Being interested in constructing a chain of approximations to $z(t, \varepsilon)$ one needs excellent estimates for the domain of validity for verifying the overlapping of two adjacent approximations.) On the other hand, different members in the sequence of approximations may satisfy different error estimates. For these reasons and in particular for the derivation of our formal approach it is convenient to stick to the weak definition (1.3). However, it might be worthwhile mentioning that when solving a singularly perturbed IVP the way we suggest in this paper one usually obtains a better approximation than required in (1.3).

—Of course, one could also wish to approximate the solution $z(t, \varepsilon)$ of the IVP (1.1), (1.2) to an order $\varepsilon$ or higher. Such a *higher order approximation* is (in general) not obtained by just solving reduced systems where $\varepsilon$ was put equal to 0.

    To find a higher order approximation to $z(t, \varepsilon)$ (on the subinterval $J_1$ of $J$) it is natural to substitute an ansatz of the form

$$(1.8) \qquad \begin{aligned} X^N(t, \varepsilon) &= X(t) + \sum_{i=1}^{N} X^i(t)\varepsilon^i \\[2mm] Y^N(t, \varepsilon) &= Y(t) + \sum_{i=1}^{N} Y^i(t)\varepsilon^i \end{aligned} \qquad , \qquad N \in \mathbb{N}$$

into Eq. (1.1), where $(X(t), Y(t))$ is the solution of the reduced problem (1.4), (1.5). Equating coefficients of like powers of $\varepsilon$ one obtains reduced (linear) differential algebraic systems for the vector functions $X^i(t)$, $Y^i(t)$ with initial conditions $X^i(0) = x^{0,i}$, if

$$x^0(\varepsilon) = \sum_{i=0}^{N} x^{0,i}\varepsilon^i + \hat{x}_N^0(\varepsilon).$$

For convenience, although it might be somehow misleading, we call (1.8) a *higher order solution of the reduced system* (1.4). In fact, under the same hypotheses a result completely analogous to Theorem 1.1 holds (error estimate $O(\varepsilon^{N+1})$) for a higher order solution of (1.4), (1.2) (see e.g. [14]). (If Eq. (1.1) and the initial conditions (1.2) were of more general form with respect to $\varepsilon$ (cf. Eq. (2.1)) the powers of $\varepsilon$ in (1.8) had to be replaced by appropriate order functions $\beta_i(\varepsilon)$ of more general type but still with the property $\beta_{i+1} = o(\beta_i)$ as $\varepsilon \to 0$.)

In the same way one would construct a higher order solution of Eq. (1.7) and thus a higher order approximation to $z(t, \varepsilon)$ on the second subinterval of $J$.

—In what follows we will restrict ourselves to finding an approximation of order unity to $z(t, \varepsilon)$. This is really the basic problem. Having solved this problem there is usually no theoretical difficulty, although it might be very costly, to really compute a higher order approximation. However, as we shall see in Section 4, to construct an approximation to order unity to $z(t, \varepsilon)$ it might be necessary to compute higher order solutions of reduced systems, i.e. in fact local higher order approximations to the corresponding full system.

## 2 THE SYSTEMATIC APPROACH

We consider the following singularly perturbed IVP of dimension $N$ which is slightly more general than (1.1), (1.2)

$$(2.1) \qquad\qquad \mathbf{\Psi}(\varepsilon)\,\frac{\mathrm{d}z}{\mathrm{d}t} = h(z, \varepsilon)$$

$$(2.2) \qquad\qquad z(0, \varepsilon) = z^0(\varepsilon)$$

where $\mathbf{\Psi}(\varepsilon) = \mathrm{diag}(\psi_1(\varepsilon), \dots, \psi_N(\varepsilon))$ and $\psi_j(\varepsilon)$ are order functions having finite limits as $\varepsilon \to 0$, at least one of them being zero. As before, all functions are supposed to be defined on appropriate domains and to be as regular as needed. For the solution $z(t, \varepsilon)$ of (2.1), (2.2) we assume that it exists for $t$ in the interval $J = [0, T]$. Moreover, let $D$ be a domain such that $z(t, \varepsilon): J \times (0, \varepsilon_0) \to D$. ($D$ may depend on $\varepsilon$, and may blow up as $\varepsilon \to 0$.)

To solve the IVP (2.1), (2.2) for $t \in J$ and $\varepsilon$ small we proceed as follows: we try to find, in a systematic way, a *sequence of approximating systems* (reduced systems) whose (if necessary higher order) solutions form a chain of approximations to $z(t,\varepsilon)$ with overlapping domains of validity.


## 2.1  Basic concepts

In this section we try to outline the ideas on which our approach will be based.


### 2.1.1  Transformations

In principle, approximating systems are obtained from the original system (2.1) by transformations. We consider the following type of transformations (*shift scaling transformation*)

$$
\begin{aligned}
z &= q(\varepsilon) + \Phi(\varepsilon)\bar{z} \\
t &= s(\varepsilon) + \nu(\varepsilon)\bar{t}
\end{aligned}
$$
(2.3)

which is of course not the most general one but appropriate for a large class of problems. Here, $q: (0, \varepsilon_0) \to \mathbb{R}^N$, $s: (0, \varepsilon_0) \to \mathbb{R}$, $\Phi(\varepsilon) = \mathrm{diag}(\varphi_1(\varepsilon), ..., \varphi_N(\varepsilon))$ with $\varphi_j(\varepsilon)$, $\nu(\varepsilon)$ being order functions.

Subjecting Eq. (2.1) to the transformation (2.3) yields

$$
\bar{\Psi}(\varepsilon)\,\frac{d\bar{z}}{d\bar{t}} = \bar{h}(\bar{z}, \varepsilon)
$$

a system of the same form as (2.1). We want to put $\varepsilon = 0$ in this equation. However, for some of the functions $\bar{\psi}_k, \bar{h}_k$ the *formal limit* as $\varepsilon \to 0$ might not exist (formal in the sense that the variables $\bar{z}$ and $\bar{t}$ are considered to be fixed with respect to $\varepsilon$). This difficulty can easily be overcome. The transformed system may always be multiplied by an appropriate diagonal matrix of order functions such that for the resulting system

$$
\Psi^*(\varepsilon)\,\frac{d\bar{z}}{d\bar{t}} = h^*(\bar{z}, \varepsilon)
$$
(2.4)

on both sides the formal limits as $\varepsilon \to 0$ exist and are finite. Hence, there is a corresponding *reduced system* (which might not make sense, however, as a system of differential and algebraic equations)

$$
\Psi^*(0)\,\frac{d\bar{z}}{d\bar{t}} = h^*(\bar{z}, 0)
$$
(2.5)

The transformation (2.3) maps $J \to \bar{J}(\varepsilon)$ and $D \to \bar{D}(\varepsilon)$ (here $^-$ does not denote the closure), and Eq. (2.4) has the solution

$$
\bar{z}(\bar{t}, \varepsilon) = \Phi^{-1}(\varepsilon)[z(s(\varepsilon) + \nu(\varepsilon)\bar{t}, \varepsilon) - q(\varepsilon)]
$$

which exists for $\bar{t} \in \bar{J}(\varepsilon) := [-s/\nu, (T-s)/\nu], \varepsilon \in (0, \varepsilon_0)$, with respect to the domain $\bar{D}(\varepsilon)$. Let $\bar{Z}(\bar{t}, \varepsilon)$ be the (if necessary higher order) solution of the approximating system (2.5) with $\bar{Z}(\bar{t}_0, \varepsilon) = \bar{Z}^0(\varepsilon), \bar{t}_0 \in \bar{J}(\varepsilon)$. Then, if $\bar{Z}^0(\varepsilon)$ is, in some sense, close to $\bar{z}(\bar{t}_0, \varepsilon)$, $\bar{Z}(\bar{t}, \varepsilon)$ may in principle be expected to be an approximation to $\bar{z}(\bar{t}, \varepsilon)$ on some finite subinterval $\bar{J}' \subset \bar{J}(\varepsilon)$ independent of $\varepsilon$ with respect to some bounded subdomain $\bar{D}' \subset \bar{D}(\varepsilon)$ also independent of $\varepsilon$ $(\bar{Z}(\bar{t}, \varepsilon): \bar{J}' \times (0, \varepsilon_1) \to \bar{D}', \varepsilon_1 \leqslant \varepsilon_0)$. This important assertion concerning the domain of validity of an approximation is not the whole truth, since $\bar{J}'$ actually can be extended slightly, in general to an interval with moving end-points (as $\varepsilon \to 0$) and this might also influence $\bar{D}'$ (compare Section 2.1.2 and the error estimates stated in Section 5).

$\tilde{Z}(\bar{t}, \varepsilon) := q(\varepsilon) + \Phi(\varepsilon)\bar{Z}(\bar{t}, \varepsilon)$, i.e. the solution $\bar{Z}(\bar{t}, \varepsilon)$ of the approximating system expressed in terms of the original dependent variables, is called a *local approximation* (to $z(s + \nu\bar{t}, \varepsilon)$ on $\bar{J}'$). If we also introduce the original independent variable, $\tilde{Z}((t - s(\varepsilon))/\nu(\varepsilon), \varepsilon)$ will be an approximation to $z(t, \varepsilon)$ on some subinterval $J' \subset J$ with respect to some subdomain $D' \subset D$ (if $\bar{Z}(\bar{t}, \varepsilon)$ approximates $\bar{z}(\bar{t}, \varepsilon)$ 'well enough'). Hence, transforming Eq. (2.1) by means of the shift scaling transformation (2.3) and considering the approximating system (2.5) means considering some subdomain $D' \subset D$ which contains $z = q(\varepsilon)$ and introducing local dependent variables $\bar{z}$ there by blowing up or shrinking $D'$, and similarly considering a subinterval $J' \subset J$ and introducing a local time scale $\bar{t}$ by blowing up or shrinking $J'$.

### 2.1.2 Validity of a local approximation—singularities

We have seen in Section 1 that, in general, a local approximation is not valid on the whole interval of the independent variable of interest. But how do we determine the *domain of validity* of a local approximation? As already mentioned before to do this in a rigorous way, i.e. to give sharp estimates for the error of an approximation as well as for its domain of validity, one has to provide analytical results which might be quite involved. On the other hand, it is often possible to specify $\bar{t}$-values $\bar{S}^+$ and $\bar{S}^-$ where $\bar{Z}(\bar{t}, \varepsilon)$ ceases to be an approximation to $\bar{z}(\bar{t}, \varepsilon)$ (an upper and lower bound so to speak for the interval of validity). We call such a value $\bar{S}$ a *singularity* of the approximation $\bar{Z}(\bar{t}, \varepsilon)$. It is usually determined by heuristic reasonings as: $\bar{Z}(\bar{t}, \varepsilon)$ becomes infinite as $\bar{t} \to \bar{S}$, or a derivative becomes infinite, or a certain stability property is no longer satisfied, etc. The typical situation is: $\bar{Z}(\bar{t}, \varepsilon)$ may be expected to be an approximation to $\bar{z}(\bar{t}, \varepsilon)$ on every compact $\bar{t}$-interval $[c^-, c^+]$ with $c^- > \bar{S}^-$ and $c^+ < \bar{S}^+$, where $\bar{S}^-$ might be $-\infty$ and $\bar{S}^+$ might be $+\infty$. (To show that there is no singularity usually again requires a hard proof.)

Let us consider the infinite case first. We assume that $\bar{S}^+ = +\infty$, i.e. $\bar{Z}(\bar{t}, \varepsilon)$ may be expected to be an approximation to $\bar{z}(\bar{t}, \varepsilon)$ on every compact $\bar{t}$-interval $[c_0, c^+], c^+ < +\infty$, but not on $[c_0, \infty)$ (here, $c_0$ is supposed to be fixed but is not specified more precisely). Then it can be shown that there exists an order function $\eta_1(\varepsilon)$, with $\eta_1 \to +\infty$ as $\varepsilon \to 0$, such that $\bar{Z}(\bar{t}, \varepsilon)$ is even an approximation on the interval $[c_0, \eta_1(\varepsilon)]$. This is a qualitative statement which, since $\eta_1$ is not specified explicitly, does not provide a quantitative estimate for the domain of validity. (Of course, there is an analogous situation for $\bar{S}^- = -\infty$.)

If the singularity $\bar{S}$ is finite we can always achieve by a shift of the independent variable that $\bar{S} = 0$. If $\bar{S}^+ = 0$, i.e. $\bar{Z}(\bar{t}, \varepsilon)$ is expected to be an approximation on every $\bar{t}$-interval $[c_0, c_1]$, $c_1 < 0$, then it again holds that $\bar{Z}(\bar{t}, \varepsilon)$ is in fact an approximation on an extended interval $[c_0, -\eta_1(\varepsilon)]$, where the not explicitly known order function $\eta_1(\varepsilon) \to 0^+$ as $\varepsilon \to 0$. (Similarly for $\bar{S}^- = 0$.)

Although the order function $\eta_1(\varepsilon)$ cannot be specified precisely the fact that the domain of validity of a local approximation typically is 'a finite interval of the local independent variable with moving endpoints (as $\varepsilon \to 0$)' will play an important role in what follows. The fact that an approximation valid on a family of compact intervals actually extends to an $\varepsilon$-dependent interval is usually referred to as *the extension theorem* and is attributed to S. Kaplun (cf. [15]). The proof is simple and is given in all details in [22] (see also [16], [17], [3]). As already stated, this result does not provide a quantitative estimate for the domain of validity of an approximation, however.

### 2.1.3 *A necessary condition for the overlapping of two local approximations*

Let $\tilde{Z}^l(t, \varepsilon)$ and $\tilde{Z}^r(t, \varepsilon)$ be two local approximations (to the solution $z(t, \varepsilon)$ of the IVP (2.1), (2.2)) expressed in the original independent variable each obtained by solving an approximating system of the type (2.5). We assume that these two local approximations are adjacent, i.e. that they have some overlap domain $I$. Then our definition (1.3) of an approximation (to order unity) implies that

$$(2.6) \qquad \lim_{\varepsilon \to 0} |\tilde{Z}^l(t, \varepsilon) - \tilde{Z}^r(t, \varepsilon)| = 0 \qquad \text{uniformly for } t \in I.$$

This is a necessary (but not a sufficient) condition for the overlapping of two local approximations, the so-called *matching condition*. If (2.6) holds the two local approximations $\tilde{Z}^l$ and $\tilde{Z}^r$ are said to *match* for $t \in I$. Replacing the (usually difficult) verification of overlapping by the (usually much easier) verification of matching in constructing our chain of local approximations will be the principal idea for the formal considerations of the next section.

*Remark* As pointed out in Section 1 a local approximation to order unity in general approximates the solution $z(t, \varepsilon)$ of (2.1), (2.2) to some higher order in fact. This of course goes through to matching, i.e. if $\tilde{Z}^l(t, \varepsilon)$ and $\tilde{Z}^r(t, \varepsilon)$ satisfy (2.6) in an overlap domain $I$ they in general also satisfy

$$(2.7) \qquad \lim_{\varepsilon \to 0} \frac{|\tilde{Z}^l(t, \varepsilon) - \tilde{Z}^r(t, \varepsilon)|}{\zeta(\varepsilon)} = 0 \qquad \text{uniformly for } t \in I$$

for certain order functions $\zeta(\varepsilon) \ll 1$. We will also make use of this fact in the next paragraph. $\qquad \square$

## 2.2 Heuristic considerations

The basic ideas developed in Section 2.1 suggest the following kind of procedure (yet vague) for constructing our chain of approximations to the solution $z(t, \varepsilon)$ of (2.1), (2.2). Starting with an approximating system of full order (for satisfying the initial conditions (2.2) with $\varepsilon = 0$) we solve at each step a reduced system and determine where its (possibly higher order) solution ceases to be an approximation to $z(t, \varepsilon)$, i.e. a singularity to the right. In the sense of the transformation (2.3) we shift into the singularity and try to determine the appropriate scalings such that the solution of the approximating system obtained by this transformation is an adjacent approximation (i.e. with an overlapping domain of validity).

However, finding the appropriate scalings (and hence the sequence of approximating systems) in general is a difficult problem and probably cannot be solved in a rigorous way. What we will try to do instead is to develop a *heuristic* concept for determining suitable shift scaling transformations.

### 2.2.1 *Shift scaling transformation*

Let us assume that we have constructed our chain of local approximations to $z(t, \varepsilon)$ up to a certain step and let $\overline{Z}(\bar{t}, \varepsilon)$ be the (possibly higher order) solution of the approximating system of type (2.5) found last. Although we actually do not know its domain of validity as an approximation to the solution $\bar{z}(\bar{t}, \varepsilon)$ considered of the corresponding full system (i.e. $z(t, \varepsilon)$ expressed in the local variables) we assume that we were able to determine a singularity $\bar{S}^+$ of $\overline{Z}(\bar{t}, \varepsilon)$ to the right (as an approximation to $\bar{z}(\bar{t}, \varepsilon)$). As we have seen in Section 2.1.2 we then may expect $\overline{Z}(\bar{t}, \varepsilon)$ to be an approximation to $\bar{z}(\bar{t}, \varepsilon)$ on a $\bar{t}$-interval with right endpoint $\bar{c}^+(\varepsilon)$ where this (not explicitly known) function of $\varepsilon$ satisfies $\bar{c}^+(\varepsilon) \to \bar{S}^+$ as $\varepsilon \to 0$. We want to determine an adjacent local approximation with domain of validity covering this $\varepsilon$-dependent neighbourhood of $\bar{S}^+$ where $\overline{Z}(\bar{t}, \varepsilon)$ is not valid (or at least part of it).

Guided by the ideas of Section 2.1 we introduce local variables in the vicinity of the point $(\bar{S}^+, \overline{Z}^+)$, $\overline{Z}^+ := \overline{Z}(\bar{S}^+, 0)$ by means of the shift scaling transformation

$$\bar{t} = \bar{s} + \bar{\nu}(\varepsilon)\hat{t}, \qquad \bar{s} := \begin{cases} \bar{S}^+, & \text{if } \bar{S}^+ < +\infty \\ 0, & \text{if } \bar{S}^+ = +\infty \end{cases}$$

(2.8)

$$\bar{z}_j = \bar{q}_j + \bar{\varphi}_j(\varepsilon)\hat{z}_j, \quad \bar{q}_j := \begin{cases} \bar{Z}_j^+ & \text{if } |\bar{Z}_j^+| < \infty \\ 0, & \text{if } |\bar{Z}_j^+| = \infty \end{cases} \qquad j = 1, \ldots, N$$

$\bar{\nu}(\varepsilon)$, $\bar{\varphi}_j(\varepsilon)$ being some order functions. If we express our original system (2.1) in terms of these new variables we obtain the system

(2.9)
$$\hat{\Psi}(\varepsilon) \frac{\mathrm{d}\hat{z}}{\mathrm{d}\hat{t}} = \hat{h}(\hat{z}, \varepsilon)$$

and if we apply the formal limit as $\varepsilon \to 0$ in (2.9) (in the sense of Eq. (2.4)) we get an approximating system of type (2.5). Let $\hat{Z}(\hat{t}; \hat{t}_0, \hat{Z}^0; \varepsilon)$ be its general solution (if necessary of higher order). The problem we have to solve is: How does one choose *a priori* the scalings $\bar{\nu}(\varepsilon)$, $\bar{\varphi}_j(\varepsilon)$ in order that, for suitable integration constants $\hat{t}_0$, $\hat{Z}^0$, the vector function $\hat{Z}^r(\hat{t}; \hat{t}_0, \hat{Z}^0; \varepsilon)$ is a local approximation adjacent to $\tilde{Z}^l(\bar{t}, \varepsilon)$, i.e. the two functions ($\bar{Z}$ and $\hat{Z}$ expressed in the original variables) are local approximations with an overlap domain.

We try to find an answer to this problem in the following formal way: We try to derive significant properties for the scaling functions $\bar{\nu}, \bar{\varphi}_j$. If they can be specified completely by such conditions and if in addition a solution of the approximating system obtained by means of (2.8) matches with $\tilde{Z}^l(\bar{t}, \varepsilon)$ in some domain of the independent variable we expect our problem to be solved.

*Remark* The *heuristic principle 'If two local approximations match they also have an overlap domain'* plays an important role in singular perturbations. Engineers and applied mathematicians are usually satisfied with such a formal result. However, as seen in Section 2.1.3 the matching condition is only a necessary condition and although the heuristic principle holds in most cases if one wants to prove rigorously the correctness of a formal result obtained it remains to verify the overlapping of the domains of validity of the (formal) local approximations. This can be done by estimating their domains of validity by analytical results as given in Section 5. $\square$

As pointed out already, the domain of validity of a local approximation in principle may be regarded as a finite interval independent of $\varepsilon$, but which can be extended slightly. We therefore want, if possible, the scaling functions $\bar{\nu}, \bar{\varphi}_j$ to be such that the neighbourhood $\bar{U}$ of $(\bar{S}^+, \bar{Z}^+)$ where $\bar{Z}(\bar{t}, \varepsilon)$ is not valid is transformed by (2.8) into a domain $\hat{U}$ which is finite and independent of $\varepsilon$ (in principle). Hence, we try to find scalings of the following orders as $\varepsilon \to 0$.

(2.10)

(a) $\bar{\nu} \ll 1$, if $\bar{S}^+ < +\infty$ $\qquad \bar{\varphi}_j \ll 1$, if $|\bar{Z}_j^+| < \infty$

(b) $\bar{\nu} \gg 1$, if $\bar{S}^+ = +\infty$ $\qquad \bar{\varphi}_j \gg 1$, if $|\bar{Z}_j^+| = \infty$.

Of course, (a) means a blowing up and (b) a shrinking of the corresponding component of the neighbourhood $\bar{U}$. However, since we do not know $\bar{U}$ explicitly the scalings $\bar{\nu}, \bar{\varphi}_j$ are not completely specified by (2.10).

Let us now consider Eq. (2.9), which contains $\bar{\nu}, \bar{\varphi}_j$ as parameters, and multiply this system with a diagonal matrix of order functions $\bar{\lambda}_j(\varepsilon)$. Then, for any choice of the $\bar{\nu}, \bar{\varphi}_j$ there are $\bar{\lambda}_j$ such that the formal limit of the system as $\varepsilon \to 0$ exists and hence there results a reduced system (which might not make sense as a system of differential and algebraic equations). In general, for such a reduced system the order functions $\bar{\nu}, \bar{\varphi}_j$ are not uniquely determined. If one varies the $\varepsilon$-order of the $\bar{\nu}, \bar{\varphi}_j$ slightly, in general, one obtains the same reduced system. Those reduced systems where the scaling functions $\bar{\nu}, \bar{\varphi}_1, ..., \bar{\varphi}_N$ are uniquely determined, unique up to order functions of the same order, are distinguished. There is exactly one transformation (2.8) that yields such a reduced system. In singular perturbations literature this situation is called a *distinguished limit* (see e.g. [3]). Moreover, a reduced system corresponding to a distinguished limit has the following property: If one varies the order of a scaling function slightly away from the distinguished limit the resulting reduced system has less terms than the distinguished limit system if both systems are expanded with respect to all variables; or more precisely, the distinguished limit system contains all terms of that reduced system and at least one more. In [17] (where only the independent variable is scaled) a reduced system corresponding to a distinguished limit is called a 'principal equation'.

So we try to find $(2N+1)$-vectors $(\bar{\nu}, \bar{\varphi}_1, ..., \bar{\varphi}_N, \bar{\lambda}_1, ..., \bar{\lambda}_N)$ that correspond to a distinguished limit and satisfy, if possible, the conditions (2.10). (The $\bar{\lambda}_j$ are of course taken such that for given scalings $\bar{\nu}, \bar{\varphi}_j$ the number of terms retained in the reduced system is maximal.)

This solves part of the problem posed above. However, there might arise several questions:

—Is there always such a scaling vector corresponding to a distinguished limit?
—If there is more than one which one should be taken?
—How does one compute the 'distinguished scalings'?

In the very general setting of the problem (2.1), (2.2) it is difficult to answer these questions. In Section 3 where we will restrict ourselves to a specific but wide class of problems we will refine the rules for an a priori selection of appropriate scalings. The important concepts, however, are derived already in this section and might well be useful to do complex applications (maybe in combination with some specific insight in the physics of the problem).

### 2.2.2 Matching

Assume we had chosen the scalings $\bar{\nu}(\varepsilon), \bar{\varphi}_1(\varepsilon), ..., \bar{\varphi}_N(\varepsilon), \bar{\lambda}_1(\varepsilon), ..., \bar{\lambda}_N(\varepsilon)$. They completely determine our subsequent approximating system. A solution

$\hat{Z}(\hat{t}; \hat{t}_0, \hat{Z}^0; \varepsilon)$ in general will have a singularity as an approximation to $\hat{z}(\hat{t}, \varepsilon)$ to the left. Assume we had already specified the integration constants $\hat{t}_0, \hat{Z}^0$ and we had determined $\hat{S}^-$, i.e. w expect $\hat{Z}$ to be an approximation to $\hat{z}(\hat{t}, \varepsilon)$ on a $\hat{t}$-interval with left endpoint $\hat{c}^-(\varepsilon)$ where this (not explicitly known) function of $\varepsilon$ satisfies $\hat{c}^-(\varepsilon) \to \hat{S}^-$ as $\varepsilon \to 0$. $\hat{S}^-$ can be finite or infinite, and the same is of course true for the right singularity $\bar{S}^+$ of the preceding local approximation $\bar{Z}(\bar{t}, \varepsilon)$. Hence, the following situations can occur:

$$
\begin{array}{ccc}
 & \bar{S}^+ & \hat{S}^- \\
\text{(i)} & +\infty & > -\infty \\
\text{(ii)} & < +\infty & -\infty \\
\text{(iii)} & +\infty & -\infty \\
\text{(iv)} & < +\infty & > -\infty
\end{array}
$$

We now want to derive typical matching situations for these cases. As we have seen in Section 2.1.3 the matching condition is a necessary condition for the overlapping of two local approximations. We therefore assume for the moment, that $\tilde{Z}^l(\bar{t}, \varepsilon)$ and $\tilde{Z}^r(\hat{t}; \hat{t}_0, \hat{Z}^0; \varepsilon)$ (i.e. $\bar{Z}$ and $\hat{Z}$ expressed in the original dependent variables) have an overlap domain which does not contain the whole domain of validity of either of the two local approximations (otherwise this approximation would be redundant in our chain of local approximations). For simplicity only and without loss of generality suppose that if a singularity is finite it lies in 0; then due to (2.8) one has the relation $\bar{t} = \bar{\nu}(\varepsilon)\hat{t}$ between the two local independent variables.

(i) In the first case, $\tilde{Z}^l$ is supposed to be valid on a $\bar{t}$-interval $[c_0, \bar{\eta}^+(\varepsilon)]$ where $c_0 \leqslant 0$ and the (not explicitly known) order function satisfies $\bar{\eta}^+ \to +\infty$ as $\varepsilon \to 0$, and $\tilde{Z}^r$ on a $\hat{t}$-interval $[\hat{\eta}^-(\varepsilon), c_1]$ with $c_1 > 0$ and $\hat{\eta}^- \to 0^+, \varepsilon \to 0$. (Since we are only interested in one side of the domains of validity, $c_0$ and $c_1$ are assumed to be fixed but are not specified more precisely.)

According to the rule (2.10) we have $\bar{\nu}(\varepsilon) \gg 1$ as $\varepsilon \to 0$. Let $\bar{\nu} \gg \bar{\eta}^+$, then the interval of validity of $\tilde{Z}^l$ expressed in terms of the new variable $\hat{t}$ has a right boundary approaching $0^+$ as $\varepsilon \to 0$, i.e. its singularity in $\hat{t}$ can be regarded to be at a finite value independent of $\varepsilon$ and hence the crucial neighbourhood of the singularity in the new variable is of order 1. Moreover, let $\hat{\eta}^- < \bar{\eta}^+/\bar{\nu}$ as $\varepsilon \to 0$. Then the two local approximations have an overlap domain, and it holds that

there exist two order functions $\hat{\sigma}_0(\varepsilon), \hat{\sigma}_1(\varepsilon)$ with the properties

$$1/\bar{\nu}(\varepsilon) \ll \hat{\sigma}_0(\varepsilon) < \hat{\sigma}_1(\varepsilon) \ll 1 \qquad \text{as } \varepsilon \to 0$$

(2.11)    such that

$$\lim_{\varepsilon \to 0} | \tilde{Z}^l(\bar{\nu}(\varepsilon)\hat{t}, \varepsilon) - \tilde{Z}^r(\hat{t}; \hat{t}_0, \hat{Z}^0; \varepsilon)| = 0 \qquad \text{for } \hat{t} \in [\hat{\sigma}_0(\varepsilon), \hat{\sigma}_1(\varepsilon)];$$

i.e. the two local approximations match for $\hat{t} \in [\hat{\sigma}_0, \hat{\sigma}_1]$.

(ii) In the second case we have $\tilde{Z}^l$ valid on a $\bar{t}$-interval $[c_0, -\bar{\eta}^+(\varepsilon)]$ where $c_0 < 0$ and $\bar{\eta}^+ \to 0^+$ as $\varepsilon \to 0$, and $\tilde{Z}^r$ on a $\hat{t}$-interval $[-\hat{\eta}^-(\varepsilon), c_1]$ with $c_1 \geqslant 0$ and $-\hat{\eta}^- \to -\infty$ as $\varepsilon \to 0$. According to (2.10) $\bar{\nu}(\varepsilon) \ll 1$ as $\varepsilon \to 0$. (If $\bar{\nu}$ were of the same order as $\bar{\eta}^+$ as $\varepsilon \to 0$ the crucial neighbourhood of the singularity of $\tilde{Z}^l$ in the variable $\hat{t}$ were of order 1.) Let $\bar{\nu} \ll 1/\hat{\eta}^-$, then the interval of validity of the new local approximation $\tilde{Z}^r$ expressed in terms of the old variable $\bar{t}$ has a left boundary approaching $0^-$ as $\varepsilon \to 0$, i.e. its singularity in $\bar{t}$ can be regarded to be at a finite value independent of $\varepsilon$. If $\bar{\eta}^+ < \bar{\nu}\hat{\eta}^-$ as $\varepsilon \to 0$ the two local approximations have an overlap domain, and it again holds that

there exist order functions $\bar{\sigma}_0(\varepsilon), \bar{\sigma}_1(\varepsilon)$ satisfying

$$\bar{\nu}(\varepsilon) \ll \bar{\sigma}_1(\varepsilon) < \bar{\sigma}_0(\varepsilon) \ll 1 \qquad \text{as } \varepsilon \to 0$$

(2.12)  such that

$$\lim_{\varepsilon \to 0} |\tilde{Z}^l(\bar{t}, \varepsilon) - \tilde{Z}^r(\bar{t}/\bar{\nu}(\varepsilon); \hat{t}_0, \hat{Z}^0; \varepsilon)| = 0 \qquad \text{for } \bar{t} \in [-\bar{\sigma}_0(\varepsilon), -\bar{\sigma}_1(\varepsilon)].$$

(iii) In this case, $\tilde{Z}^l$ is supposed to be valid on a $\bar{t}$-interval $[c_0, \bar{\eta}^+(\varepsilon)]$ where $c_0 \leqslant 0$ and $\bar{\eta}^+ \to +\infty$ as $\varepsilon \to 0$ and $\tilde{Z}^r$ on a $\hat{t}$-interval $[-\hat{\eta}^-(\varepsilon), c_1]$ with $c_1 \geqslant 0$ and $-\hat{\eta}^- \to -\infty$ as $\varepsilon \to 0$. If we had $\bar{\nu} \gg 1$ as $\varepsilon \to 0$ according to (2.10), in general one of the two local approximations would be redundant. Hence, in general $\bar{\nu} \gg 1$ is not the appropriate scaling for this situation. So it remains only $\bar{\nu}(\varepsilon) \approx 1$ as a possible scaling. But how can we then have the singularity $\bar{S}^+$ to be $O(1)$ in the new variable $\hat{t}$? This is the case if we assume that $\bar{S}^+$ depends on $\varepsilon$ and approaches $+\infty$ as $\varepsilon \to 0$ and if we shift into this $\varepsilon$-dependent singularity, i.e.

$$\bar{t} = \bar{T}(\varepsilon) + \hat{t} \qquad \text{where } \bar{T} > \bar{\eta}^+ \text{ as } \varepsilon \to 0.$$

(Of course, the time shift $\bar{T}(\varepsilon)$ can be regarded as one of the integration constants of $\hat{Z}$.) Then, if $\bar{T} - \bar{\eta}^+ < \hat{\eta}^-$ as $\varepsilon \to 0$ the two local approximations have an overlap domain, and

there exist order functions $\hat{\sigma}_0(\varepsilon), \hat{\sigma}_1(\varepsilon)$ satisfying

$$1 \ll \hat{\sigma}_1 < \hat{\sigma}_0 \text{ and } \bar{T} - \hat{\sigma}_0 \gg 1 \qquad \text{as } \varepsilon \to 0$$

(2.13)  such that

$$\lim_{\varepsilon \to 0} |\tilde{Z}^1(\bar{T}(\varepsilon) + \hat{t}, \varepsilon) - \tilde{Z}^r(\hat{t}; \hat{t}_0, \hat{Z}^0; \varepsilon)| = 0 \qquad \text{for } \hat{t} \in [-\hat{\sigma}_0(\varepsilon), -\hat{\sigma}_1(\varepsilon)].$$

((2.13) can be formulated analogously with respect to the independent variable $\bar{t}$.)

(iv) In the last case $\tilde{Z}^l$ is supposed to be valid on a $\bar{t}$-interval $[c_0, -\bar{\eta}^+(\varepsilon)]$ with $c_0 < 0$, $\bar{\eta}^+ \to 0^+$ as $\varepsilon \to 0$, and $\tilde{Z}^r$ on a $\hat{t}$-interval $[\hat{S}^- + \hat{\eta}^-(\varepsilon), c_1]$ where $\hat{S}^- < 0, c_1 > 0$ and $\hat{\eta}^- \to 0^+$ as $\varepsilon \to 0$. If according to (2.10) $\bar{\nu}(\varepsilon) \ll 1$ as $\varepsilon \to 0$ one derives a matching situation completely analogous to (2.12). (However, one could also think of a matching situation similar to (2.13) if one had $\bar{\nu} \approx 1, \bar{S}^+ = 0, \hat{S}^- = 0$, and $\bar{t} = \bar{T}(\varepsilon) + \hat{t}$ with $\bar{T}(\varepsilon) \to 0^-$ as $\varepsilon \to 0$.)

*Remarks* —In this section, having assumed an overlap domain of two local approximations being solutions of approximating systems obtained from our original system (2.1) by means of a shift scaling transformation (2.8) we have developed typical matching situations. Although we cannot claim that these are the only possible situations they are certainly the most frequent ones and give us the insight how to handle the matching of two local approximations.

Of course, in our procedure of constructing a chain of local approximations we do not know a priori either the (precise) domain of validity of $\tilde{Z}^l$ nor the integration constants (and hence the singularity $\bar{S}^-$) of $\tilde{Z}^r$ nor an overlap domain. As already pointed out, the idea is to use matching as a verification for overlapping in the sense of a necessary condition. More precisely, having specified the scaling functions $\bar{\nu}(\varepsilon), \bar{\varphi}_1(\varepsilon), \ldots, \bar{\varphi}_N(\varepsilon)$ by the considerations of Section 2.2.1 and thus having found a candidate for a subsequent local approximation the unknown integration constants $\hat{t}_0, \hat{Z}^0$ are chosen (if possible) such that the appropriate matching condition holds ((2.11), (2.12) or (2.13), respectively). (The matching condition is chosen due to the $\varepsilon$-order of $\bar{\nu}(\varepsilon)$, i.e. the type of singularity $\bar{S}^+$, since $\hat{S}^-$ is not known a priori). Usually, the matching condition is satisfied, if at all, for all values $\sigma$ of the appropriate local independent variable with $1/\bar{\nu}(\varepsilon) \ll \sigma \ll 1, \bar{\nu}(\varepsilon) \ll \sigma \ll 1$ or $1 \ll \sigma$ and $\bar{T}(\varepsilon) - \sigma \gg 1$, respectively. Hence, often it is not even necessary to determine order functions $\hat{\sigma}_k(\varepsilon)$ or $\bar{\sigma}_k(\varepsilon)$, respectively, for specifying a so-called *formal overlap domain*, i.e. a domain of the (local) independent variable where the matching condition holds.

—From Section 2.1.3 we may conclude that a matching condition (2.11), (2.12) or (2.13), respectively, in general also holds to an order higher than 1 (in the sense of (2.7)). Hence, since the matching condition is only necessary but not sufficient for the overlapping of two local approximations one would actually like to verify an *as good as possible matching*, i.e. to an as high as possible order. If $\tilde{Z}^l$ and $\tilde{Z}^r$ in the matching condition are expanded with respect to $\varepsilon$ and to the specified local independent variable this in particular implies that as many terms as possible of the two expansions cancel out. This requirement is also often helpful (or even necessary) to specify all constants of integration $\hat{t}_0, \hat{Z}^0$ of the new local approximation. (In the next section we will even make use of it for polishing up the selection rules for the scalings $\bar{\nu}(\varepsilon), \bar{\varphi}_j(\varepsilon)$.) $\square$


## 2.3 A formal algorithm

We are now able to specify our procedure in a fairly precise way as well as to define the label formal.

DEFINITION  A *formal approximation* (to order unity) *to the solution* $z(t, \varepsilon)$ of the IVP (2.1), (2.2) *on the interval* $J = [0, T]$ is a chain of (possibly higher order) solutions of approximating systems such that the first one of these vector functions satisfies the initial conditions (2.2) with $\varepsilon = 0$, each of the subsequent ones matches with its preceding one, and the last one has a singularity which (expressed in the original variable $t$) is greater than $T$. An individual member out of this sequence is called a *local formal approximation*. $\square$

A formal approximation to $z(t, \varepsilon)$ on $J$ can be constructed in the following systematic way.

## Step 1

(a) Subject the original system (2.1) to the scaling transformation

$$t = \nu(\varepsilon)t_1$$
$$z_j = \varphi_j(\varepsilon)z_j^1, \ j = 1, \ldots, N$$

and multiply this system with a diagonal matrix of order functions $\lambda_j(\varepsilon), j = 1, \ldots, N$.

(b) Choose the $2N + 1$ order functions such that
the formal limit as $\varepsilon \to 0$ exists;
the resulting reduced system is of full order
and, if possible, corresponds to a distinguished limit;
the initial conditions (2.2) are $O(1)$ as $\varepsilon \to 0$ (if possible independent of $\varepsilon$).

(c) Determine the solution $Z^1(t_1)$ of this first approximating system satisfying the initial conditions (2.2) with $\varepsilon = 0$ (if necessary a higher order solution $Z^1(t_1, \varepsilon)$.

(d) Determine a singularity $S_1^+$ of $Z^1(t_1)$ (as an approximation to $z(t, \varepsilon)$ expressed in the new variables) and define $Z^{1+} := Z^1(S_1^+)$. $\square$

## Step i: $(i > 1)$

(1) Introduce new local variables in the system (2.1) by means of the shift scaling transformation

$$t_{i-1} = \bar{s} + \bar{\nu}(\varepsilon)t_i, \qquad \bar{s} := \begin{cases} S_{i-1}^+, & \text{if } S_{i-1}^+ < +\infty \\ 0, & \text{if } S_{i-1}^+ = +\infty \end{cases}$$

$$z_j^{i-1} = \bar{q}_j + \bar{\varphi}_j(\varepsilon)z_j^i, \ \bar{q}_j := \begin{cases} Z_j^{(i-1)+}, & \text{if } |Z_j^{(i-1)+}| < \infty \\ 0, & \text{if } |Z_j^{(i-1)+}| = \infty \end{cases}, \qquad j = 1, \ldots, N.$$

Multiply this system with a diagonal matrix of order functions $\bar{\lambda}_j(\varepsilon)$ and denote it by $(\bar{E}^i)$.

(2) Determine the set $\mathbb{V}_i$ of scaling vectors $(\bar{\nu}(\varepsilon), \bar{\varphi}_1(\varepsilon), ..., \bar{\varphi}_N(\varepsilon), \bar{\lambda}_1(\varepsilon), ..., \bar{\lambda}_N(\varepsilon))$ (up to order functions of the same order class) corresponding to a distinguished limit of $(\bar{E}^i)$.

(3) Determine the subset $\mathbb{V}_i^1$ of $\mathbb{V}_i$ with the properties

$$\bar{\nu}\begin{cases} \ll 1, & \text{if } S_{i-1}^+ < +\infty \\ \gg 1, & \text{if } S_{i-1}^+ = +\infty \end{cases}, \qquad \bar{\varphi}_j\begin{cases} \ll \approx 1, & \text{if } |Z_j^{(i-1)+}| < \infty \\ \gg \approx 1, & \text{if } |Z_j^{(i-1)+}| = \infty \end{cases}, \qquad j = 1, ..., N.$$

(4) If $\mathbb{V}_i^1 = \varnothing$ determine the subset $\mathbb{V}_i^2$ of $\mathbb{V}_i$ with the properties

$$\bar{\nu} \approx 1, \qquad \bar{\varphi}_j \text{ as in (3)}.$$

Let $\mathbb{V}_i^0 := \mathbb{V}_i^2$.

If $\mathbb{V}_i^0 = \varnothing$ Step $i$ is not successful.

(5) If $\mathbb{V}_i^0$ contains more than one scaling vector choose one with the property: The $\bar{\varphi}_j$ are of an as low as possible order as $\varepsilon \to 0$.

(6) Let $(E^i)$ be the system obtained by the scaling transformation (1) with these now specified scalings and determine the general solution $\hat{Z}^i(t_i; t_{i0}, Z^{i0}; \varepsilon)$ (if necessary of higher order) of its reduced system.

(7) Let $\tilde{Z}^{l(i-1)}(\tilde{t}_{i-1}, \varepsilon)$ be the soution of the $(i-1)$st approximating system expressed in the original dependent variables and in terms of $\tilde{t}_{i-1} := t_{i-1} - \bar{s}$ and denote $\hat{Z}^i$ expressed in the original dependent variables by $\tilde{Z}^{ri}$.

Determine the integration constants $t_{i0}$, $Z^{i0}$ by an as good as possible matching of the vector functions $\tilde{Z}^{l(i-1)}(\tilde{t}_{i-1}, \varepsilon)$ and $\tilde{Z}^{ri}(t_i; t_{i0}, Z^{i0}; \varepsilon)$ according to the matching rules (2.11) (if $\bar{\nu} \gg 1$), (2.12) (if $\bar{\nu} \ll 1$) or (2.13) (if $\bar{\nu} \approx 1$), respectively.

If matching fails Step $i$ is not successful.

(8) Determine a singularity $S_i^+$ of the chosen solution $Z^i(t_i, \varepsilon)$ (as an approximation to $z(t, \varepsilon)$ expressed in $z^i, t_i$) and define $Z^{i+} := Z^i(S_i^+, 0)$.

If $S_i^+$, expressed in the original independent variable $t$, is greater than $T$ the steps $1, ..., i$ have provided a formal approximation to $z(t, \varepsilon)$ on $J$. $\square$

*Remarks* —As already emphasized our approach is strongly based on the idea of blowing up or shrinking a certain neighbourhood of the 'singularity' $(\bar{S}^+, \bar{Z}^+)$ of a local approximation to 'essentially $O(1)$' as $\varepsilon \to 0$ (if possible independent of $\varepsilon$).

Moreover, the rules (3) and (5) for specifying the scalings $\bar{\varphi}_j(\varepsilon)$, $j = 1, ..., N$, are motivated by the fact that in general we do not know the exact $\varepsilon$-order of that neighbourhood and that a function of $\varepsilon$ which is $o(1)$ as $\varepsilon \to 0$ of course is also $O(1)$.

—If in (7) matching is not possible and if $\mathbb{V}_i^0$ contains more than one scaling vector one would try (6) and (7) with new scalings.

—The concepts and ideas of Section 2 has led us to the formulation of a systematic approach for solving singularly perturbed IVP's. Although, in this very general setting, the rules given in Section 2.3 might not be precise enough, i.e. might not in any case yield a suitable adjacent approximating system, and although we could not specify how to compute the distinguished scalings, we believe that this procedure gives the correct way of how to handle a general singularly perturbed IVP (of course also a non-autonomous one). The concepts derived might even be useful for solving singularly perturbed BVP's.

In Section 3 where we will restrict ourselves to a specific class of problems we will be able to refine the rules for an a priori selection of appropriate scalings as well as to compute the distinguished scalings in an efficient way. —We have rendered precise now what we mean by formally solving a singularly perturbed IVP. This formal approach is based on the heuristic principle 'If two local approximations match they have an overlap domain'. Although matching is only a necessary condition for overlapping, having found a formal approximation one may by good reasons assume that one indeed has found an approximation to $z(t, \varepsilon)$ on $J$. Its correctness, however, has to be proved by verifying the overlapping of the (formal) local approximations, and this is to be done by estimating their domains of validity by theoretical results as stated in Section 5. $\square$

## 3 THE POLYHEDRON ALGORITHM

In Section 2 we have derived a formal approach for approximately solving a singularly perturbed IVP in a straightforward way. In the very general setting of that section, however, we were not able to answer completely all questions arising. Therefore, we now restrict ourselves to a class of singularly perturbed systems of ODE's where it is possible to motivate a precise choise of scalings for finding an appropriate sequence of approximating systems and hence for constructing a formal approximation. This is done by introducing a correspondence between a system of ODE's containing the small parameter $\varepsilon$ and a convex polyhedron.

### 3.1 Convex polyhedrons and singularly perturbed systems

Let us consider a system of the type (2.1) where the right-hand sides are polynomials with respect to all variables. For convenience and without loss of generality we develop our ideas for a system of two scalar equations

$$\frac{\mathrm{d}x}{\mathrm{d}t} = p(x, y, \varepsilon) = \sum_{i, j, k \geqslant 0} c_{ijk}\varepsilon^i x^j y^k$$

(3.1)

$$\varepsilon\frac{\mathrm{d}y}{\mathrm{d}t} = q(x, y, \varepsilon) = \sum_{i, j, k \geqslant 0} d_{ijk}\varepsilon^i x^j y^k$$

where only a finite number of coefficients $c_{ijk}$ and $d_{ijk}$ are non-zero.

If we subject Eq. (3.1) to the scaling transformation

$$x = \varepsilon^a \bar{x}$$
$$(3.2) \qquad y = \varepsilon^b \bar{y} \qquad a, b, d \in \mathbb{R}$$
$$t = \varepsilon^d \bar{t}$$

and if we multiply the first equation by $\varepsilon^{-l}$ and the second one by $\varepsilon^{-m}$, $l, m \in \mathbb{R}$, we obtain a system of the form

$$(3.3) \qquad \varepsilon^{L(a,b,d,l,m)} \frac{d\bar{x}}{d\bar{t}} = \sum_{i,j,k \geqslant 0} \varepsilon^{L_{ijk}(a,b,d,l,m)} c_{ijk} \bar{x}^j \bar{y}^k$$

$$\varepsilon^{M(a,b,d,l,m)} \frac{d\bar{y}}{d\bar{t}} = \sum_{i,j,k \geqslant 0} \varepsilon^{M_{ijk}(a,b,d,l,m)} d_{ijk} \bar{x}^j \bar{y}^k$$

where $L$, $L_{ijk}$, $M$, $M_{ijk}$ are linear functions on $\mathbb{R}^5$ with integer coefficients.

In order for the formal limit as $\varepsilon \to 0$ to exist in (3.3) the following conditions have to be satisfied:

$$L(a,b,d,l,m) \geqslant 0$$
$$(3.4) \qquad L_{ijk}(a,b,d,l,m) \geqslant 0$$
$$M(a,b,d,l,m) \geqslant 0$$
$$M_{ijk}(a,b,d,l,m) \geqslant 0$$

These linear inequality constraints define a *convex polyhedral set P* in $\mathbb{R}^5$. ($P$ is the intersection of a finite number of closed half spaces.) Each of the equations $L = 0$, $L_{ijk} = 0$, $M = 0$, $M_{ijk} = 0$ defines a hyperplane in $\mathbb{R}^5$ and together these hyperplanes define the boundary of $P$.

To each point $p = (a, b, d, l, m) \in P$ there belongs a system of the type (3.3) satisfying (3.4) and vice versa, i.e. there is a unique correspondence between the points of $P$ and the set of systems (3.3) with property (3.4); and each of these systems yields an approximating system. The original system (3.1) corresponds to the origin in $\mathbb{R}^5$:

$$p \in P \longleftrightarrow \text{system of ODE's} \xrightarrow{\ \varepsilon = 0\ } \text{approximating system}$$

$$0 \in P \longleftrightarrow \text{system (3.1).}$$

We refer to $P$ as *the polyhedral set corresponding to Eq. (3.1) by means of the scaling transformation (3.2)*.

Let the point $p^* \in P$ be chosen, and let $(E^*)$ be the corresponding system of the type (3.3). Each term of Eq. $(E^*)$ has an $\varepsilon$-factor with exponent $L(p^*)$, $L_{ijk}(p^*)$, $M(p^*)$ or $M_{ijk}(p^*)$, respectively. If $\varepsilon \to 0$, only the terms with $\varepsilon$-exponent equal to zero remain in the resulting approximating system. Therefore, there are terms left in the approximating system only if $p^*$ is a point of the boundary of $P$. (Of course not all these approximating systems make sense.)

We will use the following notions: A $k$-dimensional *face* of $P$, $k \in \{0, 1, 2, 3, 4\}$, is a $k$-dimensional subset of the boundary of $P$ which is the intersection of at least $5 - k$ hyperplanes (e.g. a 1-dimensional face is an edge of $P$). An *open face* is the interior of a face (e.g. an edge without vertices).

Unless specifically stated we assume that $P$ is a *convex polyhedron*, i.e. the number of vertices of $P$ is greater than zero. (A necessary condition is that Eq. (3.1) consists of at least five terms.)

We are now able to state some important properties concerning $P$ and the corresponding systems of ODE's, respectively. They also motivate the following

DEFINITION   A *principal system* is a system corresponding to a vertex of $P$. A *principal approximating system* is an approximating system obtained from a principal system (by putting $\varepsilon = 0$). $\square$

In the sequel we will use PS for principal system, AS for approximating system and PAS for principal AS.

LEMMA 3.1   Each point of an open face of $P$ yields the same AS. $\square$

LEMMA 3.2   An open face of $P$ and a vertex of $P$ yield different AS's. Different open faces of $P$ yield different AS's as do different vertices of $P$. $\square$

Hence there is also a unique correspondence between the set of vertices and open faces of $P$ and the set of AS's of Eq. (3.1) obtained by means of the scaling transformation (3.2).

COROLLARY 3.1   The number of possible AS's (they of course might not all make sense) of Eq. (3.1) obtained by the scaling transformation (3.2) is equal to the number of vertices plus the number of faces of $P$. $\square$

LEMMA 3.3   Let $\mathbb{P}$ be a vertex or an open face of $P$, and let $\mathbb{F}$ be some open face of $P$ adjacent to $\mathbb{P}$ and of higher dimension. Then the AS corresponding to $\mathbb{P}$ contains all terms of the AS corresponding to $\mathbb{F}$ plus at least one more. $\square$

COROLLARY 3.2   A PAS is *distinguished*, i.e. there is exactly one scaling transformation (3.2) yielding this AS; and a PAS is *maximal*, i.e. it contains (in the sense of Lemma 3.3) all AS's corresponding to adjacent faces. $\square$

COROLLARY 3.3  PAS's corresponding to adjacent vertices have most terms in common. □

The verification of the above assertions is quite easy and therefore left to the reader.

*Remarks*  (i) No other AS's of (Eq. (3.1)) are obtained if more general functions of $\varepsilon$ (compare (2.3)) are used in the scaling transformation (3.2). The powers of $\varepsilon$ are representatives of the appropriate classes of order functions.
(ii) The coordinates $a, b, d, l, m$ of a vertex of $P$ are rational numbers.
(iii) If only the independent variable $t$ is scaled a PAS is 'distinguished' in the sense of Cole [3] or Kevorkian [16]. Moreover, a PAS then were a 'principal equation' in the sense of Lagerstrom and Casten [17]. □

From the above results it has become clear that the PAS's are the important AS's of Eq. (3.1) obtained by scaling transformations (3.2).

There is a special situation, however, where a system is principal in a certain sense although it does not correspond to a vertex of $P$. Let us now examine this special case. (It becomes important especially if a system of dimension greater than two is considered.)

Suppose we have a system (obtained from Eq. (3.1) by a transformation (3.2)) corresponding to an open face of $P$ that consists of *two* singular equations, i.e. both equations have a power of $\varepsilon$ in front of the derivative. Then the reduced system consists of two algebraic equations, say

$$(3.5) \qquad F(x, y) = 0, \qquad G(x, y) = 0$$

where $F$ and $G$ are polynomials in $x, y$. Each equation defines a curve in $\mathbb{R}^2$. Let us consider the case where the two equations are degenerated, i.e. they have a common factor, say $H(x, y)$, which is a polynomial of degree greater than 0. Then this solution ($H = 0$) of the reduced system again defines a curve in $\mathbb{R}^2$, i.e. $X$ or $Y$ satisfy *algebraic identities* (e.g. $X = Y$ or $XY = 1$), and a solution is not determinable in the form $X(t)$ and $Y(t)$. We call such a situation an *algebraic degeneration*.

Let us return to the full system which contains the expressions $F(x, y)$ and $G(x, y)$. Since it corresponds to an open face of $P$ there are $\varepsilon$-factors depending on at least one parameter out of $a, b, d, l, m$. If we now introduce higher order terms into $F$ and $G$ by means of

$$(3.6) \qquad \begin{aligned} x &= X + \varepsilon^{\bar{r}} x_1 \\ y &= Y + \varepsilon^{\bar{r}} y_1 \end{aligned} \qquad \bar{r} > 0,$$

taking into account the algebraic identities between $X$ and $Y$ and if the first equation is multiplied by $\varepsilon^{-\bar{l}}$ and the second one by $\varepsilon^{-\bar{m}}$ again in an analogous way as before a (usually lower dimensional) convex polyhedral set $\bar{P}$ is defined with analogous properties as $P$. To each point of $\bar{P}$ there corresponds a (special) system of ODE's (variables $x$, $y$, $X$, $Y$, $x_1$, $y_1$), and putting $\varepsilon = 0$ such a system yields a (special) reduced system. Again vertices of $\bar{P}$ have the property that their (special) reduced systems are distinguished and maximal (in the sense of Corollary 3.2). Now choosing a vertex of $\bar{P}$ what is the corresponding AS? It is of course still Eq. (3.5). What is the conclusion of the above considerations? By choosing a vertex of $\bar{P}$ we have chosen in a unique way a (full) system (out of the ones corresponding to the open face of $P$ and yielding the reduced system (3.5)). This system is principal in the sense that if introducing higher order terms by means of (3.6) there corresponds in a unique way a special reduced system with a maximal number of terms retained from the full system. And because of the algebraic degeneration of (3.5) one may, by good reasons, expect that the first order terms $x_1$, $y_1$ cancel out (i.e. may be expressed by means of zero order terms) and that it is possible to completely determine a solution $X(t)$ and $Y(t)$ of this special reduced system and hence also of Eq. (3.5). Hence, we call the point in the open face of $P$ corresponding to this system a *hidden vertex* of $P$. Besides vertices of $P$ also hidden vertices of $P$ will play an important role in the sequel. And in *both* cases we will call the corresponding systems *principal systems*.

*Remark* Obviously, in a *higher dimensional system* (i.e. Eq. (2.1) consisting of more than two equations and having polynomial right-hand sides) *algebraic degeneration* can occur if it contains at least two singular equations with the resulting algebraic equations having a common polynomial factor such that the corresponding solution of the reduced system is not completely determinable in the form $Z(t)$. $\square$

If we first subject Eq. (3.1) to a shift transformation and then scale the transformed system in the way of (3.2) there again will be a polyhedral set corresponding to the transformed system by means of a scaling transformation which generally will be different from $P$, however. We refer to this polyhedral set as the polyhedral set corresponding to the original system (3.1) by means of the (combined) shift scaling transformation.

We now will suggest how to make use of the polydedral sets corresponding to a singularly perturbed system of ODE's by means of shift scaling transformations for determining the appropriate scalings at each step of our procedure derived in Section 2.

## 3.2 Selection of appropriate scalings

In this subsection we will continue to refine the formal procedure given in

Section 2.3 for the special class of systems considered in this Section 3.

Let us consider Eq. (3.1) together with the initial conditions

$$(3.7) \qquad x(0, \varepsilon) = x^0, \qquad y(0, \varepsilon) = y^0$$

We want to construct a formal approximation to the solution $(x(t, \varepsilon), y(t, \varepsilon))$ of the IVP (3.1), (3.7) on some interval $J = [0, T]$. Scaling Eq. (3.1) by means of a transformation (3.2) we obtain a convex polyhedron $P_1$. (Without loss of generality we here assume that all polyhedral sets are polyhedrons, which is the generic case. If there are no vertices the lowest-dimensional faces are considered instead (compare also subsection 3.4).) We choose a vertex of $P_1$ such that the corresponding PAS is of full order (i.e. two differential equations) and the transformed initial conditions (3.7) are $O(1)$ as $\varepsilon \to 0$ (if possible independent of $\varepsilon$). The solution of this first reduced IVP ($\varepsilon = 0$) provides a local approximation to $(x(t, \varepsilon), y(t, \varepsilon))$ which in general will have a singularity. Introducing local variables in this first PS by shifting into the singularity (compare Section 2) and scaling the resulting system again by a transformation (3.2) yields a new convex polyhedron $P_2$ (which might be congruent to $P_1$). The point $0 \in P_2$ corresponds to the first PS chosen expressed in terms of the new variables.

We now consider the vertices adjacent to $0 \in P_2$. Their PAS's differ from our first PAS by at least one term but in general by less terms than PAS's corresponding to vertices that are not adjacent to 0. (The number of terms of a PAS depends on how degenerated a vertex is, i.e. on how many hyperplanes intersect in that vertex.) We therefore take one of these 'adjacent' PAS's to proceed. (In [17] this is referred to as the basic assumption that 'neighbouring equations have neighbouring solutions'.) But which adjacent vertex shall we choose?

Suppose we have constructed our chain of formal local approximations up to the $(i - 1)$st one and let $(U_{i-1}(s_{i-1}, \varepsilon), V_{i-1}(s_{i-1}, \varepsilon))$ be the (possibly higher order) solution of the last PAS expressed in terms of local variables at the singularity. It is a solution of a reduced system. The variables of the corresponding full system are denoted by $u_{i-1}, v_{i-1}, s_{i-1}$. This system corresponds to the point 0 of the new convex polyhedron $P_i$ defined by the scaling transformation

$$(3.8) \qquad \begin{aligned} u_{i-1} &= \varepsilon^a x_i \\ v_{i-1} &= \varepsilon^b y_i \\ s_{i-1} &= \varepsilon^d t_i \end{aligned}$$

We want to select an appropriate vertex for $P_i$ adjacent to the point 0.

Let $R_{i-1}^{\pm}$ be the singularity of the $(i - 1)$st formal local approximation in the local variable $s_{i-1}$ ($R_{i-1}^{\pm} = 0$ or $+\infty$, respectively). According to the rule (2.10) (or, more precisely, to the rule (3) in Section 2.3) we require

$$d \begin{cases} <0, & \text{if } R_{i-1}^+ = \infty \\ >0, & \text{if } R_{i-1}^+ = 0 \end{cases};$$

(3.9)

$$a \begin{cases} \leqslant 0, & \text{if } U_{i-1}(R_{i-1}^+, 0) \text{ is infinite} \\ \geqslant 0, & \text{if } U_{i-1}(R_{i-1}^+, 0) \text{ is finite} \end{cases}$$

and similarly for $b$.

For formulating a refinement of this rule we will need the following notion.

DEFINITION  If $U_{i-1}(s_{i-1}, 0) \sim s_{i-1}^{A_{i-1}}(\log s_{i-1})^{\alpha_{i-1}}$ as $s_{i-1} \to R_{i-1}^+$ where $A_{i-1}$ and $\alpha_{i-1}$ are rational numbers, then $A_{i-1}$ is called the *asymptotic exponent* of $U_{i-1}(s_{i-1}, \varepsilon)$. If there is no such asymptotic behaviour, $U_{i-1}(s_{i-1}, \varepsilon)$ is said to have an asymptotic exponent $A_{i-1} = 0$.
(The asymptotic exponent $B_{i-1}$ of $V_{i-1}(s_{i-1}, \varepsilon)$ is defined in the same way.) $\square$

As pointed out in Section 2.2.2 one aims at having a best possible matching of two adjacent local approximations. This means that if the two vector functions in the matching condition are expanded with respect to $\varepsilon$ and to the specified local independent variable as many terms as possible of the two expansions cancel out. In (3.9) we have $d \neq 0$. If $U_{i-1}$ is supposed to have an asymptotic exponent $A_{i-1} \neq 0$ (and for simplicity $\alpha_{i-1} = 0$) this function expressed in the new variables will have a leading term $\varepsilon^{(dA_{i-1}-a)} t_i^{A_{i-1}}$. Hence, since the $i$th local approximation may be expected to be valid and to be $O(1)$, in principle, on a finite $t$-interval extended slightly in the sense of the extension theorem an appropriate requirement in order to have an as good as possible matching according to the matching rules (2.11) (if $d < 0$) or (2.12) (if $d > 0$), respectively, is that $a = dA_{i-1}$. (Similarly for $V_{i-1}$.) If there is no vertex of $P_i$ adjacent to 0 satisfying this extended rule (3.9) we also consider the adjacent vertices with $d = 0$ and $a, b$ as in (3.9) (compare rule (4) in Section 2.3). Moreover, in that case we take into account scalings corresponding to hidden vertices of $P_i$.

The considerations of Sections 3.1 and 3.2 now allow us to reformulate and render precise the formal procedure of Section 2.3 for systems of the type (2.1) with polynomial right-hand sides.

## 3.3  The algorithm

Construct a formal approximation to the solution $(x(t, \varepsilon), y(t, \varepsilon))$ of the IVP (3.1), (3.7) on $J = [0, T]$ in the following way.

*Step 1*

(A) Scale the original system (3.1) by means of

$$x = \varepsilon^a x_1$$
$$y = \varepsilon^b y_1$$
$$t = \varepsilon^d t_1$$

and multiply the first equation by $\varepsilon^{-l}$ the second one by $\varepsilon^{-m}$.

(B) Determine the linear inequality constraints in $a, b, d, l, m$ defining the convex polyhedron $P_1$.

(C) Choose a point $p_1 \in P_1$, if possible a vertex adjacent to 0, such that the corresponding approximating system is of full order and such that the transformed initial conditions (3.7) are $O(1)$ as $\varepsilon \to 0$.

Denote the corresponding system by $(PS_1)$, its reduced system by $(PAS_1)$ and the transformed initial conditions by $x_1^0(\varepsilon), y_1^0(\varepsilon)$.

(D) Determine the solution $(X_1(t_1), Y_1(t_1))$ of $(PAS_1)$ satisfying $X_1(0) = x_1^0(0)$, $Y_1(0) = y_1^0(0)$ (if necessary a higher order solution $(X_1(t_1, \varepsilon), Y_1(t_1, \varepsilon))$).

(E) Determine a singularity $S_1^+$ of $(X_1(t_1), Y_1(t_1))$ (as an approximation to $(x(t, \varepsilon), y(t, \varepsilon))$ expressed in the new variables) and define

$$T_1^* := \begin{cases} S_1^+, & \text{if } S_1^+ < +\infty \\ 0, & \text{if } S_1^+ = +\infty \end{cases}$$

$$X_1^+ := X_1(S_1^+)$$

$$X_1^* := \begin{cases} X_1^+, & \text{if } X_1^+ \text{ is finite} \\ 0, & \text{if } X_1^+ \text{ is infinite} \end{cases}$$

$Y_1^+, Y_1^*$ similarly.


*Step i: $(i > 1)$*

(I) Introduce new local variables in $(PS_{i-1})$ by means of the shift transformation

$$x_{i-1} = X_{i-1}^* + u_{i-1}$$
$$y_{i-1} = Y_{i-1}^* + v_{i-1}$$
$$t_{i-1} = T_{i-1}^* + s_{i-1}$$

and determine the asymptotic exponent $A_{i-1}$ of $U_{i-1}(s_{i-1}, \varepsilon) := -X_{i-1}^* + X_{i-1}(T_{i-1}^* + s_{i-1}, \varepsilon)$ as $s_{i-1} \to 0^-$ (if $S_{i-1}^+ < +\infty$) or $\to +\infty$ (if $S_{i-1}^+ = +\infty$), respectively, and similarly the asymptotic exponent $B_{i-1}$ of $V_{i-1}(s_{i-1}, \varepsilon)$. Find possible algebraic identities between $U_{i-1}$ and $V_{i-1}$ as $s_{i-1} \to 0^-$ or $+\infty$, respectively.

(II) Scale $(\mathrm{PS}_{i-1})$ expressed in terms of the new variables by

$$u_{i-1} = \varepsilon^a x_i$$
$$v_{i-1} = \varepsilon^b y_i$$
$$s_{i-1} = \varepsilon^d t_i$$

and multiply the first equation by $\varepsilon^{-l}$, the second one by $\varepsilon^{-m}$.

(III) Determine the linear inequality constraints in $a, b, d, l, m$ defining the convex polyhedron $P_i$ and the set $\mathbb{V}_i$ of vertices of $P_i$ which are adjacent to $0 \in P_i$.

(IV) Determine the subset $\mathbb{V}_i^1 \subset \mathbb{V}_i$ satisfying

$$d \begin{cases} <0, & \text{if } S_{i-1}^+ \text{ is infinite} \\ >0, & \text{if } S_{i-1}^+ \text{ is finite} \end{cases}$$

and

$$a \begin{cases} \leqslant 0, & \text{if } X_{i-1}^+ \text{ is infinite} \\ \geqslant 0, & \text{if } X_{i-1}^+ \text{ is finite} \end{cases}$$
$$\text{and similarly for } b$$

and

$$a = A_{i-1}d, \qquad \text{if } A_{i-1} \neq 0$$
$$b = B_{i-1}d, \qquad \text{if } B_{i-1} \neq 0.$$

If $\mathbb{V}_i^1 \neq \varnothing$, let $\mathbb{V}_i^0 := \mathbb{V}_i^1$.

(V) If $\mathbb{V}_i^1 = \varnothing$ and $A_{i-1} = B_{i-1} = 0$ determine the subset $\mathbb{V}_i^2 \subset \mathbb{V}_i$ satisfying

$$d = 0$$

and

$$a \begin{cases} \leqslant 0, & \text{if } X_{i-1}^+ \text{ is infinite} \\ \geqslant 0, & \text{if } X_{i-1}^+ \text{ is finite} \end{cases}$$
$$\text{and similarly for } b.$$

If $\mathbb{V}_i^2 \neq \varnothing$, let $\mathbb{V}_i^0 := \mathbb{V}_i^2$.

(VI) If $\mathbb{V}_i^1 = \varnothing$ and $(A_{i-1} \neq 0$ or $B_{i-1} \neq 0)$ or if $\mathbb{V}_i^2 = \varnothing$ search for a hidden vertex of $P_i$ in the following way.
(1) Determine an open linear subset $\mathbb{F}_i$ of the boundary of $P_i$ with the properties:
   (a) There results a single reduced system that admits algebraic degeneration with algebraic identities analogous to those found in (I).
   (b) The coordinates $a, b, d$ satisfy the conditions (IV) or if (IV) cannot be fulfilled and $A_{i-1} = B_{i-1} = 0$ the conditions (V).

(c) As many terms as possible remain in the resulting reduced system.

If $\mathbb{F}_i = \varnothing$, let $\mathbb{V}_i^0 := \varnothing$.

(2) If $\mathbb{F}_i \neq \varnothing$ it is defined by linear inequality and equality constraints in $a, b, d, l, m$. Hence, the $\varepsilon$-exponents of the corresponding system are linear functions of $K$ of these variables, $0 < K < 5$.
Denote the variables of this system by $\bar{x}_i, \bar{y}_i, \bar{t}_i$ and proceed as follows.

(a) Introduce higher order terms in the system by substituting

$$\bar{x}_i = \bar{X}_i + \varepsilon^{\bar{r}}\bar{x}_{i,1}$$
$$\bar{y}_i = \bar{Y}_i + \varepsilon^{\bar{r}}\bar{y}_{i,1}$$

into the terms with $\varepsilon$-exponent 0 taking into account the equations for $\bar{X}_i$ and $\bar{Y}_i$, and multiplying the first equation by $\varepsilon^{-\bar{l}}$ and the second one by $\varepsilon^{-\bar{m}}$.
This defines the polyhedron $\bar{P}_i \subset \mathbb{R}^{K+3}$.

(b) Determine the set $\bar{\mathbb{V}}_i$ of vertices of $\bar{P}_i$ satisfying

$$\bar{r} > 0$$

and

      the $K$ components without $\phantom{x}^{-}$ of the elements of $\bar{\mathbb{V}}_i$
      satisfy the constraints defining $\mathbb{F}_i$

and

      the corresponding (special) reduced systems are such that
      the higher order terms cancel out.

If $\bar{\mathbb{V}}_i = \varnothing$, let $\mathbb{V}_i^0 := \varnothing$.
If $\bar{\mathbb{V}}_i \neq \varnothing$, let $\mathbb{V}_i^0$ be the corresponding points in $\mathbb{F}_i \subset P_i$.

(VII) If $\mathbb{V}_i^0 = \varnothing$ Step $i$ is not successful.

(VIII) If $\mathbb{V}_i^0 \neq \varnothing$ choose $p_i \in \mathbb{V}_i^0$ by the rules
(1) The coordinates $a$ and $b$ are smallest possible.
(2) The number of terms in the corresponding reduced system is greatest possible.
(3) The resulting reduced system has most terms in common with the one corresponding to $0 \in P_i$.
Denote the principal system corresponding to $p_i$ by $(PS_i)$ and its reduced system by $(PAS_i)$.

(IX) Determine the general solution $(\hat{X}_i(t_i; t_{i0}, X_{i0}; \varepsilon), \hat{Y}_i(t_i; t_{i0}, Y_{i0}; \varepsilon))$ of $(PAS_i)$ (if necessary of higher order).

(X) Let $(\tilde{X}_{i-1}^l(t_{i-1}, \varepsilon), \tilde{Y}_{i-1}^l(t_{-1}, \varepsilon))$ be the solution of $(PAS_{i-1})$ expressed in the original dependent variables and denote $(\hat{X}_i, \hat{Y}_i)$ expressed in the original dependent variables by $(\tilde{X}_i^l, \tilde{Y}_i^l)$.

Determine the integration constants $t_{i0}$, $X_{i0}$, $Y_{i0}$ by a best possible matching of the functions $\tilde{X}^l_{i-1}(T^*_{i-1} + s_{i-1}, \varepsilon)$ with $\tilde{X}^r_{i-1}(t_i; t_{i0}, X_{i0}; \varepsilon)$ and $\tilde{Y}^l_{i-1}(T^*_{i-1} + s_{i-1}, \varepsilon)$ with $\tilde{Y}^r_{i-1}(t_i; t_{i0}, Y_{i0}; \varepsilon)$ according to the matching rules (2.11) (if $d < 0$), (2.12) (if $d > 0$) or (2.13) (if $d = 0$), respectively. If matching fails Step i is not successful.

(XI) Determine a singularity $S^+_i$ of the chosen solution $(X_i(t_i, \varepsilon), Y_i(t_i, \varepsilon))$ of $(PAS_i)$ (as an approximation to $(x(t, \varepsilon), y(t, \varepsilon))$ expressed in the variables $x_i$, $y_i$, $t_i$) and define

$$T^*_i := \begin{cases} S^+_i, & \text{if } S^+_i < +\infty \\ 0, & \text{if } S^+_i = +\infty \end{cases}$$

$$X^+_i := X_i(S^+_i, 0)$$

$$X^*_i := \begin{cases} X_i^+, & \text{if } X^+_i \text{ is finite} \\ 0, & \text{if } X^+_i \text{ is infinite} \end{cases}$$

$$Y^+_i, Y^*_i \text{ similarly.}$$

If $S^+_i$ expressed in the original independent variable $t$ is greater than $T$ the steps $1, \ldots, i$ have provided a formal approximation to $(x(t, \varepsilon), y(t, \varepsilon))$ on $J$. $\square$

## 3.4 Remarks and comments

(1) For convenience only we have formulated our algorithm for systems of the form (3.1) which are of dimension 2. For $n$-dimensional problems, $n > 2$, the convex polyhedral sets are of dimension $N = 2n + 1$.

The algorithm is based on the requirement of having a system with polynomial right-hand sides since for such a system scalings with powers of $\varepsilon$ (and hence convex polyhedral sets) yield all possible AS's that can be obtained by shift scaling transformations. The most general type of problem for which the algorithm is in principle applicable is one of the form (compare (2.1), (2.2))

$$\Psi(\varepsilon) \frac{dz}{dt} = h(t, z, \varepsilon), \qquad z(0, \varepsilon) = z^0(\varepsilon)$$

where the elements $\psi_i(\varepsilon)$ of the $n \times n$-diagonal matrix $\Psi(\varepsilon)$ and the components of $h(t, z, \varepsilon)$ and of $z^0(\varepsilon)$ can be expanded into generalized power series (fractional exponents) with respect to all arguments.

(2) The set of vertices adjacent to the point 0 of a convex polyhedron (or all vertices) are computed by linear programming methods (cf. [4]). There exists such an algorithm (also implemented) using pivoting (based on [2]) which takes care of the fact that the polyhedrons in our method are usually not bounded and degenerated (i.e. vertices are the intersection of more than $N$ hyperplanes).

If a $P_i$ is a polyhedral set without vertices the faces with minimal dimension are principal. They satisfy the lemmata and corollaries of Section 3.1 except that they are not distinguished. The algorithm analogously carries over to this case if $k$-dimensional faces of $P_i$, $k$ minimal, adjacent to the face containing 0 are chosen with the property that $d$ is fixed (i.e. the corresponding reduced system is distinguished with respect to the independent variable). Such faces cannot be determined by pivoting but by other methods of linear programming (cf. [11]).

(3) The procedure outlined in Section 3.3 provides a systematic method to solve singularly perturbed IVP's, yet it might not be able to handle all possible situations in a completely automatic way. In particular, we have not included all possible stops and all difficulties and subtleties which might be encountered in specific applications. The reason is twofold. On the one hand the details are somewhat cumbersome, on the other hand the procedure is a prototype of approach for a very large class of problems. In fact, the algorithm is so general that it seems hardly possible to prove its correctness even in the formal sense, nor to decide a priori if it will or will not succeed in a specific situation. It is well motivated, however, and has proved successful in several involved applications (cf. Section 4, [22], [26]), and in all cases the choice of an adjacent point in a polyhedron $P_i$ was unique. It is suited particularly for such complex (higher dimensional) singularly perturbed IVP's where there are more than two local approximations needed and where it is not obvious how (or even impossible) to guess the appropriate scalings. Also if the algorithm does not succeed, i.e. if there is a step where the selection rules do not determine an adjacent (formal) local approximation, the polyhedral set provides a survey of all possible AS's obtained by scaling transformations with powers of $\varepsilon$. (This might also be useful for solving singularly perturbed BVP's.)

(4) Due to the fact that the dependent variables are scaled, too, or due to so-called switchback phenomena (compare [17]) it might be necessary to compute a higher order solution of a PAS to be able to match two (formal) local approximations. For examples the reader is referred to the application in Section 4.

In order to obtain a higher order approximation to the solution of an IVP one has to determine an appropriate higher order solution of the PAS$_i$ at each step.

(5) The formal algorithm (if successful) yields a formal approximation. Since instead of overlapping of the adjacent local approximations only the necessary condition of matching is required it is not guaranteed that the formal approximation obtained is an approximation in a rigorous sense. Although it is indeed the case in general, to make sure one has to prove it by verifying the overlapping. This is done by deriving best possible estimates for the domains of validity of the local approximations. Results providing such estimates are

stated in Section 5. Applications (of the rigorous algorithm, so to speak) may be found in [23], [26].

## 4  APPLICATION OF THE POLYHEDRON ALGORITHM TO THE FIELD–NOYES MODEL OF THE BELOUSOV–ZHABOTINSKII REACTION

In this section we apply the formal algorithm to a rather involved three-dimensional problem exhibiting a stable limit-cycle solution. As we shall see, fifteen (formal) local approximations are needed for the construction of a formal approximation to this periodic solution. The fifteen scalings (approximating systems) were found by our systematic procedure and without making use of any other information from, say, numerical computations etc. And we want to stress that we would not have been able to solve this problem just by intuition.

We have not taken the effort to prove our result rigorously, i.e. to verify the overlapping of the fifteen local approximations. It was verified, however, by comparison with numerical computations. Moreover, the formal approximation obtained allows to estimate the period of the limit-cycle, and this estimate again agrees very well with numerical studies. (This comparison is illustrated in Tables 2 and 3.)

We consider the non-linear autonomous system of three ordinary differential equations

$$\varepsilon \frac{dx}{dt} = x + y - xy - \varepsilon^3 q x^2$$

(4.1)
$$\frac{dy}{dt} = \varepsilon(-y - xy + fz)$$

$$\frac{dz}{dt} = p(x - z)$$

used by Field and Noyes [8] to model the chemical oscillations of the Belousov–Zhabotinskii reaction.

There exists a wide literature on this chemical reaction, see e.g. Tyson [31], Murray [20], Field, Körös and Noyes [7]. The system (4.1) represents a simplified model, referred to as the Oregonator. The variables $x$, $y$ and $z$ are dimensionless measures of the concentrations of bromous acid ($HBrO_2$), bromide ion ($Br^-$) and ceric ion ($Ce^{4+}$), respectively. In [8] Field and Noyes demonstrated numerically, for certain plausible values of the parameters that the Oregonator model has a stable limit-cycle solution (of relaxation oscillation type). An existence proof was given by Hastings and Murray [12].

The equations (4.1) are those considered by Stanshine and Howard in [28] who give the leading terms of a formal asymptotic expansion of the limit cycle solution as $\varepsilon \to 0$, $p, q$ of order unity for the cases $f = 1$ and $1/2 < f < 1$. In order to find the scalings of the variables in the twelve different regions they make use of the numerical calculations by Field and Noyes for the parameter values $f = 1, p = 0.161, q = 3.864$ and $\varepsilon^{-1} = 77.27$. Their approach is somewhat unsatisfactory, however, since the reader is not given any advice how to find the many scalings. Their approximate period of the limit-cycle solution for the case $f = 1$

$$T(\varepsilon) \sim -\frac{\varepsilon^{-1}}{2} \log[\varepsilon^3 4q(\varepsilon^{-1}p - 2)] =$$

$$\varepsilon^{-1} \log(\varepsilon^{-1}) + \varepsilon^{-1} \left( -\frac{1}{2} \log(4pq) \right) + O(1)$$

as we will find out gives the two leading terms of the period as $\varepsilon \to 0$.

We restrict ourselves to the case $f = 1$, $p$ and $q$ of order unity. The polyhedron algorithm yields fifteen (formal) local approximations which in most cases agree with those given by Stanshine and Howard with the exception of the third and the twelfth ones. Their formal approximation in region 3 seems to be incorrect. Instead of their composite region 12 we have found four different scalings. The fifteen shift-scaling transformations relating the chosen principal systems to the system (4.1) are shown in Table 1.

Let $l(t, \varepsilon) := (x(t, \varepsilon), y(t, \varepsilon), z(t, \varepsilon))$ be the limit-cycle solution of the system (4.1), and let $P(\varepsilon)$ be its period. Since its amplitude is unbounded as $\varepsilon \to 0$, higher order solutions of the chosen PAS's are needed to obtain a formal approximation to order unity. However, since it can be very costly to compute higher order solutions we shall restrict ourselves in most cases to determine the zero order solutions of the chosen PAS's assuming that these are the leading terms (with respect to $\varepsilon$) of the local approximations to order unity if they match with preceding ones (possibly only to an order lower than one, but at least such that the leading terms in the expansions of those zero order solutions cancel out; what we will call '*matching to leading order*'). Therefore, however, the period of the limit-cycle solution cannot be computed up to terms of order unity. The leading terms of the period $P(\varepsilon)$ will be obtained by matching the fifteenth formal approximation with the first one.

Since, actually, we do not have an IVP, how do we start our procedure, i.e. what is a good PAS to start with so that we may assume by good reasons that its solution is close to the limit-cycle solution $l(t, \varepsilon)$?

We know that the limit-cycle we want to approximate is attractive. Hence, guided by the knowledge about the Tikhonov case (cf. Theorems 1.1 and 5.4) we try to start with a PAS containing two algebraic equations and having a slowest possible time-scale. Then we may hope that those algebraic equations define a curve in $\mathbb{R}^3$ (a reduced manifold) which is attractive and gives a large

**Table 1**  *The shift scaling transformations relating the chosen principal systems to the original system (4.1)*

| $(PS_i)$ | $x =$ | $y =$ | $z =$ | $t =$ |
|---|---|---|---|---|
| 1 | $x_1$ | $y_1$ | $z_1$ | $\mathbf{T}_1(t_1, \varepsilon) = \varepsilon^{-1} t_1$ |
| 2 | $\varepsilon^{-1/2} x_2$ | $1 + \varepsilon^{1/2} y_2$ | $\varepsilon^{-1/2} z_2$ | $\mathbf{T}_2(t_2, \varepsilon) = \varepsilon^{-1/2} t_2$ |
| 3 | $\varepsilon^{-1/2}\sqrt{(p)} + \varepsilon^{-1/3} x_3$ | $1 + \varepsilon^{1/2}/\sqrt{(p)} + \varepsilon^{2/3} y_3$ | $\varepsilon^{-1/2}\sqrt{(p)} + \varepsilon^{-1/3} z_3$ | $\mathbf{T}_3(t_3, \varepsilon) = \varepsilon^{-1/2} T_2^* + \varepsilon^{-1/6} t_3$ |
| 4 | $\varepsilon^{-1/2}\sqrt{(p)} + \varepsilon^{-1/2} x_4$ | $1 + \varepsilon^{1/2}/\sqrt{(p)} + \varepsilon^{1/2} y_4$ | $\varepsilon^{-1/2}\sqrt{(p)} + \varepsilon^{-1/2} z_4$ | $\mathbf{T}_4(t_4, \varepsilon) = \mathbf{T}_3(T_3^*, \varepsilon) + t_4$ |
| 5 | $\varepsilon^{-1/2}\sqrt{(p)} + \varepsilon^{-2/3} x_5$ | $1 + \varepsilon^{2/3} y_5$ | $\varepsilon^{-1/2} 2\sqrt{(p)} + \varepsilon^{-1/3} z_5$ | $\mathbf{T}_5(t_5, \varepsilon) = \mathbf{T}_4(T_4^*, \varepsilon) + \varepsilon^{1/3} t_5$ |
| 6 | $\varepsilon^{-1/2}\sqrt{(p)} + \varepsilon^{-2} x_6$ | $1 + y_6$ | $\varepsilon^{-1/2} 2\sqrt{(p)} + \varepsilon^{-1} z_6$ | $\mathbf{T}_6(t_6, \varepsilon) = \mathbf{T}_5(T_5^*, \varepsilon) + \varepsilon t_6$ |
| 7 | $\varepsilon^{-1/2}\sqrt{(p)} + \varepsilon^{-3} x_7$ | $y_7$ | $\varepsilon^{-1/2} 2\sqrt{(p)} + \varepsilon^{-2} z_7$ | $\mathbf{T}_7(t_7, \varepsilon) = \mathbf{T}_6(T_6^*, \varepsilon) + \varepsilon t_7$ |
| 8 | $\varepsilon^{-3}/q + \varepsilon^{-1/2}\sqrt{(p)} + \varepsilon^{-3} x_8$ | $y_8$ | $\varepsilon^{-1/2} 2\sqrt{(p)} + \varepsilon^{-3} z_8$ | $\mathbf{T}_8(t_8, \varepsilon) = \mathbf{T}_6(T_6^*, \varepsilon) + t_8$ |
| 9 | $\varepsilon^{-3}/2q + \varepsilon^{-1/2}\sqrt{(p)} + \varepsilon^{-8/3} x_9$ | $1/2 + \varepsilon^{1/3} y_9$ | $\varepsilon^{-3}/4q + \varepsilon^{-1/2} 2\sqrt{(p)} + \varepsilon^{-7/3} z_9$ | $\mathbf{T}_9(t_9, \varepsilon) = \mathbf{T}_8(T_8^*, \varepsilon) + \varepsilon^{2/3} t_9$ |
| 10 | $\varepsilon^{-3}/2q + \varepsilon^{-1/2}\sqrt{(p)} + \varepsilon^{-3} x_{10}$ | $1/2 + y_{10}$ | $\varepsilon^{-3}/4q + \varepsilon^{-7/3} Z_9^* + \varepsilon^{-1/2} 2\sqrt{(p)} + \varepsilon^{-2} z_{10}$ | $\mathbf{T}_{10}(t_{10}, \varepsilon) = \mathbf{T}_9(T_9^*, \varepsilon) + \varepsilon t_{10}$ |
| 11 | $\varepsilon^{-1/2}\sqrt{(p)} + \varepsilon^{-5/2} x_{11}$ | $1/2 + \varepsilon^{-1/2} y_{11}$ | $\varepsilon^{-3}/4q + \varepsilon^{-7/3} Z_9^* + \varepsilon^{-2} Z_{10}^* + \varepsilon^{-1/2} 2\sqrt{(p)} + \varepsilon^{-3/2} z_{11}$ | $\mathbf{T}_{11}(t_{11}, \varepsilon) = \mathbf{T}_{10}(T_{10}^*, \varepsilon) + \varepsilon^{3/2} t_{11}$ |
| 12 | $\varepsilon^{-1/2}\sqrt{(p)} + \varepsilon^{-1} x_{12}$ | $1/2 + \varepsilon^{-2} y_{12}$ | $\varepsilon^{-3}/4q + \varepsilon^{-7/3} Z_9^* + \varepsilon^{-2} Z_{10}^* + \varepsilon^{-1/2} 2\sqrt{(p)} + \varepsilon^{-3} z_{12}$ | $\mathbf{T}_{12}(t_{12}, \varepsilon) = \mathbf{T}_{10}(T_{10}^*, \varepsilon) + t_{12}$ |
| 13 | $\varepsilon^{-1/2}\sqrt{(p)} + \varepsilon^{-1/2} x_{13}$ | $\varepsilon^{-2}/4pq + 1/2 + \varepsilon^{-2} y_{13}$ | $\varepsilon^{-7/3} Z_9^* + \varepsilon^{-2} Z_{10}^* + \varepsilon^{-1/2} 2\sqrt{(p)} + \varepsilon^{-5/2} z_{13}$ | $\mathbf{T}_{13}(t_{13}, \varepsilon) = \mathbf{T}_{10}(T_{10}^*, \varepsilon) + \varepsilon^{-1/2} t_{13}$ |
| 14 | $\varepsilon^{-1/3} x_{14}$ | $\varepsilon^{-2}/4pq + 1/2 + \varepsilon^{-2} y_{14}$ | $\varepsilon^{-7/3} Z_9^* + \varepsilon^{-2} Z_{10}^* + \varepsilon^{-1/2} 2\sqrt{(p)} + \varepsilon^{-7/3} z_{14}$ | $\mathbf{T}_{14}(t_{14}, \varepsilon) = \mathbf{T}_{10}(T_{10}^*, \varepsilon) + \varepsilon^{-2/3} t_{14}$ |
| 15 | $x_{15}$ | $\varepsilon^{-2}/4pq + 1/2 + \varepsilon^{-2} y_{15}$ | $\varepsilon^{-2} Z_{10}^* + \varepsilon^{-1/2} 2\sqrt{(p)} + \varepsilon^{-2} z_{15}$ | $\mathbf{T}_{15}(t_{15}, \varepsilon) = \mathbf{T}_{10}(T_{10}^*, \varepsilon) + \varepsilon^{-1} t_{15}$ |

$T_2^* = -1/\sqrt{(p)},\ T_3^* = 2^{-1/3} p^{-5/6}\hat{t},\ T_4^* = -1/p,\ T_5^* = 2^{1/3}\hat{t},\ T_6^* = \log(\varepsilon^{-1}),\ T_8^* = [2\log(2) - 1]/p,\ T_9^* = (4/p)^{1/3}\hat{t},\ T_{10}^* = -2;$
$Z_9^* = (p/4)^{2/3}\hat{t}/q,\ Z_{10}^* = -p[1 - 2\log(2q)]/2q;\ \hat{t} = 2.338\ldots\,.$

contribution to the approximation of the limit-cycle solution. (Of course, the Tikhonov stability assumption may be checked; and a successful termination of our procedure, i.e. there is a local approximation matching to the first one, will justify this choice of a first PAS.)

*Step 1*

Applying the scaling transformation

$$x = \varepsilon^a x_1$$
$$y = \varepsilon^b y_1$$
$$z = \varepsilon^c z_1$$
$$t = \varepsilon^d t_1$$

to the system (4.1) and multiplying the equations by $\varepsilon^{-l}, \varepsilon^{-m}, \varepsilon^{-n}$, respectively, we obtain

$$\varepsilon^{1+a-d-l} \frac{\mathrm{d}x_1}{\mathrm{d}t_1} = \varepsilon^{a-l} x_1 + \varepsilon^{b-l} y_1 - \varepsilon^{a+b-l} x_1 y_1 - \varepsilon^{3+2a-l} q x_1^2$$

$$\varepsilon^{b-d-m} \frac{\mathrm{d}y_1}{\mathrm{d}t_1} = -\varepsilon^{1+b-m} y_1 - \varepsilon^{1+a+b-m} x_1 y_1 + \varepsilon^{1+c-m} z_1$$

$$\varepsilon^{c-d-n} \frac{\mathrm{d}z_1}{\mathrm{d}t_1} = \varepsilon^{a-n} p x_1 - \varepsilon^{c-n} p z_1$$

The linear functions in $a, b, c, d, l, m, n$ required to be non-negative define the convex polyhedron $P_1 \subset \mathbb{R}^7$ corresponding to the system (4.1).

Using a code based on [2] we have found the following set of vertices adjacent to $0 \in P_1$

|  | $a$ | $b$ | $c$ | $d$ | $l$ | $m$ | $n$ |
|---|---|---|---|---|---|---|---|
| (1) | $-1$ | 0 | $-1$ | 0 | $-1$ | 0 | $-1$ |
| (2) | 0 | 0 | 1 | 1 | 0 | $-1$ | 0 |
| (3) | 0 | 0 | $-1$ | 0 | 0 | 0 | $-1$ |
| (4) | 0 | 0 | 0 | $-1$ | 0 | 1 | 0 |

As motivated above we choose

$$p_1 = (0, 0, 0, -1, 0, 1, 0)$$

with the corresponding PS

$$\varepsilon^2 \frac{\mathrm{d}x_1}{\mathrm{d}t_1} = x_1 + y_1 - x_1 y_1 - \varepsilon^3 q x_1^2$$

$$(\mathrm{PS}_1) \qquad \frac{\mathrm{d}y_1}{\mathrm{d}t_1} = -y_1 - x_1 y_1 + z_1$$

$$\varepsilon \frac{\mathrm{d}z_1}{\mathrm{d}t_1} = p(x_1 - z_1)$$

where

$$x = x_1$$
$$y = y_1$$
$$z = z_1$$
$$t = \mathbf{T}_1(t_1, \varepsilon) := \varepsilon^{-1} t_1$$

Its reduced system

$$0 = X_1 + Y_1 - X_1 Y_1$$

$$(\mathrm{PAS}_1) \qquad \frac{\mathrm{d}Y_1}{\mathrm{d}t_1} = -Y_1 - X_1 Y_1 + Z_1$$

$$0 = X_1 - Z_1$$

has the solution

$$\hat{Y}_1(t_1) = e^{-2(t_1 - t_1^0)}$$
$$\hat{X}_1(t_1) = \hat{Z}_1(t_1) = \hat{Y}_1(t_1)/(\hat{Y}_1(t_1) - 1)$$

where $t_1^0$ is an integration constant. For convenience we choose this arbitrary time shift $t_1^0 = 0$. This solution is denoted by $(X_1(t_1), Y_1(t_1), Z_1(t_1))$. We suppose that it indeed approximates the limit-cycle solution on some interval of the independent variable, a hypothesis which will be verified in the final step of the algorithm.

Since $x \geqslant 0, y \geqslant 0, z \geqslant 0$ is required, and since $X_1(t_1)$ and $Z_1(t_1)$ become infinite at $Y_1(t_1) = 1$, this solution of $(\mathrm{PAS}_1)$ cannot approximate $\mathbf{l}(\varepsilon^{-1}t_1, \varepsilon)$ for $t_1 \geqslant 0$; i.e. $t_1 = 0$ is a singularity of the first local approximation. (The stability assumptions of Theorem 5.6 hold on every compact $t_1$-interval $[-c_2, -c_1]$, where $0 < c_1 < c_2 < \infty$.)

We have

$$Y_1(0) = 1, \qquad X_1(0) = Z_1(0) = \infty.$$

More precisely, the first local approximation (to $\mathbf{l}(t, \varepsilon)$) has the following asymptotic behaviour near the singularity:

$$Y_1(t_1) = Y_1^* + 2(-t_1) + O(t_1^2)$$

$$X_1(t_1) = Z_1(t_1) = \frac{1}{2(-t_1)} + \frac{1}{2} + O((-t_1)), \qquad t_1 \to 0^-$$

where $Y_1^* := 1$. Hence, the asymptotic exponents are $A_1 = C_1 = -1$ and $B_1 = 1$.

**Step 2**

The shift transformation

$$x_1 = u_1$$
$$y_1 = 1 + v_1$$
$$z_1 = w_1$$
$$t = s_1$$

takes the system (PS$_1$) into

$$\varepsilon^2 \frac{du_1}{ds_1} = 1 + v_1 - u_1 v_1 - \varepsilon^3 q u_1{}^2$$

$$\frac{dv_1}{ds_1} = -1 - u_1 - v_1 - u_1 v_1 + w_1$$

$$\varepsilon \frac{dw_1}{ds_1} = p(u_1 - w_1)$$

The first local approximation expressed in the new variables satisfies the following algebraic identities as $s_1 \to 0^-$:

(4.2) $$U_1 = W_1, \qquad U_1 V_1 = 1.$$

Applying the scaling transformation

$$u_1 = \varepsilon^a x_2$$
$$v_1 = \varepsilon^b y_2$$
$$w_1 = \varepsilon^c z_2$$
$$s_1 = \varepsilon^d t_2$$

to the preceding system and multiplying the equations by $\varepsilon^{-l}, \varepsilon^{-m}, \varepsilon^{-n}$, respectively, yields

$$\varepsilon^{2+a-d-l} \frac{dx_2}{dt_2} = \varepsilon^{-l} 1 + \varepsilon^{b-l} y_2 - \varepsilon^{a+b-l} x_2 y_2 - \varepsilon^{3+2a-l} q x_2{}^2$$

(4.3) $$\varepsilon^{b-d-m} \frac{dy_2}{dt_2} = -\varepsilon^{-m} 1 - \varepsilon^{a-m} x_2 - \varepsilon^{b-m} y_2 - \varepsilon^{a+b-m} x_2 y_2 + \varepsilon^{c-m} z_2$$

$$\varepsilon^{1+c-d-n} \frac{dz_2}{dt_2} = \varepsilon^{a-n} p x_2 - \varepsilon^{c-n} p z_2$$

The convex polyhedron $P_2 \subset \mathbb{R}^7$ contains the following set of vertices adjacent to $0 \in P_2$

|      | $a$ | $b$ | $c$ | $d$ | $l$ | $m$ | $n$ |
|------|------|------|------|------|------|------|------|
| (1)  | $-1$ | $0$ | $-1$ | $1$ | $-1$ | $-1$ | $-1$ |
| (2)  | $0$ | $1$ | $0$ | $1$ | $0$ | $0$ | $0$ |
| (3)  | $-1/2$ | $1/2$ | $-1/2$ | $1$ | $0$ | $-1/2$ | $-1/2$ |
| (4)  | $0$ | $0$ | $-1$ | $1$ | $0$ | $-1$ | $-1$ |
| (5)  | $0$ | $0$ | $0$ | $1$ | $0$ | $-1$ | $0$ |

Since none of these five vertices satisfies

$$(4.4) \qquad\qquad d > 0 \qquad a = c = -d \qquad b = d$$

required by the selection rule (IV) of the algorithm, we have to look for a *hidden vertex* according to the rule (VI).

Substituting (4.4) into the system (4.3) we obtain

$$\varepsilon^{2-2d-l}\frac{\mathrm{d}x_2}{\mathrm{d}t_2} = \varepsilon^{-l}\mathbf{1} + \varepsilon^{d-l}y_2 - \varepsilon^{-l}x_2 y_2 - \varepsilon^{3-2d-l}qx_2{}^2$$

$$\varepsilon^{-m}\frac{\mathrm{d}y_2}{\mathrm{d}t_2} = -\varepsilon^{-m}\mathbf{1} - \varepsilon^{-d-m}x_2 - \varepsilon^{d-m}y_2 - \varepsilon^{-m}x_2 y_2 + \varepsilon^{-d-m}\mathbf{z}_2$$

$$\varepsilon^{1-2d-n}\frac{\mathrm{d}z_2}{\mathrm{d}t_2} = \varepsilon^{-d-n}\mathbf{p}x_2 - \varepsilon^{-d-n}\mathbf{p}z_2$$

where the terms in bold letters admit algebraic identities analogous to those stated in (4.2). Algebraic degeneration can only occur for these terms of the second and the third equation and can only be achieved if

$$(4.5) \qquad\qquad m = n = -d < 0 \text{ and } d < 1.$$

If, in addition,

$$(4.6) \qquad\qquad\qquad l = 0$$

a maximum number of terms remain in the resulting reduced system. Thus we get the system

$$\varepsilon^{2-2d}\frac{\mathrm{d}\bar{x}_2}{\mathrm{d}\bar{t}_{-}} = 1 + \varepsilon^{d}\bar{y}_2 - \bar{x}_2\bar{y}_2 - \varepsilon^{3-2d}q\bar{x}_2{}^2$$

$$\varepsilon^{d}\frac{\mathrm{d}\bar{y}_2}{\mathrm{d}\bar{t}_2} = -\varepsilon^{d}1 - \bar{x}_2 - \varepsilon^{2d}\bar{y}_2 - \varepsilon^{d}\bar{x}_2\bar{y}_2 + \bar{z}_2$$

$$\varepsilon^{1-d}\frac{\mathrm{d}\bar{z}_2}{\mathrm{d}\bar{t}_2} = p(\bar{x}_2 - \bar{z}_2)$$

with $\varepsilon$-exponents depending on $d$.

Introducing higher order terms by substituting

$$
\begin{aligned}
\bar{x}_2 &= \bar{X}_2 + \varepsilon^{\bar{r}}\bar{x}_{2,1} \\
\bar{y}_2 &= \bar{Y}_2 + \varepsilon^{\bar{r}}\bar{y}_{2,1} \\
\bar{z}_2 &= \bar{Z}_2 + \varepsilon^{\bar{r}}\bar{z}_{2,1}
\end{aligned}
$$

(4.7)

into the terms with $\varepsilon$-exponent 0 and multiplying the equations by $\varepsilon^{-\bar{l}}, \varepsilon^{-\bar{m}}, \varepsilon^{-\bar{n}}$ yields the system

$$
\varepsilon^{2-2d-\bar{l}}\frac{d\bar{x}_2}{d\bar{t}_2} = -\varepsilon^{\bar{r}-\bar{l}}(\bar{X}_2\bar{y}_{2,1} + \bar{Y}_2\bar{x}_{2,1}) - \varepsilon^{2\bar{r}-\bar{l}}\bar{x}_{2,1}\bar{y}_{2,1} + \varepsilon^{d-\bar{l}}\bar{y}_2 - \varepsilon^{3-2d-\bar{l}}q\bar{x}_2^2
$$

$$
\varepsilon^{d-\bar{m}}\frac{d\bar{y}_2}{d\bar{t}_2} = -\varepsilon^{d-\bar{m}}1 - \varepsilon^{2d-\bar{m}}\bar{y}_2 - \varepsilon^{d-\bar{m}}\bar{x}_2\bar{y}_2 - \varepsilon^{\bar{r}-\bar{m}}(\bar{x}_{2,1} - \bar{z}_{2,1})
$$

$$
\varepsilon^{1-d-\bar{n}}\frac{d\bar{z}_2}{d\bar{t}_2} = \varepsilon^{\bar{r}-\bar{n}}p(\bar{x}_{2,1} - \bar{z}_{2,1})
$$

This defines the convex polyhedron $\bar{P}_2 \subset \mathbb{R}^5$ having two vertices

| | $d$ | $\bar{r}$ | $\bar{l}$ | $\bar{m}$ | $\bar{n}$ |
|---|---|---|---|---|---|
| (1) | 1/2 | 1/2 | 1/2 | 1/2 | 1/2 |
| (2) | 2/3 | 2/3 | 2/3 | 2/3 | 1/3 |

satisfying $\bar{r} > 0$ and $0 < d < 1$. Only the vertex (1) can be chosen since in the (special) reduced system corresponding to the vertex (2) the terms $\bar{x}_{2,1}, \bar{z}_{2,1}$ do not cancel out.

Hence, taking into account (4.4), (4.5) and (4.6) we have determined a unique point

$$
p_2 = \left(-\tfrac{1}{2}, \tfrac{1}{2}, -\tfrac{1}{2}, \tfrac{1}{2}, 0, -\tfrac{1}{2}, -\tfrac{1}{2}\right)
$$

with the corresponding system

$$
\varepsilon\frac{dx_2}{dt_2} = 1 - x_2y_2 + \varepsilon^{1/2}y_2 - \varepsilon^2qx_2^2
$$

(PS$_2$)

$$
\varepsilon^{1/2}\frac{dy_2}{dt_2} = -x_2 + z_2 - \varepsilon^{1/2}(1 + x_2y_2) - \varepsilon y_2
$$

$$
\varepsilon^{1/2}\frac{dz_2}{dt_2} = p(x_2 - z_2)
$$

where

$$
\begin{aligned}
x_1 &= \varepsilon^{-1/2}x_2 & x &= \varepsilon^{-1/2}x_2 \\
y_1 &= 1 + \varepsilon^{1/2}y_2 & y &= 1 + \varepsilon^{1/2}y_2 \\
z_1 &= \varepsilon^{-1/2}z_2 & z &= \varepsilon^{-1/2}z_2 \\
t_1 &= \varepsilon^{1/2}t_2 & t &= \mathbf{T}_2(t_2, \varepsilon) := \varepsilon^{-1/2}t_2
\end{aligned}
$$

The reduced system (PAS$_2$) consists of the two algebraic equations

$$(4.8) \qquad X_2 Y_2 = 1 \qquad X_2 = Z_2$$

Introducing higher-order terms (as in (4.7), with $\bar{r} = 1/2$) into (PS$_2$) the following non-linear differential equation for the $y$-component is obtained

$$Y_2'\left(1 - \frac{1}{p Y_2^2}\right) = -2$$

with the one parameter family of 'solutions'

$$\hat{Y}_2(t_2, C) = -(t_2 + C) \pm \sqrt{\left[(t_2 + C)^2 - \frac{1}{p}\right]}.$$

(Actually, this describes a family of hyperbolas in the $(t_2, y_2)$-plane, each hyperbola representing four different solution branches which all exist only on a one-sided unbounded $t_2$-interval.)

It is obvious that for matching with the preceding local approximation the branch with $(t_2 + C) < 0$ and the positive square root is needed. It has the following asymptotic behaviour as $t_2 \to -\infty$:

$$\hat{Y}_2(t_2, C) = -2(t_2 + C) - \frac{1}{2p[-(t_2 + C)]} + O([-(t_2 + C)]^{-3}).$$

Hence, this function expressed in the original $y$-variable and in terms of $t_1$ satisfies

$$\tilde{Y}_2^r(\varepsilon^{-1/2} t_1, C, \varepsilon) = 1 + 2(-t_1) - \varepsilon^{1/2} 2C - \frac{\varepsilon}{2p(-t_1)} \, [1 + O(\varepsilon^{1/2} C/(-t_1))]$$

$$+ O(\varepsilon^2/(-t_1)^3), \qquad \text{for } (-t_1) \gg \varepsilon^{1/2}, \, \varepsilon \to 0.$$

So, $Y_1$ and $\tilde{Y}_2^r$ match in the sense of (2.12) in the formal overlap domain $\varepsilon^{1/2} \ll (-t_1) \ll 1$. Although they actually match to order $\zeta(\varepsilon)$ for every $\varepsilon^{1/2} \ll \zeta \ll \approx 1$, the integration constant $C$ cannot be determined by this matching. It is easy to see, however, that for matching to an order higher than $\varepsilon^{1/2}$ we had to take $C = 0$ since neither in a higher order solution of (PAS$_1$) nor of (PAS$_2$) another term $\varepsilon^{1/2} \cdot \text{const}$ will turn up.

The $x$- and $z$-components ($X_2$ and $Z_2$ defined by (4.8) and expressed in the original dependent variables and in $t_1$) also match to leading order as can easily be verified (precisely: to order $\varepsilon^{-1/2}$ in $\varepsilon^{1/2} \ll (-t_1) \ll 1$, or even to every order $\zeta(\varepsilon) \gg 1$ in $\varepsilon^{1/3} \ll (-t_1) \ll 1$).

Thus, the following solution of (PAS$_2$)

$$Y_2(t_2) = -t_2 + \sqrt{\left(t_2^2 - \frac{1}{p}\right)}$$

$$X_2(t_2) = Z_2(t_2) = 1/Y_2(t_2)$$

provides the second local approximation (to $l(t, \varepsilon)$). Since $Y_2(t_2)$ does not exist for $t_2 > -1/\sqrt{p}$ this second local approximation ceases to be valid in a (left) neighbourhood of the singularity $t_2 = T_2^* := -1\sqrt{p}$. Moreover, one finds the following asymptotic behaviour:

$$X_2(T_2^* + s_2) = X_2^* - p^{3/4}\sqrt{(-2s_2)} + O((-s_2))$$
$$Y_2(T_2^* + s_2) = Y_2^* + p^{-1/4}\sqrt{(-2s_2)} + O((-s_2)), \qquad s_2 \to 0^-$$
$$Z_2(T_2^* + s_2) = Z_2^* - p^{3/4}\sqrt{(-2s_2)} + O((-s_2))$$

where $X_2^* = Z_2^* = \sqrt{p}$, $Y_2^* = 1/\sqrt{p}$. Hence, the asymptotic exponents are

$$A_2 = B_2 = C_2 = 1/2.$$

## Step 3

Introducing new variables by means of the shift transformation

$$x_2 = X_2^* + u_2$$
$$y_2 = Y_2^* + v_2$$
$$z_2 = Z_2^* + w_2$$
$$t_2 = T_2^* + s_2$$

into $(PS_2)$ we obtain the system

$$\varepsilon \frac{du_2}{ds_2} = -\frac{u_2}{\sqrt{p}} - \sqrt{(p)}v_2 - u_2 v_2 + \varepsilon^{1/2}\left(\frac{1}{\sqrt{p}} + v_2\right) + \varepsilon^2(-pq - 2q\sqrt{(p)}u_2 - qu_2{}^2)$$

$$\varepsilon^{1/2}\frac{dv_2}{ds_2} = -u_2 + w_2 + \varepsilon^{1/2}\left(-2 - \frac{u_2}{\sqrt{p}} - \sqrt{(p)}v_2 - u_2 v_2\right) + \varepsilon\left(-\frac{1}{\sqrt{p}} - v_2\right)$$

$$\varepsilon^{1/2}\frac{dw_2}{ds_2} = p(u_2 - w_2)$$

The second local approximation expressed in the new variables satisfies the algebraic identities

$$U_2 = W_2 = -pV_2 \qquad \text{as } s_2 \to 0^-.$$

Scaling the above system by means of

$$u_2 = \varepsilon^a x_3$$
$$v_2 = \varepsilon^b y_3$$
$$w_2 = \varepsilon^c z_3$$
$$s_2 = \varepsilon^d t_3$$

and multiplying the equations by $\varepsilon^{-l}, \varepsilon^{-m}, \varepsilon^{-n}$, respectively, defines the convex polyhedron $P_3$.

The selection rule (IV) requires

$$a = b = c = d/2 > 0.$$

There is no vertex adjacent to $0 \in P_3$ having this property. Therefore, we again look for a *hidden vertex*.

The rule (VI) yields a unique point

$$p_3 = (\tfrac{1}{6}, \tfrac{1}{6}, \tfrac{1}{6}, \tfrac{1}{3}, \tfrac{1}{6}, \tfrac{1}{6}, \tfrac{1}{6})$$

in $P_3$ with the corresponding system

$$\varepsilon^{2/3} \frac{\mathrm{d}x_3}{\mathrm{d}t_3} = -\frac{x_3}{\sqrt{p}} - \sqrt{(p)}\, y_3 - \varepsilon^{1/6} x_3 y_3 + \varepsilon^{1/3} \frac{1}{\sqrt{p}} + \varepsilon^{1/2} y_3 - \varepsilon^{11/16} pq$$

$$- \varepsilon^2 2q\sqrt{(p)}\, x_3 - \varepsilon^{13/6} q x_3{}^2$$

$$(\mathrm{PS}_3) \quad \varepsilon^{1/6} \frac{\mathrm{d}y_3}{\mathrm{d}t_3} = -x_3 + z_3 - \varepsilon^{1/3} 2 + \varepsilon^{1/2} \left( -\frac{x_3}{\sqrt{p}} - \sqrt{(p)}\, y_3 \right)$$

$$- \varepsilon^{2/3} x_3 y_3 - \varepsilon^{5/6} \frac{1}{\sqrt{p}} - \varepsilon y_3$$

$$\varepsilon^{1/6} \frac{\mathrm{d}z_3}{\mathrm{d}t_3} = p(x_3 - z_3)$$

where

$$
\begin{aligned}
x_2 &= \sqrt{p} + \varepsilon^{1/6} x_3 & x &= \varepsilon^{-1/2}\sqrt{p} + \varepsilon^{-1/3} x_3 \\
y_2 &= 1/\sqrt{p} + \varepsilon^{1/6} y_3 & y &= 1 + \varepsilon^{1/2}/\sqrt{p} + \varepsilon^{2/3} y_3 \\
z_2 &= \sqrt{p} + \varepsilon^{1/6} z_3 & z &= \varepsilon^{-1/2}\sqrt{p} + \varepsilon^{-1/3} z_3 \\
t_2 &= T_2^* + \varepsilon^{1/3} t_3 & t &= \mathbf{T}_3(t_3, \varepsilon) := \varepsilon^{-1/2} T_2^* + \varepsilon^{-1/6} t_3
\end{aligned}
$$

A solution of $(\mathrm{PAS}_3)$ satisfies

$$\hat{X}_3 = \hat{Z}_3 = -p\hat{Y}_3$$

and

$$(4.9) \qquad \frac{\mathrm{d}\hat{Y}_3}{\mathrm{d}t_3} + p^{3/2} \hat{Y}_3{}^2 + 2p(t_3 + C) = 0$$

where $C$ is an integration constant. By means of the transformation

$$\hat{Y}_3 = p^{-3/2} \frac{\mathrm{d}}{\mathrm{d}t_3} (\log \hat{U})$$

the non-linear differential equation (4.9) can be transformed into a linear equation of second order:

$$\frac{\mathrm{d}^2}{\mathrm{d}t_3{}^2} \hat{U} + 2p^{5/2}(t_3 + C)\hat{U} = 0.$$

This is Airy's equation which has as linearly independent solutions the standard Airy functions $\text{Ai}(-2^{1/3}p^{5/6}(t_3 + C))$ and $\text{Bi}(-2^{1/3}p^{5/6}(t_3 + C))$ (cf. [1]). From the asymptotic formulas for the Airy functions (see again [1]) we find that we need $C = 0$ and the special solution

$$Y_3(t_3) = p^{-3/2} \frac{\mathrm{d}}{\mathrm{d}t_3} \log[\text{Ai}(-2^{1/3}p^{5/6}t_3)]$$

of Eq. (4.9) for the matching with the preceding local approximation. Since

$$Y_3(t_3) = p^{-1/4}\sqrt{(-2t_3)} + O((-t_3)^{-1}) \qquad \text{as } t_3 \to -\infty$$

the functions

$$\tilde{X}_3^r(\varepsilon^{-1/3}s_2, \varepsilon) := \varepsilon^{-1/2}\sqrt{p} - \varepsilon^{-1/3}p Y_3(\varepsilon^{-1/3}s_2)$$
$$\tilde{Y}_3^r(\varepsilon^{-1/3}s_2, \varepsilon) := 1 + \varepsilon^{1/2}/\sqrt{p} + \varepsilon^{2/3}Y_3(\varepsilon^{-1/3}s_2)$$
$$\tilde{Z}_3^r(\varepsilon^{-1/3}s_2, \varepsilon) := \varepsilon^{-1/2}\sqrt{p} - \varepsilon^{-1/3}p Y_3(\varepsilon^{-1/3}s_2)$$

match to leading order (in the sense of (2.12)) with the functions $\tilde{X}_2^l(T_2^* + s_2, \varepsilon)$, $\tilde{Y}_2^l(T_2^* + s_2, \varepsilon)$, $\tilde{Z}_2^l(T_2^* + s_2, \varepsilon)$ in the formal overlap domain $\varepsilon^{1/3} \ll (-s_2) \ll 1$.

$Y_3(t_3)$ has a pole at $t_3 = T_3^* := 2^{-1/3}p^{-5/6}\hat{t}_3$, where $\hat{t}_3$ is the first zero of $\text{Ai}(-s)$, $\hat{t}_3 = 2.338\ldots$. Hence, the third local approximation (to $1(t, \varepsilon)$) ceases to be valid in a (left) neighbourhood of the singularity $t_3 = T_3^*$. Expanding $\text{Ai}(-s)$ in a Taylor series near $-s = \hat{t}_3$ we find the following asymptotic behaviour:

$$X_3(T_3^* + s_3) = \frac{1}{\sqrt{(p)(-s_3)}} + O((-s_3))$$

$$Y_3(T_3^* + s_3) = -\frac{1}{p^{3/2}(-s_3)} + O((-s_3)), \qquad s_3 \to 0^-.$$

$$Z_3(T_3^* + s_3) = \frac{1}{\sqrt{(p)(-s_3)}} + O((-s_3))$$

Hence, the asymptotic exponents are

$$A_3 = B_3 = C_3 = -1.$$

*Step 4*

Subjecting (PS$_3$) to the shift scaling transformation

$$x_3 = \varepsilon^a x_4$$
$$y_3 = \varepsilon^b y_4$$
$$z_3 = \varepsilon^c z_4$$
$$t_3 = T_3^* + \varepsilon^d t_4$$

and multiplication by $\varepsilon^{-l}, \varepsilon^{-m}, \varepsilon^{-n}$ yields the convex polyhedron $P_4$ (which is congruent to $P_3$).

The selection rule (IV) requires

$$a = b = c = -d < 0.$$

There is exactly one vertex of $P_4$ adjacent to $0$ having this property:

$$p_4 = (-\tfrac{1}{6}, -\tfrac{1}{6}, -\tfrac{1}{6}, \tfrac{1}{6}, -\tfrac{1}{6}, -\tfrac{1}{6}, -\tfrac{1}{6})$$

with the corresponding system

$$\varepsilon^{1/2}\frac{\mathrm{d}x_4}{\mathrm{d}t_4} = -\frac{x_4}{\sqrt{p}} - \sqrt{(p)}\,y_4 - x_4 y_4 + \varepsilon^{1/2}\left(\frac{1}{\sqrt{p}} + y_4\right)$$
$$+ \varepsilon^2(-pq - 2q\sqrt{(p)}\,x_4 - qx_4{}^2$$

$\text{(PS}_4)\quad \dfrac{\mathrm{d}y_4}{\mathrm{d}t_4} = -x_4 + z_4 + \varepsilon^{1/2}\left(-2 - \dfrac{x_4}{\sqrt{p}} - \sqrt{(p)}\,y_4 - x_4 y_4\right) + \varepsilon\left(-\dfrac{1}{\sqrt{p}} - y_4\right)$

$$\frac{\mathrm{d}z_4}{\mathrm{d}t_4} = p(x_4 - z_4)$$

where

$$x_3 = \varepsilon^{-1/6}x_4 \qquad\qquad x = \varepsilon^{-1/2}\sqrt{p} + \varepsilon^{-1/2}x_4$$
$$y_3 = \varepsilon^{-1/6}y_4 \qquad\qquad y = 1 + \varepsilon^{1/2}/\sqrt{p} + \varepsilon^{1/2}y_4$$
$$z_3 = \varepsilon^{-1/6}z_4 \qquad\qquad z = \varepsilon^{-1/2}\sqrt{p} + \varepsilon^{-1/2}z_4$$
$$t_3 = T_3^* + \varepsilon^{1/6}t_4 \qquad\quad t = T_4(t_4, \varepsilon) := \varepsilon^{-1/2}T_2^* + \varepsilon^{-1/6}T_3^* + t_4$$

The appropriate solutions of $(\text{PAS}_4)$ (for the matching with the preceding local approximation) satisfy

$$\hat{X}_4(t_4) = -\frac{p\hat{Y}_4(t_4)}{1 + \sqrt{(p)}\,\hat{Y}_4(t_4)}$$

$$-p(t_4 + C) = \log(-\sqrt{(p)}\,\hat{Y}_4(t_4)) - \frac{1}{\sqrt{(p)}\,\hat{Y}_4(t_4)}$$

$$\hat{Z}_4(t_4) = -p\hat{Y}_4(t_4)$$

and

$$\hat{X}_4(t_4) = \frac{p^{-1/2}}{-(t_4 + C)} + O((t_4 + C)^{-2})$$

$$\hat{Y}_4(t_4) = -\frac{p^{-3/2}}{-(t_4 + C)} + O\left(\frac{\log[-(t_4 + C)]}{(t_4 + C)^2}\right), \qquad t_4 \to -\infty.$$

$$\hat{Z}_4(t_4) = \frac{p^{-1/2}}{-(t_4 + C)} + O((t_4 + C)^{-2})$$

Taking $C = 0$ the functions

$$\tilde{X}_4^r(\varepsilon^{-1/6}s_3, \varepsilon) := \varepsilon^{-1/2}\sqrt{p} + \varepsilon^{-1/2}X_4(\varepsilon^{-1/6}s_3)$$

$$\tilde{Y}_4^r(\varepsilon^{-1/6}s_3, \varepsilon) := 1 + \varepsilon^{1/2}/\sqrt{p} + \varepsilon^{1/2}Y_4(\varepsilon^{-1/6}s_3)$$

$$\tilde{Z}_4^r(\varepsilon^{-1/6}s_3, \varepsilon) := \varepsilon^{-1/2}\sqrt{p} + \varepsilon^{-1/2}Z_4(\varepsilon^{-1/2}s_3)$$

match to leading order (in the sense of (2.12)) with the functions $\tilde{X}_3^l(T_3^* + s_3, \varepsilon)$, $\tilde{Y}_3^l(T_3^* + s_3, \varepsilon)$, $\tilde{Z}_3^l(T_3^* + s_3, \varepsilon)$. (Actually $C$ cannot be determined by this matching to leading order. Since it is a time shift which anyway is neglected in our approximation of the period of $l(t, \varepsilon)$ we choose, for simplicity, $C = 0$.)

The fourth local approximation (to $l(t, \varepsilon)$) ceases to be valid as $Y_4(t_4)$ approaches $-1/\sqrt{(p)}$, i.e. has a singularity at $t_4 = T_4^* := -1/p$ with the asymptotic behaviour

$$X_4(T_4^* + s_4) = \frac{1}{\sqrt{(-2s_4)}} + O(1)$$

$$Y_4(T_4^* + s_4) = Y_4^* + \sqrt{(-2s_4)} + O((-s_4)), \qquad s_4 \to 0^-$$

$$Z_4(T_4^* + s_4) = Z_4^* - p\sqrt{(-2s_4)} + O((-s_4))$$

where $Y_4^* := -\dfrac{1}{\sqrt{p}}$ and $Z_4^* := \sqrt{p}$.

Hence

$$A_4 = -1/2, \qquad B_4 = C_4 = 1/2.$$

*Step 5*

Subjecting (PS$_4$) to the shift scaling transformation

$$x_4 = \varepsilon^a x_5$$
$$y_4 = Y_4^* + \varepsilon^b y_5$$
$$z_4 = Z_4^* + \varepsilon^c z_5$$
$$t_4 = T_4^* + \varepsilon^d t_5$$

and multiplication by $\varepsilon^{-l}, \varepsilon^{-m}, \varepsilon^{-n}$ yields the convex polyhedron $P_5$. The selection rule (IV) requires

$$-a = b = c = d/2 > 0.$$

There is exactly one vertex of $P_5$ adjacent to 0 having this property:

$$p_5 = (-\tfrac{1}{6}, \tfrac{1}{6}, \tfrac{1}{6}, \tfrac{1}{3}, 0, -\tfrac{1}{6}, -\tfrac{1}{6})$$

with the corresponding system

$$\frac{dx_5}{dt_5} = 1 - x_5 y_5 - \varepsilon^{1/6}\sqrt{(p)}y_5 + \varepsilon^{2/3}y_5 - \varepsilon^{5/3}q x_2^5$$

$$- \varepsilon^{11/6}2q\sqrt{(p)}x_5 - \varepsilon^2 pq$$

$$(\text{PS}_5) \quad \frac{dy_5}{dt_5} = -x_5 + \varepsilon^{1/6}\sqrt{p} + \varepsilon^{1/3}z_5 + \varepsilon^{2/3}(-1 - x_5 y_5)$$

$$-\varepsilon^{5/6}\sqrt{(p)}y_5 - \varepsilon^{4/3}y_5$$

$$\frac{dz_5}{dt_5} = px_5 - \varepsilon^{1/6}p^{3/2} - \varepsilon^{1/3}pz_5$$

where

$$x_4 = \varepsilon^{-1/6}x_5 \qquad\qquad x = \varepsilon^{-1/2}\sqrt{p} + \varepsilon^{-2/3}x_5$$

$$y_4 = -\frac{1}{\sqrt{p}} + \varepsilon^{1/6}y_5 \qquad y = 1 + \varepsilon^{2/3}y_5$$

$$z_4 = \sqrt{p} + \varepsilon^{1/6}z_5 \qquad z = \varepsilon^{-1/2}2\sqrt{p} + \varepsilon^{-1/3}z_5$$

$$t_4 = T_4^* + \varepsilon^{1/3}t_5 \qquad t = \mathbf{T}_5(t_5,\varepsilon) := \varepsilon^{-1/2}T_2^* + \varepsilon^{-1/6}T_3^* + T_4^* + \varepsilon^{1/3}t_5$$

The solutions of $(\text{PAS}_5)$ satisfy

$$\frac{d\hat{Y}_5}{dt_5} + \frac{\hat{Y}_5^{\,2}}{2} + t_5 + C = 0$$

$$\hat{X}_5(t_5) = t_5 + C + \frac{\hat{Y}_5(t_5)^2}{2}$$

$$\hat{Z}_5(t_5) = -p\hat{Y}_5(t_5) + D.$$

If we take the integration constants $C = D = 0$ and

$$Y_5(t_5) = 2\,\frac{d}{dt_5}\,\log[\text{Ai}(-2^{-1/3}t_5)]$$

we find the asymptotic behaviour

$$X_5(t_5) = \frac{1}{\sqrt{(-2t_5)}} + O((-t_5)^{-2})$$

$$Y_5(t_5) = \sqrt{(-2t_5)} + O((-t_5)^{-1}) \quad , \qquad t_5 \to -\infty,$$

$$Z_5(t_5) = -p\sqrt{(-2t_5)} + O((-t_5)^{-1})$$

and the functions

$$\tilde{X}_5^r(\varepsilon^{-1/3}s_4,\varepsilon) := \varepsilon^{-1/2}\sqrt{p} + \varepsilon^{-2/3}X_5(\varepsilon^{-1/3}s_4)$$

$$\tilde{Y}_5^r(\varepsilon^{-1/3}s_4,\varepsilon) := 1 + \varepsilon^{2/3}Y_5(\varepsilon^{-1/3}s_4)$$

$$\tilde{Z}_5^r(\varepsilon^{-1/3}s_4,\varepsilon) := \varepsilon^{-1/2}2\sqrt{p} + \varepsilon^{-1/3}Z_5(\varepsilon^{-1/3}s_4)$$

match to leading order (in the sense of (2.12)) with the functions $\tilde{X}_4^l(T_4^* + s_4,\varepsilon)$, $\tilde{Y}_4^l(T_4^* + s_4,\varepsilon)$, $\tilde{Z}_4^l(T_4^* + s_4,\varepsilon)$.

As seen in Step 3, $Y_5(t_5)$ has a pole at $t_5 = T_5^* := 2^{1/3}\hat{t}_5$, where $\hat{t}_5$ is the first zero of Airy's function $\mathrm{Ai}(-s)$; i.e. the fifth local approximation (to $\mathbf{I}(t, \varepsilon)$) has a singularity at $t_5 = T_5^*$ with the asymptotic behaviour

$$X_5(T_5^* + s_5) = \frac{2}{s_5^2} + \frac{T_5^*}{3} + O((-s_5))$$

$$Y_5(T_5^* + s_5) = -\frac{2}{(-s_5)} + O((-s_5)), \qquad s_5 \to 0^-.$$

$$Z_5(T_5^* + s_5) = \frac{2p}{(-s_5)} + O((-s_5))$$

Hence

$$A_5 = -2, \qquad B_5 = C_5 = -1.$$

*Step 6*

Subjecting (PS$_5$) to the shift scaling transformation

$$x_5 = \varepsilon^a x_6$$
$$y_5 = \varepsilon^b y_6$$
$$z_5 = \varepsilon^c z_6$$
$$t_5 = T_5^* + \varepsilon^d t_6$$

and multiplication by $\varepsilon^{-l}, \varepsilon^{-m}, \varepsilon^{-n}$ yields the convex polyhedron $P_6$ (which is congruent to $P_5$).
The selection rule (IV) requires

$$a/2 = b = c = -d < 0.$$

There is exactly one vertex of $P_6$ adjacent to 0 having this property:

$$p_6 = (-\tfrac{4}{3}, -\tfrac{2}{3}, -\tfrac{2}{3}, \tfrac{2}{3}, -2, -\tfrac{4}{3}, -\tfrac{4}{3})$$

with the corresponding system

$$\frac{\mathrm{d}x_6}{\mathrm{d}t_6} = -x_6 y_6 - \varepsilon q x_6^2 - \varepsilon^{3/2}\sqrt{(p)}y_6 + \varepsilon^2(1 + y_6)$$

$$- \varepsilon^{5/2} 2q\sqrt{(p)}x_6 - \varepsilon^4 pq$$

$$\text{(PS}_6) \quad \frac{\mathrm{d}y_6}{\mathrm{d}t_6} = -x_6 - x_6 y_6 + \varepsilon z_6 + \varepsilon^{3/2}(\sqrt{p} - \sqrt{(p)}y_6) + \varepsilon^2(-1 - y_6)$$

$$\frac{\mathrm{d}z_6}{\mathrm{d}t_6} = px_6 - \varepsilon p z_6 - \varepsilon^{3/2} p^{3/2}$$

where

$$x_5 = \varepsilon^{-4/3} x_6 \qquad\qquad x = \varepsilon^{-1/2}\sqrt{p} + \varepsilon^{-2} x_6$$
$$y_5 = \varepsilon^{-2/3} y_6 \qquad\qquad y = 1 + y_6$$
$$z_5 = \varepsilon^{-2/3} z_6 \qquad\qquad z = \varepsilon^{-1/2} 2\sqrt{p} + \varepsilon^{-1} z_6$$
$$t_5 = T_5^* + \varepsilon^{2/3} t_6 \qquad\quad t = \mathbf{T}_6(t_6, \varepsilon) := \varepsilon^{-1/2} T_2^* + \varepsilon^{-1/6} T_3^* + T_4^* + \varepsilon^{1/3} T_5^* + \varepsilon t_6$$

The solutions of (PAS$_6$) satisfy

(4.10)
$$\frac{\mathrm{d}\hat{Y}_6}{\mathrm{d}t_6} = -(1 + \hat{Y}_6)[\hat{Y}_6 - \log(1 + \hat{Y}_6) + C]$$

$$\hat{X}_6(t_6) = \hat{Y}_6(t_6) - \log(1 + \hat{Y}_6(t_6)) + C$$
$$\hat{Z}_6(t_6) = p[\hat{X}_6(t_6) - \hat{Y}_6(t_6)] + D.$$

If we take the integration constants $C = D = 0$ and a solution $Y_6(t_6)$ of (4.10) with $Y_6(0) \in (-1, 0)$ we find the asymptotic behaviour

$$X_6(t_6) = \frac{2}{t_6^2} + O((-t_6)^{-3})$$

$$Y_6(t_6) = -\frac{2}{(-t_6)} + O((-t_6)^{-2}), \qquad t_6 \to -\infty,$$

$$Z_6(t_6) = \frac{2p}{(-t_6)} + O((-t_6)^{-2})$$

and the functions

$$\tilde{X}_6^r(\varepsilon^{-2/3} s_5, \varepsilon) := \varepsilon^{-1/2}\sqrt{p} + \varepsilon^{-2} X_6(\varepsilon^{-2/3} s_5)$$
$$\tilde{Y}_6^r(\varepsilon^{-2/3} s_5, \varepsilon) := 1 + Y_6(\varepsilon^{-2/3} s_5)$$
$$\tilde{Z}_6^r(\varepsilon^{-2/3} s_5, \varepsilon) := \varepsilon^{-1/2} 2\sqrt{p} + \varepsilon^{-1} Z_6(\varepsilon^{-2/3} s_5)$$

match to leading order (in the sense of (2.12)) with the functions $\tilde{X}_5^l(T_5^* + s_5, \varepsilon)$, $\tilde{Y}_5^l(T_5^* + s_5, \varepsilon)$, $\tilde{Z}_5^l(T_5^* + s_5, \varepsilon)$.

$X_6(t_6)$ (and hence also $Z_6(t_6)$) becomes infinite as $Y_6(t_6)$ approaches $-1$. From this we find that the sixth local approximation (to $\mathbf{l}(t, \varepsilon)$) has a singularity at $t_6 = +\infty$ with the asymptotic behaviour

$$X_6(t_6) = e^{t_6 + E} + \mathrm{h.o.t.}$$
$$Y_6(t_6) = Y_6^* + e^{-e^{t_6 + E}} + \mathrm{h.o.t.}, \qquad t_6 \to +\infty$$
$$Z_6(t_6) = p e^{t_6 + E} + \mathrm{h.o.t.}$$

where $Y_6^* := -1$. Hence

$$A_6 = B_6 = C_6 = 0.$$

For convenience, we take the time shift $E = 0$, i.e. we fix the third integration constant of (4.10) which could not be determined by matching to leading order.

## Step 7

Subjecting (PS$_6$) to the shift scaling transformation

$$x_6 = \varepsilon^a x_7$$
$$y_6 = Y_6^* + \varepsilon^b y_7$$
$$z_6 = \varepsilon^c z_7$$
$$t_6 = \varepsilon^d t_7$$

and multiplication by $\varepsilon^{-l}, \varepsilon^{-m}, \varepsilon^{-n}$ yields the convex polyhedron $P_7$. The selection rule (IV) requires

$$a, c \leqslant 0, b \geqslant 0, d < 0.$$

There is no such vertex adjacent to $0 \in P_7$. The selection rule (V) yields three vertices:

|       | $a$ | $b$ | $c$ | $d$ | $l$ | $m$ | $n$ |
|-------|-----|-----|-----|-----|-----|-----|-----|
| (1)   | 0   | 0   | $-1$ | 0  | 0   | 0   | $-1$ |
| (2)   | 0   | 1   | 0   | 0   | 0   | 1   | 0   |
| (3)   | $-1$ | 0  | $-1$ | 0  | $-1$ | $-1$ | $-1$ |

By means of the rule (VIII(1)) we choose the vertex (3). Hence

$$p_7 = (-1, 0, -1, 0, -1, -1, -1).$$

$d = 0$ means that we have the same time scale as before. Since $S_6^+ = +\infty$, we expect to be in the situation (iii) of Section 2.2.2. So we put $t_6 = T_6^*(\varepsilon) + t_7$ with $T_6^*(\varepsilon) \to +\infty$ as $\varepsilon \to 0$. The yet unknown time shift $T_6^*$ will be determined by matching.

To the point $p_7$ there corresponds the system

$$\frac{dx_7}{dt_7} = x_7 - x_7 y_7 - qx_7{}^2 + \varepsilon^{5/2}(\sqrt{p} - \sqrt{(p)}y_7 - 2q\sqrt{(p)}x_7) + \varepsilon^3 y_7 - \varepsilon^5 pq$$

$$\text{(PS}_7) \qquad \varepsilon \frac{dy_7}{dt_7} = -x_7 y_7 + \varepsilon z_7 + \varepsilon^{5/2}(2\sqrt{p} - \sqrt{(p)}y_7) - \varepsilon^3 y_7$$

$$\frac{dz_7}{dt_7} = px_7 - \varepsilon pz_7 - \varepsilon^{5/2} p^{3/2}$$

where

$$x_6 = \varepsilon^{-1}x_7 \qquad x = \varepsilon^{-1/2}\sqrt{p} + \varepsilon^{-3}x_7$$
$$y_6 = -1 + y_7 \qquad y = y_7$$
$$z_6 = \varepsilon^{-1}z_7 \qquad z = \varepsilon^{-1/2}2\sqrt{p} + \varepsilon^{-2}z_7$$
$$t_6 = T_6^* + t_7 \qquad t = \mathbf{T}_7(t_7, \varepsilon) := \varepsilon^{-1/2}T_2^* + \varepsilon^{-1/6}T_3^* + T_4^*$$
$$+ \varepsilon^{1/3}T_5^* + \varepsilon T_6^* + \varepsilon t_7$$

The appropriate solutions of $(\mathrm{PAS}_7)$ (as one easily finds from a phase-plane analysis in the $(x_7, t_7)$-plane) are of the form

$$\hat{X}_7(t_7) = \frac{1}{q + Ce^{-t_7}}$$

$$Y_7(t_7) = 0$$

$$\hat{Z}_7(t_7) = \frac{p}{q}[t_7 + \log(q + Ce^{-t_7})] + D$$

where $C$ and $D$ are integration constants. We may put $C = 1$, i.e. we take this integration constant to the time shift $T_6^*$. If we take $D = 0$ and $T_6^*(\varepsilon) := \log \varepsilon^{-1}$ the functions

$$\tilde{X}_7^r(t_7, \varepsilon) := \varepsilon^{-1/2}\sqrt{p} + \varepsilon^{-3}X_7(t_7)$$
$$\tilde{Y}_7^r(t_7, \varepsilon) := 0$$
$$\tilde{Z}_7^r(t_7, \varepsilon) := \varepsilon^{-1/2}2\sqrt{p} + \varepsilon^{-2}Z_7(t_7)$$

match to leading order with the functions $\tilde{X}_6^l(T_6^* + t_7, \varepsilon)$, $\tilde{Y}_6^l(T_6^* + t_7, \varepsilon)$, $\tilde{Z}_6^l(T_6^* + t_7, \varepsilon)$ in the sense of (2.13) in the formal overlap domain $1 \ll (-t_7)$ and $T_6^* + t_7 \gg 1$ as $\varepsilon \to 0$. Since $Z_7(t_7)$ becomes infinite as $t_7 \to +\infty$ the seventh local approximation (to $\mathbf{l}(t, \varepsilon)$) cannot be valid for all times, i.e. we expect to have a singularity at $t_7 = +\infty$ with the asymptotic behaviour

$$X_7(t_7) = X_7^* + O(e^{-t_7})$$
$$Y_7(t_7) = 0 \qquad\qquad , \qquad t_7 \to +\infty$$

$$Z_7(t_7) = \frac{p}{q}\,t_7 + O(1)$$

where $X_7^* := 1/q$.
Hence

$$A_7 = B_7 = 0, \qquad C_7 = 1.$$

## Step 8

Subjecting (PS$_7$) to the shift scaling transformation

$$x_7 = X_7^* + \varepsilon^a x_8$$
$$y_7 = \varepsilon^b y_8$$
$$z_7 = \varepsilon^c z_8$$
$$t_7 = \varepsilon^d t_8$$

and multiplication by $\varepsilon^{-l}, \varepsilon^{-m}, \varepsilon^{-n}$ yields the convex polyhedron $P_8$. The selection rule (IV) requires

$$a, b \geqslant 0 \qquad \text{and} \qquad c = d < 0.$$

There is exactly one vertex of $P_8$ adjacent to 0 having this property:

$$p_8 = (0, 0, -1, -1, 0, 0, 0)$$

with the corresponding system

$$\varepsilon \frac{dx_8}{dt_8} = -x_8 - x_8 y_8 - \frac{y_8}{q} - qx_8^2 + \varepsilon^{5/2}(-\sqrt{p} - \sqrt{(p)}y_8 - 2q\sqrt{(p)}x_8)$$
$$+ \varepsilon^3 y_8 - \varepsilon^5 pq$$

(PS$_8$) $\quad \varepsilon^2 \dfrac{dy_8}{dt_8} = -\dfrac{y_8}{q} - x_8 y_8 + z_8 + \varepsilon^{5/2}(2\sqrt{p} - \sqrt{(p)}y_8) - \varepsilon^3 y_8$

$$\frac{dz_8}{dt_8} = \frac{p}{q} + px_8 - pz_8 - \varepsilon^{5/2}p^{3/2}$$

where

$$x_7 = \frac{1}{q} + x_8 \qquad x = \varepsilon^{-1/2}\sqrt{(p)} + \varepsilon^{-3}\frac{1}{q} + \varepsilon^{-3}x_8$$
$$y_7 = y_8 \qquad y = y_8$$
$$z_7 = \varepsilon^{-1}z_8 \qquad z = \varepsilon^{-1/2}2\sqrt{(p)} + \varepsilon^{-3}z_8$$
$$t_7 = \varepsilon^{-1}t_8 \qquad t = \mathbf{T}_8(t_8, \varepsilon) := \varepsilon^{-1/2}T_2^* + \varepsilon^{-1/6}T_3^* + T_4^* + \varepsilon^{1/3}T_5^* + \varepsilon T_6^* + t_8$$

The solutions of (PAS$_8$) satisfy

$$\frac{\hat{X}_8}{dt_8} = -\frac{p}{q}\frac{(1 + q\hat{X}_8)^2}{1 + 2q\hat{X}_8}$$

$$\hat{Y}_8(t_8) = \frac{q\hat{Z}_8(t_8)}{1 + q\hat{X}_8(t_8)}$$

$$\hat{Z}_8(t_8) = -\hat{X}_8(t_8) - q\hat{X}_8(t_8)^2$$

Since the $x$-component of the preceding local approximation (expressed in the

$x_8$-variable) approaches $0^-$ as $t_7 \to +\infty$ it is obvious that we need the initial condition $X_8(0) = 0$. (This is precisely the Tikhonov situation; compare Theorem 5.6.) Hence we have

$$pt_8 = 1 - 2 \log[1 + qX_8(t_8)] - \frac{1}{1 + qX_8(t_8)}$$

Then the solutions have the asymptotic behaviour

$$X_8(t_8) = -\frac{p}{q} t_8 + O(t_8{}^3)$$

$$Y_8(t_8) = pt_8 + O(t_8{}^3) \quad , \qquad t_8 \to 0^+,$$

$$Z_8(t_8) = \frac{p}{q} t_8 + O(t_8{}^2)$$

and the functions

$$\tilde{X}_8^r(t_8, \varepsilon) := \varepsilon^{-1/2}\sqrt{p} + \varepsilon^{-3}\frac{1}{q} + \varepsilon^{-3}X_8(t_8)$$

$$\tilde{Y}_8^r(t_8, \varepsilon) := Y_8(t_8)$$

$$\tilde{Z}_8^r(t_8, \varepsilon) := \varepsilon^{-1/2}2\sqrt{p} + \varepsilon^{-3}Z_8(t_8)$$

match to leading order with the functions $\tilde{X}_7^l(\varepsilon^{-1}t_8, \varepsilon)$, $\tilde{Y}_7^l(\varepsilon^{-1}t_8, \varepsilon)$, $\tilde{Z}_7^l(\varepsilon^{-1}t_8, \varepsilon)$ in the sense of (2.11) in the formal overlap domain $\varepsilon \ll t_8 \ll 1, \varepsilon \to 0$.

The eighth local approximation (to $I(t, \varepsilon)$) ceases to be valid as $X_8(t_8)$ approaches $-1/2q$, i.e. has a singularity at $t_8 = T_8^* := (2 \log 2 - 1)/p$ with the asymptotic behaviour

$$X_8(T_8^* + s_8) = X_8^* + \frac{1}{2q} \sqrt{(-ps_8)} + O((-s_8))$$

$$Y_8(T_8^* + s_8) = Y_8^* - \frac{1}{2} \sqrt{(-ps_8)} + O((-s_8)) \quad , \qquad s_8 \to 0^-$$

$$Z_8(T_8^* + s_8) = Z_8^* + \frac{p}{4q} s_8 + O((-s_8)^{3/2})$$

where $X_8^* := -1/2q$, $Y_8^* := 1/2$, $Z_8^* := 1/4q$.
Hence

$$A_8 = B_8 = 1/2, \qquad C_8 = 1;$$

and the eighth local approximation expressed in the local variables near the singularity satisfies the algebraic identities

$$V_8 = -qU_8, \quad W_8 = -qU_8{}^2, \qquad \text{as } s_8 \to 0^-.$$

**Step 9**

Subjecting (PS$_8$) to the shift scaling transformation

$$x_8 = X_8^* + \varepsilon^a x_9$$
$$y_8 = Y_8^* + \varepsilon^b y_9$$
$$z_8 = Z_8^* + \varepsilon^c z_9$$
$$t_8 = T_8^* + \varepsilon^d t_9$$

and multiplication by $\varepsilon^{-l}, \varepsilon^{-m}, \varepsilon^{-n}$ yields the convex polyhedron $P_9$. The selection rule (IV) requires

$$2a = 2b = c = d > 0.$$

There is no vertex adjacent to $0 \in P_9$ having this property. We therefore look for a *hidden vertex*.
The rule (VI) of the algorithm yields a unique point

$$p_9 = (\tfrac{1}{3}, \tfrac{1}{3}, \tfrac{2}{3}, \tfrac{2}{3}, \tfrac{1}{3}, \tfrac{1}{3}, 0)$$

in $P_9$ with the corresponding system

$$\varepsilon^{1/3}\frac{dx_9}{dt_9} = -\frac{x_9}{2} - \frac{y_9}{2q} + \varepsilon^{1/3}(-qx_9^2 - x_9y_9) - \varepsilon^{13/6}\frac{\sqrt{p}}{2}$$
$$+ \varepsilon^{5/2}(-2q\sqrt{(p)}x_9 - \sqrt{(p)}y_9) + \varepsilon^{8/3}\frac{1}{2} + \varepsilon^3 y_9 - \varepsilon^{14/3}pq$$

(PS$_9$)  $\varepsilon^{4/3}\dfrac{dy_9}{dt_9} = -\dfrac{x_9}{2} - \dfrac{y_9}{2q} + \varepsilon^{1/3}(-x_9y_9 + z_9) + \varepsilon^{13/6}\dfrac{3\sqrt{p}}{2}$

$$- \varepsilon^{5/2}\sqrt{(p)}y_9 - \varepsilon^{8/3}\frac{1}{2} - \varepsilon^3 y_9$$

$$\frac{dz_9}{dt_9} = \frac{p}{4q} + \varepsilon^{1/3}px_9 - \varepsilon^{2/3}pz_9 - \varepsilon^{5/2}p^{3/2}$$

where

$$x_8 = -\frac{1}{2q} + \varepsilon^{1/3}x_9 \qquad x = \varepsilon^{-1/2}\sqrt{p} + \varepsilon^{-3}\frac{1}{2q} + \varepsilon^{-8/3}x_9$$

$$y_8 = \frac{1}{2} + \varepsilon^{1/3}y_9 \qquad y = \frac{1}{2} + \varepsilon^{1/3}y_9$$

$$z_8 = \frac{1}{4q} + \varepsilon^{2/3}z_9 \qquad z = \varepsilon^{-1/2}2\sqrt{p} + \varepsilon^{-3}\frac{1}{4q} + \varepsilon^{-7/3}z_9$$

$$t_8 = T_8^* + \varepsilon^{2/3}t_9 \qquad t = \mathbf{T}_9(t_9, \varepsilon) := \varepsilon^{-1/2}T_2^* + \varepsilon^{-1/6}T_3^* + T_4^* + \varepsilon^{1/3}T_5^*$$
$$+ \varepsilon T_6^* + T_8^* + \varepsilon^{2/3}t_9$$

The solutions of (PAS$_9$) satisfy

$$\frac{\mathrm{d}\hat{X}_9}{\mathrm{d}t_9} = -\frac{p}{4q}\,t_9 - q\hat{X}_9{}^2 - D$$

$$\hat{Y}_9(t_9) = -q\hat{X}_9(t_9)$$

$$\hat{Z}_9(t_9) = \frac{p}{4q}\,t_9 + D$$

If we take $D = 0$ and (see Step 3)

$$X_9(t_9) = \frac{1}{q}\frac{\mathrm{d}}{\mathrm{d}t_9}\,\log[\,\mathrm{Ai}(-(p/4)^{1/3}t_9)\,]$$

we find the asymptotic behaviour

$$X_9(t_9) = \frac{1}{2q}\,\sqrt{(-pt_9)} + O((-t_9)^{-1})$$

$$Y_9(t_9) = -\frac{1}{2}\sqrt{(-pt_9)} + O((-t_9)^{-1}), \qquad t_9 \to -\infty,$$

$$Z_9(t_9) = \frac{p}{4q}\,t_9$$

and the functions

$$\tilde{X}^r_9(\varepsilon^{-2/3}s_8, \varepsilon) := \varepsilon^{-1/2}\sqrt{p} + \varepsilon^{-3}\,\frac{1}{2q} + \varepsilon^{-8/3}X_9(\varepsilon^{-2/3}s_8)$$

$$\tilde{Y}^r_9(\varepsilon^{-2/3}s_8, \varepsilon) := \frac{1}{2} + \varepsilon^{1/3}Y_9(\varepsilon^{-2/3}s_8)$$

$$\tilde{Z}^r_9(\varepsilon^{-2/3}s_8, \varepsilon) := \varepsilon^{-1/2}2\sqrt{p} + \varepsilon^{-3}\,\frac{1}{4q} + \varepsilon^{-7/3}Z_9(\varepsilon^{-2/3}s_8)$$

match to leading order (in the sense of (2.12)) with the functions $\tilde{X}^l_8(T^*_8 + s_8, \varepsilon)$, $\tilde{Y}^l_8(T^*_8 + s_8, \varepsilon)$, $\tilde{Z}^l_8(T^*_8 + s_8, \varepsilon)$.

As seen in Step 3, $X_9(t_9)$ has a pole at $t_9 = T^*_9 := (4/p)^{1/3}\hat{t}_9$, where $\hat{t}_9$ is the first zero of Airy's function $\mathrm{Ai}(-s)$; i.e. the ninth local approximation (to $\mathbf{l}(t, \varepsilon)$) has a singularity at $t_9 = T^*_9$ with the asymptotic behaviour

$$X_9(T^*_9 + s_9) = -\frac{1}{q(-s_9)} + O((-s_9))$$

$$Y_9(T^*_9 + s_9) = \frac{1}{(-s_9)} + O((-s_9)) \qquad , \qquad s_9 \to 0^-$$

$$Z_9(T^*_9 + s_9) = Z^*_9 - \frac{p}{4q}\,(-s_9)$$

where $Z^*_9 := \left(\dfrac{p}{4}\right)^{2/3}\dfrac{\hat{t}_9}{q}$.

Hence

$$A_9 = B_9 = -1, \qquad C_9 = 1.$$

*Step 10*

Subjecting (PS$_9$) to the shift scaling transformation

$$x_9 = \varepsilon^a x_{10}$$
$$y_9 = \varepsilon^b y_{10}$$
$$z_9 = Z_9^* + \varepsilon^c z_{10}$$
$$t_9 = T_9^* + \varepsilon^d t_{10}$$

and multiplication by $\varepsilon^{-l}, \varepsilon^{-m}, \varepsilon^{-n}$ yields the convex polyhedron $P_{10}$. The selection rule (IV) requires

$$a = b = -c = -d < 0.$$

There is exactly one vertex of $P_{10}$ adjacent to 0 having this property:

$$p_{10} = (-\tfrac{1}{3}, -\tfrac{1}{3}, \tfrac{1}{3}, \tfrac{1}{3}, -\tfrac{1}{3}, -\tfrac{1}{3}, 0)$$

with the corresponding system

$$\frac{\mathrm{d}x_{10}}{\mathrm{d}t_{10}} = -\frac{x_{10}}{2} - \frac{y_{10}}{2q} - qx_{10}^2 - x_{10}y_{10}$$

$$+ \varepsilon^{5/2}\left(-\frac{\sqrt{p}}{2} - \sqrt{(p)}\,y_{10} - 2q\sqrt{(p)}\,x_{10}\right) + \varepsilon^3\left(\frac{1}{2} + y_{10}\right) - \varepsilon^5 pq$$

(PS$_{10}$)    $$\varepsilon\frac{\mathrm{d}y_{10}}{\mathrm{d}t_{10}} = -\frac{x_{10}}{2} - \frac{y_{10}}{2q} - x_{10}y_{10} + \varepsilon^{2/3} Z_9^* + \varepsilon z_{10} + \varepsilon^{5/2}\left(\frac{3\sqrt{p}}{2} - \sqrt{(p)}\,y_{10}\right)$$

$$+ \varepsilon^3\left(-\frac{1}{2} - y_{10}\right)$$

$$\frac{\mathrm{d}z_{10}}{\mathrm{d}t_{10}} = \frac{p}{4q} + px_{10} - \varepsilon^{2/3} pZ_9^* - \varepsilon p z_{10} - \varepsilon^{5/2} p^{3/2}$$

where

$$x_9 = \varepsilon^{-1/3} x_{10} \qquad\qquad x = \varepsilon^{-1/2}\sqrt{p} + \varepsilon^{-3}\frac{1}{2q} + \varepsilon^{-3} x_{10}$$

$$y_9 = \varepsilon^{-1/3} y_{10} \qquad\qquad y = \frac{1}{2} + y_{10}$$

$$z_9 = Z_9^* + \varepsilon^{1/3} z_{10} \qquad\qquad z = \varepsilon^{-1/2} 2\sqrt{p} + \varepsilon^{-3}\frac{1}{4q} + \varepsilon^{-7/3} Z_9^* + \varepsilon^{-2} z_{10}$$

$$t_9 = T_9^* + \varepsilon^{1/3} t_{10} \qquad\qquad t = \mathbf{T}_{10}(t_{10}, \varepsilon) := \varepsilon^{-1/2} T_2^* + \varepsilon^{-1/6} T_3^* + T_4^* + \varepsilon^{1/3} T_5^*$$

$$+ \varepsilon T_6^* + T_8^* + \varepsilon^{2/3} T_9^* + \varepsilon t_{10}$$

The general solution of $(PAS_{10})$ is

$$\hat{X}_{10}(t_{10}) = \frac{1}{qt_{10} + C}$$

$$\hat{Y}_{10}(t_{10}) = - \frac{1}{2 + t_{10} + C/q}$$

$$\hat{Z}_{10}(t_{10}) = \frac{p}{4q}\, t_{10} + \frac{p}{q}\, \log(-qt_{10} - C) + D$$

If we take the integration constants $C = D = 0$ the functions

$$\tilde{X}^r_{10}(\varepsilon^{-1/3}s_9, \varepsilon) := \varepsilon^{-1/2}\sqrt{(p)} + \varepsilon^{-3}\,\frac{1}{2q} + \varepsilon^{-3}X_{10}(\varepsilon^{-1/3}s_9)$$

$$\tilde{Y}^r_{10}(\varepsilon^{-1/3}s_9, \varepsilon) := \frac{1}{2} + Y_{10}(\varepsilon^{-1/3}s_9)$$

$$\tilde{Z}^r_{10}(\varepsilon^{-1/3}s_9, \varepsilon) := \varepsilon^{-1/2}2\sqrt{(p)} + \varepsilon^{-3}\,\frac{1}{4q} + \varepsilon^{-7/3}Z^*_9 + \varepsilon^{-2}Z_{10}(\varepsilon^{-1/3}s_9)$$

match to leading order (in the sense of (2.12)) with the functions $\tilde{X}^l_9(T^*_9 + s_9, \varepsilon)$, $\tilde{Y}^l_9(T^*_9 + s_9, \varepsilon)$, $\tilde{Z}^l_9(T^*_9 + s_9, \varepsilon)$.

$Y_{10}(t_{10})$ has a pole at $t_{10} = T^*_{10} := -2$, i.e. the tenth local approximation (to $l(t, \varepsilon)$) has a singularity at $t_{10} = T^*_{10}$ with the asymptotic behaviour

$$X_{10}(T^*_{10} + s_{10}) = X^*_{10} + \frac{(-s_{10})}{4q} + O(s_{10}^2)$$

$$Y_{10}(T^*_{10} + s_{10}) = \frac{1}{(-s_{10})} \qquad , \qquad s_{10} \to 0^-$$

$$Z_{10}(T^*_{10} + s_{10}) = Z^*_{10} + \frac{p}{4q}\,(-s_{10}) + O(s_{10}^2)$$

where $X^*_{10} := -1/2q$ and $Z^*_{10} := -p/2q + (p/q)\log 2q$. Hence

$$A_{10} = C_{10} = 1, \quad B_{10} = -1.$$

*Step 11*

Subjecting $(PS_{10})$ to the shift scaling transformation

$$x_{10} = X^*_{10} + \varepsilon^a x_{11}$$
$$y_{10} = \varepsilon^b y_{11}$$
$$z_{10} = Z^*_{10} + \varepsilon^c z_{11}$$
$$t_{10} = T^*_{10} + \varepsilon^d t_{11}$$

and multiplication by $\varepsilon^{-l}, \varepsilon^{-m}, \varepsilon^{-n}$ yields the convex polyhedron $P_{11}$. The selection rule (IV) requires

$$a = c = -b = d > 0.$$

There is exactly one vertex of $P_{11}$ adjacent to 0 having this property:

$$p_{11} = (\tfrac{1}{2}, -\tfrac{1}{2}, \tfrac{1}{2}, \tfrac{1}{2}, 0, 0, 0)$$

with the corresponding system

$$\frac{dx_{11}}{dt_{11}} = -x_{11}y_{11} + \varepsilon^{1/2}\frac{x_{11}}{2} - \varepsilon q x_{11}^2 - \varepsilon^2 \sqrt{(p)}y_{11} + \varepsilon^{5/2}\left(\frac{\sqrt{p}}{2} + y_{11}\right)$$

$$+ \varepsilon^3\left(\frac{1}{2} - 2q\sqrt{(p)}x_{11}\right) - \varepsilon^5 pq$$

(PS$_{11}$) $\quad \dfrac{dy_{11}}{dt_{11}} = \dfrac{1}{4q} - x_{11}y_{11} - \varepsilon^{1/2}\dfrac{x_{11}}{2} + \varepsilon Z_{10}^* + \varepsilon^{2/3}Z_9^* + \varepsilon^{3/2}z_{11} - \varepsilon^2\sqrt{(p)}y_{11}$

$$+ \varepsilon^{5/2}\left(\frac{3\sqrt{p}}{2} - y_{11}\right) - \varepsilon^3\frac{1}{2}$$

$$\frac{dz_{11}}{dt_{11}} = -\frac{p}{4q} + \varepsilon^{1/2}px_{11} - \varepsilon^{2/3}pZ_9^* - \varepsilon pZ_{10}^* - \varepsilon^{3/2}pz_{11} - \varepsilon^{5/2}p^{3/2}$$

where

$$x_{10} = -\frac{1}{2q} + \varepsilon^{1/2}x_{11} \qquad x = \varepsilon^{-1/2}\sqrt{(p)} + \varepsilon^{-5/2}x_{11}$$

$$y_{10} = \varepsilon^{-1/2}y_{11} \qquad\qquad y = \frac{1}{2} + \varepsilon^{-1/2}y_{11}$$

$$z_{10} = Z_{10}^* + \varepsilon^{1/2}z_{11} \qquad z = \varepsilon^{-3}\frac{1}{4q} + \varepsilon^{-7/3}Z_9^* + \varepsilon^{-2}Z_{10}^* + \varepsilon^{-1/2}2\sqrt{p}$$

$$+ \varepsilon^{-3/2}z_{11}$$

$$t_{10} = T_{10}^* + \varepsilon^{1/2}t_{11} \qquad t = T_{11}(t_{11}, \varepsilon) = \varepsilon^{-1/2}T_2^* + \varepsilon^{-1/6}T_3^* + T_4^* + \varepsilon^{1/3}T_5^*$$

$$+ \varepsilon T_6^* + T_8^* + \varepsilon^{2/3}T_9^* + \varepsilon T_{10}^*$$

$$+ \varepsilon^{3/2}t_{11}$$

If we take the following solutions of (PAS$_{11}$)

$$X_{11}(t_{11}) = \varepsilon^{-t_{11}^2/8q}\left[\int_{-\infty}^{t_{11}} e^{-s^2/8q}\, ds\right]^{-1}$$

$$Y_{11}(t_{11}) = \frac{t_{11}}{4q} + X_{11}(t_{11})$$

$$Z_{11}(t_{11}) = -\frac{p}{4q}t_{11}$$

we find the asymptotic behaviour

$$X_{11}(t_{11}) = \frac{(-t_{11})}{4q} + \frac{1}{(-t_{11})} + O(t_{11}^{-2})$$

$$Y_{11}(t_{11}) = \frac{1}{(-t_{11})} + O(t_{11}^{-2}) \qquad , \qquad t_{11} \to -\infty,$$

$$Z_{11}(t_{11}) = \frac{p}{4q}\,(-t_{11})$$

and the functions

$$\tilde{X}_{11}^r(\varepsilon^{-1/2}s_{10}, \varepsilon) := \varepsilon^{-1/2}\sqrt{(p)} + \varepsilon^{-5/2}X_{11}(\varepsilon^{-1/2}s_{10})$$

$$\tilde{Y}_{11}^r(\varepsilon^{-1/2}s_{10}, \varepsilon) := \frac{1}{2} + \varepsilon^{-1/2}Y_{11}(\varepsilon^{-1/2}s_{10})$$

$$\tilde{Z}_{11}^r(\varepsilon^{-1/2}s_{10}, \varepsilon) := \varepsilon^{-3}\,\frac{1}{4q} + \varepsilon^{-7/3}Z_9^* + \varepsilon^{-2}Z_{10}^* + \varepsilon^{-1/2}2\sqrt{(p)}$$

$$+ \varepsilon^{-3/2}Z_{11}(\varepsilon^{-1/2}s_{10})$$

match to leading order (in the sense of (2.12)) with the functions $\tilde{X}_{10}^l(T_{10}^* + s_{10}, \varepsilon)$, $\tilde{Y}_{10}^l(T_{10}^* + s_{10}, \varepsilon)$, $\tilde{Z}_{10}^l(T_{10}^* + s_{10}, \varepsilon)$.

Since $Y_{11}(t_{11})$ and $Z_{11}(t_{11})$ become infinite as $t_{11} \to +\infty$ the eleventh local approximation (to $\mathbf{l}(t, \varepsilon)$) cannot be valid for all times, i.e. we expect to have a singularity at $t_{11} = +\infty$ with the asymptotic behaviour

$$X_{11}(t_{11}) = O(e^{-t_{11}^2/8q})$$

$$Y_{11}(t_{11}) = \frac{t_{11}}{4q} + O(e^{-t_{11}^2/8q}), \qquad t_{11} \to +\infty.$$

$$Z_{11}(t_{11}) = -\frac{p}{4q}\,t_{11}$$

Hence

$$A_{11} = 0, \quad B_{11} = C_{11} = 1.$$

*Step 12*

Subjecting $(PS_{11})$ to the shift scaling transformation

$$x_{11} = \varepsilon^a x_{12}$$
$$y_{11} = \varepsilon^b y_{12}$$
$$z_{11} = \varepsilon^c z_{12}$$
$$t_{11} = \varepsilon^d t_{12}$$

and multiplication by $\varepsilon^{-l}, \varepsilon^{-m}, \varepsilon^{-n}$ yields the convex polyhedron $P_{12}$. The selection rule (IV) requires

$$a \geqslant 0, \quad b = c = d < 0.$$

There is exactly one vertex of $P_{12}$ adjacent to 0 having this property:

$$p_{12} = (\tfrac{3}{2}, -\tfrac{3}{2}, -\tfrac{3}{2}, -\tfrac{3}{2}, 0, 0, 0)$$

with the corresponding system

$$\varepsilon^3 \frac{dx_{12}}{dt_{12}} = -x_{12}y_{12} - \varepsilon^{1/2}\sqrt{(p)}y_{12} + \varepsilon y_{12} + \varepsilon^2 \frac{x_{12}}{2} + \varepsilon^{5/2} \frac{\sqrt{p}}{2} + \varepsilon^3 \frac{1}{2}$$

$$- \varepsilon^4 q x_{12}^2 - \varepsilon^{9/2} 2q\sqrt{(p)}x_{12} - \varepsilon^5 pq$$

(PS$_{12}$) $\quad \dfrac{dy_{12}}{dt_{12}} = \dfrac{1}{4q} - x_{12}y_{12} + z_{12} - \varepsilon^{1/2}\sqrt{(p)}y_{12} + \varepsilon^{2/3} Z_9^*$

$$+ \varepsilon(Z_{10}^* - y_{12}) - \varepsilon^2 \frac{x_{12}}{2} + \varepsilon^{5/2} \frac{3\sqrt{p}}{2} - \varepsilon^3 \frac{1}{2}$$

$$\frac{dz_{12}}{dt_{12}} = -\frac{p}{4q} - pz_{12} - \varepsilon^{2/3} pZ_9^* - \varepsilon pZ_{10}^* + \varepsilon^2 px_{12} - \varepsilon^{5/2} p^{3/2}$$

where

$$x_{11} = \varepsilon^{3/2} x_{12} \qquad\qquad x = \varepsilon^{-1/2}\sqrt{p} + \varepsilon^{-1} x_{12}$$

$$y_{11} = \varepsilon^{-3/2} y_{12} \qquad\qquad y = \frac{1}{2} + \varepsilon^{-2} y_{12}$$

$$z_{11} = \varepsilon^{-3/2} z_{12} \qquad\qquad z = \varepsilon^{-3}\frac{1}{4q} + \varepsilon^{-7/3} Z_9^* + \varepsilon^{-2} Z_{10}^* + \varepsilon^{-1/2} 2\sqrt{p} + \varepsilon^{-3} z_{12}$$

$$t_{11} = \varepsilon^{-3/2} t_{12} \qquad\qquad t = \mathbf{T}_{12}(t_{12}, \varepsilon) := \varepsilon^{-1/2} T_2^* + \varepsilon^{-1/6} T_3^* + T_4^* + \varepsilon^{1/3} T_5^*$$

$$+ \varepsilon T_6^* + T_8^* + \varepsilon^{2/3} T_9^* + \varepsilon T_{10}^* + t_{12}$$

(PAS$_{12}$) has the (higher order) solutions

$$\hat{X}_{12}(t_{12}, \varepsilon) = 0 - \varepsilon^{1/2}\sqrt{p} + \varepsilon 1 + \varepsilon^3 \frac{1}{\hat{Y}_{12}(t_{12})}$$

$$\hat{Y}_{12}(t_{12}) = \frac{1}{4pq} + Ce^{-pt_{12}} + D$$

$$\hat{Z}_{12}(t_{12}, \varepsilon) = -\frac{1}{4q} - pCe^{-pt_{12}} - \varepsilon^{2/3} Z_9^* - \varepsilon Z_{10}^* - \varepsilon^{5/2} 2\sqrt{p} + \varepsilon^3 1$$

$$+ [\varepsilon^{1/2} D_1 + \varepsilon D_2 + \varepsilon^{3/2} D_3 + \varepsilon^2 D_4 + \varepsilon^{5/2} D_5 + \varepsilon^3 D_6]e^{-pt_{12}}$$

where $C, D$ and $D_1, \ldots, D_6$ are integration constants.

If we choose $C = -1/4pq$ and $D = 0$ the functions

$$\tilde{X}^r_{12}(t_{12}, \varepsilon) := 1 + \varepsilon^2 4pq/(1 - e^{-pt_{12}})$$

$$\tilde{Y}^r_{12}(t_{12}, \varepsilon) := \frac{1}{2} + \varepsilon^{-2} Y_{12}(t_{12})$$

$$\tilde{Z}^r_{12}(t_{12}, \varepsilon) := \varepsilon^{-3} \frac{1}{4q} + \varepsilon^{-7/3} Z_9^* + \varepsilon^{-2} Z_{10}^*$$

$$+ \varepsilon^{-1/2} 2\sqrt{(p)} + \varepsilon^{-3} Z_{12}(t_{12}, \varepsilon) = O(\varepsilon^{-3}) \cdot e^{-pt_{12}} + 1$$

match to leading order (in the sense of (2.11)) with the functions $\tilde{X}^l_{11}(\varepsilon^{-3/2} t_{12}, \varepsilon)$, $\tilde{Y}^l_{11}(\varepsilon^{-3/2} t_{12}, \varepsilon)$, $\tilde{Z}^l_{11}(\varepsilon^{-3/2} t_{12}, \varepsilon)$. ($D_1, ..., D_6$ cannot be determined by this matching.) Although it exists for all times the twelfth local approximation obviously cannot approximate $l(t, \varepsilon)$ for all times since it approaches constant values as $t_{12} \to +\infty$. Hence we expect to have a singularity at $t_{12} = +\infty$. One finds the asymptotic behaviour

$$X_{12}(t_{12}, \varepsilon) = -\varepsilon^{1/2}\sqrt{p} + \varepsilon + O(\varepsilon^3)$$

$$Y_{12}(t_{12}) = Y_{12}^* - \frac{1}{4pq} e^{-pt_{12}} \qquad , \qquad t_{12} \to +\infty.$$

$$Z_{12}(t_{12}, \varepsilon) = Z_{12}^* - \frac{1}{4q} e^{-pt_{12}} + O(\varepsilon^{1/2} e^{-pt_{12}} + \varepsilon^{2/3})$$

where $Y_{12}^* := 1/4pq$ and $Z_{12}^* := -1/4q$. Hence

$$A_{12} = B_{12} = C_{12} = 0.$$

*Step 13*

Subjecting (PS$_{12}$) to the shift scaling transformation

$$x_{12} = \varepsilon^a x_{13}$$
$$y_{12} = Y_{12}^* + \varepsilon^b y_{13}$$
$$z_{12} = Z_{12}^* + \varepsilon^c z_{13}$$
$$t_{12} = \varepsilon^d t_{13}$$

and multiplication by $\varepsilon^{-l}, \varepsilon^{-m}, \varepsilon^{-n}$ yields the convex polyhedron $P_{13}$. The selection rule (IV) requires

$$a, b, c \geqslant 0, \qquad d < 0.$$

There is exactly one vertex of $P_{13}$ adjacent to 0 having this property:

$$p_{13} = (\tfrac{1}{2}, 0, \tfrac{1}{2}, -\tfrac{1}{2}, \tfrac{1}{2}, \tfrac{1}{2}, \tfrac{1}{2})$$

with the corresponding system

$$\varepsilon^{7/2}\frac{dx_{13}}{dt_{13}} = -\frac{1}{4\sqrt{(p)}q} - \frac{x_{13}}{4pq} - \sqrt{(p)}y_{13} - x_{13}y_{13} + \varepsilon^{1/2}\left(\frac{1}{4pq} + y_{13}\right)$$

$$+ \varepsilon^2\left(\frac{\sqrt{p}}{2} + \frac{x_{13}}{2}\right) + \varepsilon^{5/2}\frac{1}{2} + \varepsilon^{9/2}(-pq - 2q\sqrt{(p)}x_{13} - qx_{13}^2)$$

$$(PS_{13}) \qquad \frac{dy_{13}}{dt_{13}} = -\frac{1}{4\sqrt{(p)}q} - \frac{x_{13}}{4pq} - \sqrt{(p)}y_{13} - x_{13}y_{13} + z_{13} + \varepsilon^{1/6}Z_9^*$$

$$+ \varepsilon^{1/2}\left(Z_{10}^* - \frac{1}{4pq} - y_{13}\right) + \varepsilon^2\left(\frac{3\sqrt{p}}{2} - \frac{x_{13}}{2}\right) - \varepsilon^{5/2}\frac{1}{2}$$

$$\varepsilon^{1/2}\frac{dz_{13}}{dt_{13}} = -pz_{13} - \varepsilon^{1/6}pZ_9^* - \varepsilon^{1/2}pZ_{10}^* + \varepsilon^2(-p^{3/2} + px_{13})$$

where

$$x_{12} = \varepsilon^{1/2}x_{13} \qquad\qquad x = \varepsilon^{-1/2}\sqrt{p} + \varepsilon^{-1/2}x_{13}$$

$$y_{12} = \frac{1}{4pq} + y_{13} \qquad\qquad y = \varepsilon^{-2}\frac{1}{4pq} + \frac{1}{2} + \varepsilon^{-2}y_{13}$$

$$z_{12} = -\frac{1}{4q} + \varepsilon^{1/2}z_{13} \qquad\qquad z = \varepsilon^{-7/3}Z_9^* + \varepsilon^{-2}Z_{10}^* + \varepsilon^{-1/2}2\sqrt{p} + \varepsilon^{-5/2}z_{13}$$

$$t_{12} = \varepsilon^{-1/2}t_{13} \qquad\qquad t = \mathbf{T}_{13}(t_{13}, \varepsilon) := \varepsilon^{-1/2}T_2^* + \varepsilon^{-1/6}T_3^* + T_4^*$$
$$+ \varepsilon^{1/3}T_5^* + \varepsilon T_6^* + T_8^* + \varepsilon^{2/3}T_9^*$$
$$+ \varepsilon T_{10}^* + \varepsilon^{-1/2}t_{13}$$

If we take the following (higher order) solutions of $(PAS_{13})$

$$X_{13}(t_{13}, \varepsilon) = -\sqrt{p} + \varepsilon^{1/2}1 + \varepsilon^{5/2}4pq$$
$$Y_{13}(t_{13}) = 0$$
$$Z_{13}(t_{13}, \varepsilon) = 0 - \varepsilon^{1/6}Z_9^* - \varepsilon^{1/2}Z_{10}^* - \varepsilon^2 2\sqrt{p} + \varepsilon^{5/2}1$$

the functions

$$\tilde{X}_{13}^r(t_{13}, \varepsilon) := 1 + \varepsilon^2 4pq$$

$$\tilde{Y}_{13}^r(t_{13}, \varepsilon) := \frac{1}{2} + \varepsilon^{-2}\frac{1}{4pq}$$

$$\tilde{Z}_{13}^r(t_{13}, \varepsilon) := \varepsilon^{-7/3}Z_9^* + \varepsilon^{-2}Z_{10}^* + \varepsilon^{-1/2}2\sqrt{p}$$
$$+ \varepsilon^{-5/2}Z_{13}(t_{13}, \varepsilon) = 1$$

match to leading order (in the sense of (2.11)) with the functions $\tilde{X}'_{12}(\varepsilon^{-1/2}t_{13}, \varepsilon)$, $\tilde{Y}'_{12}(\varepsilon^{-1/2}t_{13}, \varepsilon)$, $\tilde{Z}'_{12}(\varepsilon^{-1/2}t_{13}, \varepsilon)$.

Again by the same argument as in Step 12, the thirteenth local approximation may be expected to have a singularity at $t_{13} = +\infty$. We have $X^*_{13} = -\sqrt{p}$ and of course $A_{13} = B_{13} = C_{13} = 0$.

## Step 14

Subjecting $(\mathrm{PS}_{13})$ to the shift scaling transformation

$$x_{13} = X^*_{13} + \varepsilon^a x_{14}$$
$$y_{13} = \varepsilon^b y_{14}$$
$$z_{13} = \varepsilon^c z_{14}$$
$$t_{13} = \varepsilon^d t_{14}$$

and multiplication by $\varepsilon^{-l}, \varepsilon^{-m}, \varepsilon^{-n}$ yields the convex polyhedron $P_{14}$. The selection rule (IV) requires

$$a, b, c \geqslant 0, \qquad d < 0.$$

There is exactly one vertex of $P_{14}$ adjacent to 0 having this property:

$$p_{14} = (\tfrac{1}{6}, 0, \tfrac{1}{6}, -\tfrac{1}{6}, \tfrac{1}{6}, \tfrac{1}{6}, \tfrac{1}{6})$$

with the corresponding system

$$\varepsilon^{11/3}\frac{\mathrm{d}x_{14}}{\mathrm{d}t_{14}} = -\frac{x_{14}}{4pq} - x_{14}y_{14} + \varepsilon^{1/3}\left(\frac{1}{4pq} + y_{14}\right) + \varepsilon^2\frac{x_{14}}{2} + \varepsilon^{7/3}\frac{1}{2}$$
$$- \varepsilon^{14/3}px_{14}{}^2$$

$$(\mathrm{PS}_{14}) \qquad \frac{\mathrm{d}y_{14}}{\mathrm{d}t_{14}} = Z^*_9 - \frac{x_{14}}{4pq} - x_{14}y_{14} + z_{14} + \varepsilon^{1/3}\left(Z^*_{10} - \frac{1}{4pq} - y_{14}\right)$$

$$+ \varepsilon^{11/6}2\sqrt{p} - \varepsilon^2\frac{x_{14}}{2} - \varepsilon^{7/3}\frac{1}{2}$$

$$\varepsilon^{2/3}\frac{\mathrm{d}z_{14}}{\mathrm{d}t_{14}} = -pZ^*_9 - pz_{14} - \varepsilon^{1/3}pZ^*_{10} - \varepsilon^{11/6}2p^{3/2} + \varepsilon^2px_{14}$$

where

$$x_{13} = \sqrt{p} + \varepsilon^{1/6}x_{14} \qquad x = \varepsilon^{-1/3}x_{14}$$

$$y_{13} = y_{14} \qquad y = \varepsilon^{-2}\frac{1}{4pq} + \frac{1}{2} + \varepsilon^{-2}y_{14}$$

$$z_{13} = \varepsilon^{1/6}z_{14} \qquad z = \varepsilon^{-7/3}Z^*_9 + \varepsilon^{-2}Z^*_{10} + \varepsilon^{-1/2}2\sqrt{p} + \varepsilon^{-7/3}z_{14}$$

$$t_{13} = \varepsilon^{-1/6} t_{14} \qquad t = \mathbf{T}_{14}(t_{14}, \varepsilon) := \varepsilon^{-1/2} T_2^* + \varepsilon^{-1/6} T_3^* + T_4^*$$
$$+ \varepsilon^{1/3} T_5^* + \varepsilon T_6^* + T_8^*$$
$$+ \varepsilon^{2/3} T_9^* + \varepsilon T_{10}^* + \varepsilon^{-2/3} t_{14}$$

If we take the following higher order solutions of (PAS$_{14}$)

$$X_{14}(t_{14}, \varepsilon) = 0 + \varepsilon^{1/3} 1 + \varepsilon^{7/3} 4pq$$

$$Y_{14}(t_{14}, \varepsilon) = 0 - \varepsilon^{1/3} \frac{t_{14}}{2pq}$$

$$Z_{14}(t_{14}, \varepsilon) = -Z_9^* - \varepsilon^{1/3} Z_{10}^* - \varepsilon^{11/6} 2\sqrt{p} + \varepsilon^{7/3} 1$$

the functions

$$\tilde{X}_{14}^r(t_{14}, \varepsilon) := 1 + \varepsilon^2 4pq$$

$$\tilde{Y}_{14}^r(t_{14}, \varepsilon) := \varepsilon^{-2} \frac{1}{4pq} + \frac{1}{2} + \varepsilon^{-2} Y_{14}(t_{14}, \varepsilon)$$

$$\tilde{Z}_{14}^r(t_{14}, \varepsilon) := \varepsilon^{-7/3} Z_9^* + \varepsilon^{-2} Z_{10}^* + \varepsilon^{-1/2} 2\sqrt{p}$$
$$+ \varepsilon^{-7/3} Z_{14}(t_{14}, \varepsilon) = 1$$

match to leading order (in the sense of (2.11)) with the functions $\tilde{X}_{13}^l(\varepsilon^{-1/6} t_{14}, \varepsilon)$, $\tilde{Y}_{13}^l(\varepsilon^{-1/6} t_{14}, \varepsilon)$, $\tilde{Z}_{13}^l(\varepsilon^{-1/6} t_{14}, \varepsilon)$.

Again, the fourteenth local approximation may be expected to have a singularity at $t_{14} = +\infty$, and we have $Z_{14}^* = -Z_9^*$ and $A_{14} = B_{14} = C_{14} = 0$.

*Step 15*

Subjecting (PS$_{14}$) to the shift scaling transformation

$$x_{14} = \varepsilon^a x_{15}$$
$$y_{14} = \varepsilon^b y_{15}$$
$$z_{14} = Z_{14}^* + \varepsilon^c z_{15}$$
$$t_{14} = \varepsilon^d t_{15}$$

and multiplication by $\varepsilon^{-l}, \varepsilon^{-m}, \varepsilon^{-n}$ yields the convex polyhedron $P_{15}$. The selection rule (IV) requires

$$a, b, c \geqslant 0, \qquad d < 0.$$

There is exactly one vertex of $P_{15}$ adjacent to 0 having this property:

$$p_{15} = (\tfrac{1}{3}, 0, \tfrac{1}{3}, -\tfrac{1}{3}, \tfrac{1}{3}, \tfrac{1}{3}, \tfrac{1}{3})$$

with the corresponding system

$$\varepsilon^4 \frac{dx_{15}}{dt_{15}} = \frac{1}{4pq} - \frac{x_{15}}{4pq} + y_{15} - x_{15}y_{15} + \varepsilon^2\left(\frac{1}{2} + \frac{x_{15}}{2}\right) - \varepsilon^5 q x_{15}{}^2$$

(PS$_{15}$)
$$\frac{dy_{15}}{dt_{15}} = -\frac{1}{4pq} + Z_{10}^* - \frac{x_{15}}{4pq} - y_{15} - x_{15}y_{15} + z_{15} + \varepsilon^{3/2}2\sqrt{p}$$

$$+ \varepsilon^2\left(-\frac{1}{2} - \frac{x_{15}}{2}\right)$$

$$\varepsilon \frac{dz_{15}}{dt_{15}} = -pZ_{10}^* - pz_{15} - \varepsilon^{3/2}2p^{3/2} + \varepsilon^2 px_{15}$$

where

$$x_{14} = \varepsilon^{1/3}x_{15} \qquad\qquad x = x_{15}$$

$$y_{14} = y_{15} \qquad\qquad y = \varepsilon^{-2}\frac{1}{4pq} + \frac{1}{2} + \varepsilon^{-2}y_{15}$$

$$z_{14} = -Z_9^* + \varepsilon^{1/3}z_{15} \qquad z = \varepsilon^{-2}Z_{10}^* + \varepsilon^{-1/2}2\sqrt{p} + \varepsilon^{-2}z_{15}$$

$$t_{14} = \varepsilon^{-1/3}t_{15} \qquad t = \mathbf{T}_{15}(t_{15}, \varepsilon) := \varepsilon^{-1/2}T_2^* + \varepsilon^{-1/6}T_3^* + T_4^*$$

$$+ \varepsilon^{1/3}T_5^* + \varepsilon T_6^* + T_8^* + \varepsilon^{2/3}T_9^*$$

$$+ \varepsilon T_{10}^* + \varepsilon^{-1}t_{15}$$

(PAS$_{15}$) has the higher order solutions

$$\hat{X}_{15}(t_{15}, \varepsilon) = 1 + \varepsilon^2\frac{1}{C}e^{2t_{15}}$$

$$\hat{Y}_{15}(t_{15}, \varepsilon) = -\frac{1}{4pq} + Ce^{-2t_{15}} + \varepsilon C_1 e^{-2t_{15}} + \varepsilon^2\left(C_2 e^{-2t_{15}} - \frac{1}{2}\right)$$

$$Z_{15}(t_{15}, \varepsilon) = -Z_{10}^* - \varepsilon^{3/2}2\sqrt{p} + \varepsilon^2 1$$

with integration constants $C, C_1, C_2$.
The functions

$$\tilde{X}_{15}^l(t_{15}, \varepsilon) := \hat{X}_{15}(t_{15}, \varepsilon)$$

$$\tilde{Y}_{15}^l(t_{15}, \varepsilon) := \varepsilon^{-2}\frac{1}{4pq} + \frac{1}{2} + \varepsilon^{-2}\hat{Y}_{15}(y_{15}, \varepsilon)$$

$$\tilde{Z}_{15}^l(t_{15}, \varepsilon) := \varepsilon^{-2}Z_{10}^* + \varepsilon^{-1/2}2\sqrt{p} + \varepsilon^{-2}Z_{15}(t_{15}, \varepsilon) = 1$$

match to leading order (in the sense of (2.11)) with the functions $\tilde{X}_{14}^l(\varepsilon^{-1/3}t_{15}, \varepsilon)$, $\tilde{Y}_{14}^l(\varepsilon^{-1/3}t_{15}, \varepsilon), \tilde{Z}_{14}^l(\varepsilon^{-1/3}t_{15}, \varepsilon)$, if we choose $C = 1/4pq$. The integration constants $C_1$ and $C_2$ cannot be determined by this matching. Hence, we have found that the fifteenth local approximation (to order unity) to $l(t, \varepsilon)$ is of the form

$$\tilde{X}_{15}(t_{15}, \varepsilon) = 1 + \varepsilon^2 4pq \; e^{2t_{15}}$$

$$\tilde{Y}_{15}(t_{15}, \varepsilon) = \varepsilon^{-2} \frac{1}{4pq} \; e^{-2t_{15}} + \varepsilon^{-1} C_1 e^{-2t_{15}} + C_2 e^{-2t_{15}} + O(\varepsilon)$$

$$\tilde{Z}_{15}(t_{15}, \varepsilon) = 1 + O(\varepsilon), \qquad \varepsilon \to 0.$$

It has, up to a shift, the same time-scale as the first local approximation, and the two local approximations exhibit the same type of exponential behaviour. We therefore assume that we have approximated one period of the limit-cycle solution $l(t, \varepsilon)$ of (4.1) and try to verify this by matching (in a special way) the fifteenth with the first local approximation.

We know that (Tikhonov situation)

(4.11)    $(\tilde{X}_{15}(s, \varepsilon), \tilde{Y}_{15}(s, \varepsilon), \tilde{Z}_{15}(s, \varepsilon))$ approximates (to leading order) $l(\mathbf{T}^* + \varepsilon^{-1}s, \varepsilon)$ for $s$ in some finite interval $J_{15} := [\delta_0, \delta_1]$ with $0 < \delta_0 < \delta_1$, where $\mathbf{T}^*(\varepsilon) := \mathbf{T}_{10}(T_{10}^*, \varepsilon)$.

In Step 1 we have seen that

(4.12)    $(X_1(t_1), Y_1(t_1), Z_1(t_1))$ approximates $l(\varepsilon^{-1}t_1, \varepsilon)$ for $t_1$ in some interval $J_1 := [-\rho_1, -\rho_0]$ with $0 < \rho_0 < \rho_1$.

We now suppose that (compare Fig. 1)

(4.13)    there is $\hat{T}(\varepsilon) > 0$ such that $(X_1(s - \hat{T}), Y_1(s - \hat{T}), Z_1(s - \hat{T}))$ also approximates (to leading order) $l(\mathbf{T}^* + \varepsilon^{-1}s, \varepsilon)$ for $s$ in some finite interval $\hat{J}_{15} := [\hat{\delta}_0, \hat{\delta}_1]$ with $\hat{\delta}_0 > \delta_0$, $\hat{\delta}_1 < \delta_1$ and $\hat{T} > \hat{\delta}_1$.

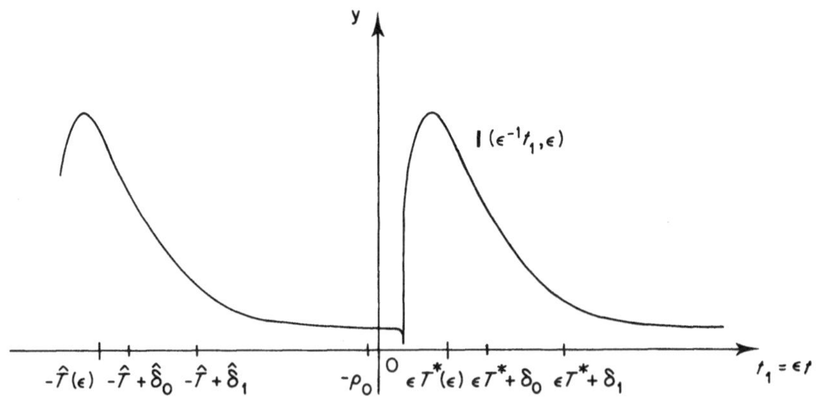


**Fig. 1.** Sketch to assumption (4.13)

Under this hypothesis the period $P(\varepsilon)$ of the limit-cycle solution has to satisfy

$$(4.14) \qquad P(\varepsilon) = \mathbf{T}^*(\varepsilon) + \varepsilon^{-1}\hat{T}(\varepsilon).$$

A necessary condition for (4.13) to hold is that

$$(4.15) \qquad \lim_{\varepsilon \to 0} [X_1(s - \hat{T}) - \tilde{X}_{15}(s, \varepsilon)] = 0 \qquad \text{for } s \in \hat{J}_{15},$$

and similar for the $y$- and $z$-components.

Therefore, in our formal approach we now use this matching condition (4.15) to verify the hypothesis (4.13), i.e. that we have indeed approximated one cycle of a periodic solution of (4.1). And this will also allow us to determine (to leading orders) the unknown quantity $\hat{T}(\varepsilon)$ in (4.14) and hence the leading order terms of the period $P(\varepsilon)$.

We have

$$Y_1(s - \hat{T}) = e^{2\hat{T}}e^{-2s}$$

$$\tilde{Y}_{15}(s, \varepsilon) = \varepsilon^{-2}\frac{1}{4pq}\,e^{-2s} + \varepsilon^{-1}C_1 e^{-2s} + C_2 e^{-2s} + O(\varepsilon), \qquad \varepsilon \to 0.$$

Hence, these two functions match for $s \in \hat{J}_{15}$ to an order $\zeta(\varepsilon)$ with $\varepsilon^{-1} \ll \zeta \ll \varepsilon^{-2}$, if we take

$$(4.16) \qquad \hat{T}(\varepsilon) = \log \varepsilon^{-1} - \tfrac{1}{2}\log(4pq).$$

Thus also the $x$- and $z$-components of the two local approximations match perfectly (to leading order).

A higher order solution of $(\text{PAS}_1)$ suggests that $\hat{T}(\varepsilon)$ is correct at least up to terms $O(\varepsilon \log \varepsilon^{-1})$.

Adding up all time-shifts in the fifteen steps we find that

$$(4.17) \qquad \mathbf{T}^*(\varepsilon) = \varepsilon^{-1/2}(-1/\sqrt{p}) + \varepsilon^{-1/6}2^{-1/3}p^{-5/6}\hat{i} + 2[\log(2) - 1]/p$$
$$+ \varepsilon^{1/3}2^{1/3}\hat{i} + \varepsilon^{2/3}(4/p)^{1/3}\hat{i} + \varepsilon \log \varepsilon^{-1} + \varepsilon(-2)$$

where $\hat{i} = 2.338...$ is the first zero of Airy's function $\text{Ai}(-t)$, and hence inserting (4.16) and (4.17) into (4.14) we get

$$(4.18) \qquad \bar{P}(\varepsilon) = \varepsilon^{-1}\log \varepsilon^{-1} + \varepsilon^{-1}(-\tfrac{1}{2}\log 4pq) + \varepsilon^{-1/2}(-1/\sqrt{p})$$
$$+ \varepsilon^{-1/6}2^{-1/3}p^{-5/6}\hat{i} + \text{h.o.t.} \qquad \text{as } \varepsilon \to 0.$$

The $^{-}$ indicates that we do not know how many terms in this expansion are correct since we have carried out the matching only to leading orders.

A comparison, given in Tables 2 and 3 below, of $\bar{P}(\varepsilon)$ with $\tilde{P}(\varepsilon)$ computed numerically by an algorithm for stiff systems suggests that the difference

$|\bar{P}(\varepsilon) - \tilde{P}(\varepsilon)|$ is $O(\varepsilon^{-1/3})$, which confirms the first three terms of our approximate period $\bar{P}(\varepsilon)$. Hence, we have

$$(4.19) \qquad P(\varepsilon) = \varepsilon^{-1} \log \varepsilon^{-1} + \varepsilon^{-1}(-\tfrac{1}{2} \log 4pq) + \varepsilon^{-1/2}(-1/\sqrt{p})$$
$$+ O(\varepsilon^{-1/3}) \qquad \text{as } \varepsilon \to 0.$$

**Table 2**   $p = 0.161,\ q = 3.864$

| $\varepsilon^{-1}$ | $\bar{P}(\varepsilon)$ | $\bar{P}(\varepsilon) - \tilde{P}(\varepsilon)$ |
|---|---|---|
| 77.27 | 296.332 162 8 | − 6.52 |
| 1 000 | 6 400.004 175 | − 10.37 |
| 10 000 | 87 335.408 53 | − 18.23 |
| 100 000 | 1 104 980.037 | − 33.42 |
| 1 000 000 | 13 357 280.17 | − 64.05 |

**Table 3**   $p = 3.864,\ q = 0.161$

| $\varepsilon^{-1}$ | $\bar{P}(\varepsilon)$ | $\bar{P}(\varepsilon) - \tilde{P}(\varepsilon)$ |
|---|---|---|
| 77.27 | 297.464 586 4 | − 10.31 |
| 1 000 | 6 437.747 272 | − 31.58 |
| 10 000 | 87 497.091 94 | − 79.36 |
| 100 000 | 1 105 553.456 | − 187.7 |
| 1 000 000 | 13 359 184.68 | − 427.0 |

## 5   ESTIMATES

As already stressed several times before if one wants to verify a formal approximation to a singularly perturbed initial value problem obtained by the algorithmic approach developed in Sections 2 and 3 one needs best possible estimates for the domains of validity of the local approximations. We therefore state such results for typical situations arising.

The theorems given in Sections 5.1, 5.2 (except for Theorem 5.3) are, in principle, well known standard results (see e.g. [29]), but they cannot be found in the literature in this sharp form needed for the purposes of this paper. For this reason and since it seems difficult to refer to neat proofs even for the standard results we present complete proofs. We think that these proofs also provide insight and understanding and should be generalizable to analogous situations which might arise in specific applications. This in particular also seems to be true for the proof of Theorem 5.7 given in [25], [26].

For simplicity we state the results for the autonomous case. The analogous theorems in Sections 5.1, 5.2 for the non-autonomous case are obtained from the autonomous ones by adding the trivial equation $\dot{t} = 1$. Moreover, we again restrict ourselves to approximations of order unity. It is a matter of routine to get correspondingly sharper error estimates for higher order approximations.

## 5.1  The regular perturbation case

In constructing a chain of local approximations to the solution of a singularly perturbed IVP by means of the systematic approach derived in Sections 2 and 3 some of the approximating systems may be of full order (compare the application in Section 4). Hence in that case one has, locally, a regular perturbation problem, and the domain of validity of such a local approximation is estimated by regular perturbation results. For completeness, we state the standard regular perturbation theorem and also give the simple proof since this will enable us to deduce directly a trivial extension which is important, however, in the framework of singular perturbations.

We consider the autonomous regular perturbation problem

$$(5.1) \qquad \frac{\mathrm{d}z}{\mathrm{d}t} = f(z, \varepsilon), \qquad z(0, \varepsilon) = z^0(\varepsilon)$$

where $f$ is defined for $z$ in some domain $D \subset \mathbb{R}^N$ and $\varepsilon \in (-\varepsilon_0, \varepsilon_0), 0 < \varepsilon_0 < 1$, $z^0(\varepsilon)$ in $D$ for all $\varepsilon$ considered, all functions being as regular as needed; and the problem

$$(5.2) \qquad \frac{\mathrm{d}z}{\mathrm{d}t} = f(z, 0), \qquad z(0) = Z^0 \quad \text{for } Z^0 \in D.$$

We make the *assumptions*

(A) There is $M_0 > 0$ and an order function $\varphi(\varepsilon)$, with $\varphi = o(1)$ as $\varepsilon \to 0$, such that

$$| z^0(\varepsilon) - Z^0 | < M_0\varphi(\varepsilon) \qquad \text{for } \varepsilon \in (0, \varepsilon_0).$$

(B) The solution $Z(t)$ of the reduced problem (5.2) exists for $t \in [0, T], T > 0$, with respect to $D$.

Then the following result holds.

THEOREM 5.1   There are positive constants $C$ and $\varepsilon_1 \leqslant \varepsilon_0$ such that the solution $z(t, \varepsilon)$ of the IVP (5.1) exists at least for $t \in [0, T]$ and satisfies

$$| z(t, \varepsilon) - Z(t) | < C\psi(\varepsilon) \qquad \text{for } t \in [0, T] \text{ and } \varepsilon \in (0, \varepsilon_1)$$

where

$$\psi(\varepsilon) := \begin{cases} \varepsilon, & \text{if } \varphi \ll \approx \varepsilon \\ \varphi(\varepsilon), & \text{if } \varphi \gg \varepsilon \end{cases}.$$

$\quad\square$

*Proof*   Let $G$ be a bounded subdomain (independent of $\varepsilon$) of $D$ with $\bar{G} \subset D$ such that $z(t) \in G$ for $t \in [0, T]$, and consider a domain $\Omega_r := \{ w \mid |w| < r \} \subset D$ where $r$ is such that $\{ z + w \mid z \in \bar{G}, w \in \bar{\Omega}_r \} \subset D$. We consider the IVP

$$\dot{w} = f_z(Z(t), 0)w + \hat{f}(t, w, \varepsilon), \qquad w(0, \varepsilon) = w^0(\varepsilon) := z^0(\varepsilon) - Z^0,$$

where $\hat{f}(t, w, \varepsilon) := f(Z(t) + w, \varepsilon) - f(Z(t), 0) - f_z(Z(t), 0)w$; and we suppose that for $\varepsilon$ small enough its solution $w(t, \varepsilon)$ exists for $t \in [0, m_+)$ with respect to the set $\{ w \in \Omega_r \} \times \{ t \in [0, T] \}$.

There are positive constants $M$, $\hat{M}_1$ and $\hat{M}_2$ such that on this set the following estimates hold:

$$|f_z(Z(t), 0)| < M$$
$$|\hat{f}(t, w, \varepsilon)| < \hat{M}_1 w + \hat{M}_2 \varepsilon.$$

Therefore, from the integral equation formulation for $w(t, \varepsilon)$ we derive

$$|w(t, \varepsilon)| \leqslant M_0 \varphi(\varepsilon) + (M + \hat{M}_1) \int_0^t |w(s, \varepsilon)| \, ds + \hat{M}_2 \varepsilon t \qquad \text{for } t \in [0, m_+),$$

and by means of the Gronwall Lemma

$$|w(t, \varepsilon)| \leqslant (M_0 \varphi(\varepsilon) + \hat{M}_2 \varepsilon t) e^{(M + \hat{M}_1)t} \qquad \textit{for } t \in [0, m_+).$$

For $\varepsilon$ small enough this implies $|w(t, \varepsilon)| < r/2$ for $t \in [0, m_+)$.

Hence, applying the 'global existence theorem' for ODE's and taking into account that $z(t, \varepsilon) := Z(t) + w(t, \varepsilon)$ solves the problem (5.1) completes the proof of Theorem 5.1. $\square$


If instead of (B) the stronger assumption


$(\bar{B})$ The solution $Z(t)$ of the reduced problem (5.2) exists for all $t \geqslant 0$ with respect to some bounded domain $G$ with $\bar{G} \subset D$.


holds, the above proof also yields the following modification of Theorem 5.1.


THEOREM 5.2   Let the assumptions (A) and $(\bar{B})$ be satisfied. Then there are positive constants $\bar{C}, \bar{k}$ and for every $\delta \in (0, 1)$ there is $\bar{\varepsilon}_1 \leqslant \varepsilon_0$ such that the solution $z(t)$ of (5.1) exists at least for $t \in [0, (\delta/\bar{k}) \log \psi^{-1}]$ and

$$|z(t, \varepsilon) - Z(t)| < \bar{C} \psi^{1-\delta} \log \psi^{-1} \qquad \text{for } t \in \left[0, \frac{\delta}{k} \log \psi^{-1}\right] \text{ and } \varepsilon \in (0, \bar{\varepsilon}_1),$$

$\psi(\varepsilon)$ as in Theorem 5.1. $\square$

The proof of this result is identical to the proof of Theorem 5.1 if the solution $w(t, \varepsilon)$ is assumed to exist for $t \in [0, m_+)$ with respect to the set

$$\{|w| < r\} \times \left\{ t \in \left[0, \frac{\delta}{M + \hat{M}_1} \log \psi^{-1}\right] \right\}.$$

(This in particular shows that Theorem 5.2 actually holds if $Z(t)$ can be verified to exist with respect to $G$ on a $t$-interval $[0, L \log \psi^{-1}]$ with $L \geqslant 1/(M + \hat{M}_1)$.)

Theorem 5.2 states that under the assumptions (A), $(\bar{B})$ the reduced solution $Z(t)$ is an approximation to $z(t, \varepsilon)$ on an expanding (with $\varepsilon \to 0$) $t$-interval.

The reader should notice that in this situation (where of course on every finite $t$-interval $[0, T]$ the estimate of Theorem 5.1 holds) the larger a portion of the domain of validity of $Z(t)$ is considered the weaker the error estimate on this extended interval. This is a typical phenomenon and will also be encountered in the results of the subsequent section.

## 5.2 The Tikhonov case

We consider the singularly perturbed autonomous system

$$\frac{\mathrm{d}x}{\mathrm{d}t} = f(x, y, \varepsilon)$$

(5.3)

$$\varepsilon \frac{\mathrm{d}y}{\mathrm{d}t} = g(x, y, \varepsilon)$$

where

$$f : D \to \mathbb{R}^m$$
$$g : D \to \mathbb{R}^n,$$

$D = D_1 \times D_2 \times (-\varepsilon_0, \varepsilon_0), \varepsilon_0 \in (0, 1)$, being some domain in $\mathbb{R}^{m+n+1}$, together with the initial conditions

(5.4)
$$x(0, \varepsilon) = x^0(\varepsilon)$$
$$y(0, \varepsilon) = y^0(\varepsilon),$$

$(x^0(\varepsilon), y^0(\varepsilon)) \in D_1 \times D_2$ for $\varepsilon \in (-\varepsilon_0, \varepsilon_0)$.

All functions are supposed to be as regular as needed. The Tikhonov situation is characterized by the *assumptions*:

(1) There is a smooth function $p: D_1 \to D_2$ satisfying

$$g(x, p(x), 0) = 0 \qquad \text{for } x \in D_1.$$

(2) There is a positive constant $b$ such that all eigenvalues of the Jacobian matrix $B(x) := g_y(x, p(x), 0)$ have real parts smaller than $-b$ for all $x \in D_1$.

Assumption (2) implies that for every $x^* \in D_1$ $p(x^*)$ is an asymptotically stable equilibrium solution of the system

$$(5.5) \qquad \frac{du}{d\tau} = g(x^*, u, 0).$$

(3) $y^0(0)$ lies in the domain of attraction of the asymptotically stable equilibrium solution $p(x^0(0))$ of Eq. (5.5).

(4) The solution $X(t)$ of the reduced problem

$$(5.6) \qquad \frac{dx}{dt} = f(x, p(x), 0), \qquad x(0) = x^0(0)$$

exists for $t \in [0, T]$, $T > 0$, with respect to $D_1$. $\square$

Under these assumptions the assertion of Theorem 1.1 holds, i.e. for $\varepsilon$ small enough the solution of the IVP (5.3), (5.4) approaches the solution of the reduced problem ($\varepsilon = 0$) in a very short time and stays $\varepsilon$-close to it. We now will prove this result and a refinement of it in the sense that the estimate for the domain of validity of the reduced solution can be improved to an $\varepsilon$-dependent interval.

There is another extension of the original Tikhonov result. Under the above assumptions one cannot only show that a solution of the full problem is $\varepsilon$-close to the corresponding solution of the reduced problem, but that $\varepsilon$-close to the reduced manifold $\{(x, y) \mid y = p(x), x \in D_1\}$ there exists an invariant manifold to the system (5.3) which is smooth and strongly attractive.

We first state this result which is proved in [24] since it will enable us to give a nice simple proof of the Tikhonov theorem.

THEOREM 5.3    Let $G = G_1 \times G_2 \times (-\varepsilon_0, \varepsilon_0)$ be a bounded domain in $\mathbb{R}^{m+n+1}$, $G_1$ with $C^{\nu+1}$-boundary, and $f \in C^{\nu+1}(G)$, $g \in C^{\nu+3}(G)$, $\nu \geq 1$, and let the assumptions (1) and (2) be satisfied with respect to $G$. Then there is a (arbitrarily close) subdomain $G_1' \subset G_1$, a positive constant $\varepsilon_0' \leq \varepsilon_0$ and a vector function $s(x, \varepsilon)$: $G_1' \times (0, \varepsilon_0') \to G_2, C^\nu$, with the following properties:

(a) The set $M := \{(x, y) \mid y = s(x, \varepsilon), x \in G_1'\}$ is invariant under the system (5.3), i.e. if $(x^0, y^0) \in M$ then also $(x(t), y(t)) \in M$ for all $t$ such that $x(t) \in G_1'$, $(x(t), y(t))$ being the solution of (5.3) with $(x(0), y(0)) = (x^0, y^0)$.
(b) $s(x, \varepsilon) = p(x) + O(\varepsilon)$, $\varepsilon \in (0, \varepsilon_0')$, uniformly for all $x \in G_1'$.
(c) There are positive constants $\chi, \gamma, \lambda$ such that for $\varepsilon(0, \varepsilon_0')$ every solution $(x(t), y(t))$ of (5.3) with $|y(0) - s(x(0), \varepsilon)| \leqslant \gamma$ satisfies

$$|y(t) - s(x(t), \varepsilon)| \leqslant \chi |y(0) - s(x(0), \varepsilon)| e^{-\lambda t / \varepsilon}$$

for all $t \geqslant 0$ such that $x(t) \in G_1'$.
(d) There is a positive constant $\zeta$ such that for $\varepsilon \in (0, \varepsilon_0')$ every solution $(x(t), y(t))$ of (5.3) satisfying $x(t) \in G_1'$ and $|y(t) - p(x(t))| < \zeta$ for all $t \in \mathbb{R}$ lies in $M$, i.e. $y(t) = s(x(t), \varepsilon)$ for all $t$. $\square$

In [24] we required $G_1$ to be star-shaped. The extension of the right-hand sides of (5.3) and of the matrix $B(x)$ to all $x \in \mathbb{R}^m$ needed for the proof of Theorem 5.3 given in [24], however, can also be achieved without this technical assumption. Moreover, the constant $\lambda$ in Assertion (c) is smaller than $b$ but can be chosen arbitrarily close to $b$ in the proof of Theorem 5.3.

We now state and prove the Tikhonov result in a sharp form as needed for the purposes of our algorithmic approach to singularly perturbed IVP's which was developed in Sections 2 and 3.

THEOREM 5.4   Let the assumptions (1), (2), (3), (4) be satisfied, and let $X(t)$ be the solution of the reduced problem (5.6) which exists for $t \in [0, T]$ and $Y(t) := p(X(t))$. Then there are positive constants $C, \tau^*$ and $\varepsilon^* \leqslant \varepsilon_0$ such that the following assertions hold for $\varepsilon \in (0, \varepsilon^*)$:
The solution $(x(t, \varepsilon), y(t, \varepsilon))$ of the IVP (5.3), (5.4) exists at least for $t \in [0, T]$ and satisfies

$$|x(t, \varepsilon) - X(t)| < C\varepsilon \qquad \text{for } t \in [0, T]$$

and for every $\rho \in (0, 1]$

$$|y(t, \varepsilon) - Y(t)| < C\varepsilon^\rho \qquad \text{for } t \in \left[\varepsilon\left(\tau^* + \rho \frac{2}{b} \log \varepsilon^{-1}\right), T\right].$$

$\square$

*Proof* (i) Let $G_1$ be a bounded subdomain of $D_1$ with a smooth boundary and with $\bar{G}_1 \subset D_1$ such that the solution $X(t)$ of (5.6) is in $G_1$ for all $t \in [0, T]$

and $x^0(\varepsilon) \in G_1$ for $\varepsilon \in (-\varepsilon_0, \varepsilon_0)$. We consider the IVP

$$\frac{dx}{d\tau} = \varepsilon f(x, y, \varepsilon), \qquad x(0, \varepsilon) = x^0(\varepsilon)$$

(5.7)

$$\frac{dy}{d\tau} = g(x, y, \varepsilon), \qquad y(0, \varepsilon) = y^0(\varepsilon)$$

with respect to the domain $G_1 \times D_2 \times (-\varepsilon_0, \varepsilon_0)$.

Note that if $((\tilde{x}(\tau), \tilde{y}(\tau))$ (to save writing we suppress the $\varepsilon$) is its solution for $\tau \in [0, L], \varepsilon \in (0, \tilde{\varepsilon})$, then $((\tilde{x}(t/\varepsilon), \tilde{y}(t/\varepsilon))$ solves the IVP (5.3), (5.4) for $t \in [0, \varepsilon L], \varepsilon \in (0, \tilde{\varepsilon})$.

Assumption (3) implies that the solution $\tilde{Y}(\tau)$ of the reduced problem

(5.8)
$$\frac{dy}{d\tau} = g(X^0, y, 0), \qquad y(0) = Y^0$$

where $X^0 := x^0(0)$, $Y^0 := y^0(0)$, exists in $D_2$ for all $\tau \geq 0$.

Let $G_2$ be a bounded subdomain of $D_2$ with $\bar{G}_2 \subset D_2$ such that $p(\bar{G}_1) \subset G_2$ and $\tilde{Y}(\tau) \in G_2$ for all $\tau \geq 0$, $y^0(\varepsilon) \in G_2$ for $\varepsilon \in (-\varepsilon_0, \varepsilon_0)$, and let $r > 0$ be such that $p(\bar{G}_1)$ together with its $r$-neighbourhood lies in $G_2$ ($r$ will be specified more precisely later).

Due to Assumption (3) there is $\tau^*(r) > 0$ such that

(5.9)
$$|\tilde{Y}(\tau) - p(X^0)| < \frac{r}{6} \qquad \text{for } \tau \geq \tau^*.$$

According to Theorem 5.1 there are positive constants $\tilde{C}(\tau^*)$ and $\varepsilon_1(\tau^*) \leq \varepsilon_0$ such that the solution $(\tilde{x}(\tau), \tilde{y}(\tau))$ of (5.7) exists for $\tau \in [0, \tau^*]$ and

(5.10)
$$\begin{aligned} |\tilde{x}(\tau) - X^0| &< \tilde{C}\varepsilon \\ |\tilde{y}(\tau) - \tilde{Y}(\tau)| &< \tilde{C}\varepsilon \end{aligned} \qquad \text{for } \tau \in [0, \tau^*], \varepsilon \in (0, \varepsilon_1).$$

This implies that for $\varepsilon$ small enough

(5.11)
$$|\tilde{y}(\tau^*) - \tilde{Y}(\tau^*)| < \frac{r}{6}$$

Moreover, if $L_p$ denotes the Lipschitz constant of $p(x)$ on $G_1$, we have

(5.12)
$$|p(\tilde{x}(\tau^*)) - p(X^0)| \leq L_p |(\tilde{x}(\tau^*) - (X^0)| < L_p \tilde{C}\varepsilon < \frac{r}{6}$$

for $\varepsilon$ small enough.

Combining (5.9), (5.11) and (5.12) we have shown that there are positive constants $\tau^*(r) > 0$ and $\varepsilon_2(r) \leq \varepsilon_0$ such that

(5.13)
$$|\tilde{y}(\tau^*) - p(\tilde{x}(\tau^*))| < \frac{r}{2} \qquad \text{for } \varepsilon \in (0, \varepsilon_2).$$

This means that the solution $(x(t), y(t))$ of the IVP (5.3), (5.4) comes arbitrarily close to the reduced manifold in a time $t$ which is $O(\varepsilon)$.

(ii) Under the hypotheses made Eq. (5.3) satisfies Theorem 5.3 in a subdomain $G_1' \subset G_1$ such that $X(t) \in G_1'$ for $t \in [0, T]$. Hence, for $r$ small enough the estimate (5.13) together with Assertion (b) of Theorem 5.3 implies that Assertion (c) applies.

Hence, there are positive constants $C^*, \tau^*$ and $\varepsilon_3 \leqslant \varepsilon_2$ such that for $\varepsilon \in (0, \varepsilon_3)$ the solution $(x(t), y(t))$ of (5.3), (5.4) exists and satisfies

$$(5.14) \qquad |y(t) - s(x(t), \varepsilon)| < C^* r e^{-b(t - \varepsilon\tau^*)/2\varepsilon} \qquad \text{for } t \in [\varepsilon\tau^*, t_+),$$

where $t_+$ denotes the right endpoint of the maximal interval of existence of $(x(t), y(t))$ with respect to the domain $G_1' \times G_2$.

(iii) The solution $X(t)$ of the reduced problem (5.6) satisfies the integral equation

$$(5.15) \qquad X(t) = X^0 + \int_0^t f(X(s), p(X(s)), 0) \, ds \qquad \text{for } t \in [0, T]$$

and the $x$-component of the solution $(x(t), y(t))$ of the full problem (5.3), (5.4) satisfies the integral equation

$$(5.16) \qquad x(t) = x^0(\varepsilon) + \int_0^t f(x(s), y(s), \varepsilon) \, ds \qquad \text{for } t \in [0, t_+), \varepsilon \in (0, \varepsilon_3).$$

Hence, there is $M_0^*(\tau^*) > 0$ such that

$$(5.17) \qquad |x(t) - X(t)| < M_0^* \varepsilon \qquad \text{for } t \in [0, \varepsilon\tau^*], \varepsilon \in (0, \varepsilon_3).$$

Now choose $\mu > 0$ such that the $\mu$-neighbourhood $\Omega_\mu$ of the orbit of $(X(t), p(X(t)), t \in [0, T]$, is in $G_1' \times G_2$, and let $r$ finally be also so small that $r < \mu$ and $C^* r < \mu/4$. Then, due to the estimates (5.13) and (5.17) there is $\varepsilon_4 \leqslant \varepsilon_3$ such that $(x(\varepsilon\tau^*), y(\varepsilon\tau^*)) \in \Omega_\mu$. We suppose that $(x(t), y(t))$ exists with respect to $\Omega_\mu$ for $t \in [\varepsilon\tau^*, m_+)$ (of course $m_+ \leqslant t_+$).

From the integral equations (5.15), (5.16) using (5.17) we derive that there are positive constants $M_1$ and $L$ such that

$$(5.18) \qquad |x(t) - X(t)| < M_0^* \varepsilon + M_1 \varepsilon T$$

$$+ L \int_{\varepsilon\tau^*}^t (|x(s) - X(s)| + |y(s) - p(X(s))|) \, ds$$

for $t \in [\varepsilon\tau^*, M_+)$, where $M_+ := \min(T, m_+)$, and for $\varepsilon \in (0, \varepsilon_4)$.

The second term under the integral satisfies

$$(5.19) \quad |y(t) - p(X(t))| \leqslant |y(t) - s(x(t), \varepsilon)| + |s(x(t), \varepsilon) - s(X(t), \varepsilon)|$$
$$+ |s(X(t), \varepsilon) - p(X(t))|$$

$$< \frac{\mu}{4} e^{-b(t - \varepsilon \tau^*)/2\varepsilon} + L_s |x(t) - X(t)| + K\varepsilon$$

$$\text{for } t \in [\varepsilon \tau^*, M_+),$$

where we have used the estimate (5.14) and Assertion (b) of Theorem 5.3; $L_s$ denotes the Lipschitz constant of $s(x, \varepsilon)$.

Hence, combining (5.18) and (5.19) we obtain

$$|x(t) - X(t)| < \bar{M}_0 \varepsilon + L^* \int_{\varepsilon \tau^*}^{t} |x(s) - X(s)| \, ds + L \frac{\mu}{4} e^{b\tau^*/2} \int_{\varepsilon \tau^*}^{t} e^{-bs/2\varepsilon} \, ds$$
$$(5.20)$$
$$< \hat{M}_0 \varepsilon + L^* \int_{\varepsilon \tau^*}^{t} |x(s) - X(s)| \, ds \quad \text{for } t \in [\varepsilon \tau^*, M_+)$$

where $\bar{M}_0 := M_0^* + M_1 T + LKT$, $\hat{M}_0 := \bar{M}_0 + L\mu/2b$ and $L^* := L(1 + L_s)$; and by means of the Gronwall lemma

$$(5.21) \quad |x(t) - X(t)| < M^* \varepsilon \quad \text{for } t \in [\varepsilon \tau^*, M_+), \varepsilon \in (0, \varepsilon_4),$$

with $M^* := \hat{M}_0 e^{L^*(T - \varepsilon \tau^*)}$.

Thus, we have for the $x$-component

$$(5.22) \quad |x(t) - X(t)| < M\varepsilon \quad \text{for } t \in [0, M_+), \varepsilon \in (0, \varepsilon_4),$$

$M := \max(M_0^*, M^*)$,

On the other hand, from (5.19) using (5.22) we conclude that there is a positive constant $\bar{C}$ such that for every $\beta \in (0, 2/b]$

$$(5.23) \quad |y(t) - p(X(t))| < \bar{C} \varepsilon^{b\beta/2}$$
$$\text{for } t \in [\varepsilon(\tau^* + \beta \log \varepsilon^{-1}), M_+), \varepsilon \in (0, \varepsilon_4).$$

Using (5.19), (5.21) we have also shown that for $\varepsilon$ small enough

$$|x(t) - (X(t))| < \frac{\mu}{2}$$
$$\text{for } t \in [\varepsilon \tau^*, M_+);$$
$$|y(t) - p(X(t))| < \frac{\mu}{2}$$

and hence the 'global existence theorem' completes the proof of Theorem 5.4.

$$\square$$

The following (weaker) form of the Tikhonov result is a trivial corollary of Theorem 5.4.

THEOREM 5.4a   Let the hypotheses (1), (2), (3), (4) be satisfied. Then there is a positive constant $C$ and for every $\delta > 0$ there is $\varepsilon_\delta \leqslant \varepsilon_0$ such that the solution $(x(t, \varepsilon), y(t, \varepsilon))$ of the IVP (5.3), (5.4) exists for $t \in [0, T]$ and satisfies

$$|x(t, \varepsilon) - X(t)| < C\varepsilon \qquad \text{for } t \in [0, T]$$
$$|y(t, \varepsilon) - Y(t)| < C\varepsilon \qquad \text{for } t \in [\delta, T],\ \varepsilon \in (0, \varepsilon_\delta),$$

where $X(t)$ is the solution of the reduced problem (5.6) and $Y(t) := p(X(t))$.
$\square$

This result includes of course the assertion of Theorem 1.1. The original Tikhonov result (cf. [29], [35]) only states that the solution of the unperturbed problem approaches the solution of the perturbed problem in the limit as $\varepsilon \to 0$, although for a slightly weaker stability condition.

The sharp estimates for the domains of validity of the reduced solutions as derived in Theorems 5.2 and 5.4 now enable us to prove that in the Tikhonov case the 'initial approximation' ($\varepsilon = 0$ in Eq. (5.7)) and the 'standard approximation' ($\varepsilon = 0$ in Eq. (5.3)) have an overlap domain. We state this application of the above results as

COROLLARY 5.5   Let the assumptions (1), (2), (3), (4) be satisfied and let $(x(t, \varepsilon), (y(t, \varepsilon))$ be the solution of the IVP (5.3), (5.4), $\tilde{Y}(\tau)$ be the solution of the problem (5.8) and $X(t)$ be the solution of the problem (5.6), $Y(t) := p(X(t))$. Then there is $M > 0$ and for every $\delta \in (0, 1)$ there are positive constants $\bar{\varepsilon}, \rho \leqslant 1, \mu_1$ and $\mu_2$, with $\mu_2 < \mu_1$, such that

$$
\begin{aligned}
&|x(t, \varepsilon) - X(t)| < M\varepsilon && \text{for } t \in [0, T] \\
&|y(t, \varepsilon) - \tilde{Y}(t/\varepsilon)| < M\varepsilon^{1-\delta} \log \varepsilon^{-1} && \text{for } t \in [0, \varepsilon\mu_1 \log \varepsilon^{-1}] \\
&|y(t, \varepsilon) - Y(t)| < M\varepsilon^\rho && \text{for } t \in [\varepsilon\mu_2 \log \varepsilon^{-1}, T] \\
&|y(t, \varepsilon) - Y(t)| < M\varepsilon && \text{for } t \in [\delta, T]
\end{aligned}
\qquad \varepsilon \in (0, \bar{\varepsilon}).
$$

$\square$

*Proof*   Let $\delta \in (0, 1)$ be fixed. Theorem 5.4a implies that there is $M_1 > 0$ and $\varepsilon_1(\delta)$ such that

$$|x(t, \varepsilon) - X(t)| < M_1\varepsilon \qquad \text{for } t \in [0, T]$$
$$|y(t, \varepsilon) - Y(t)| < M_1\varepsilon \qquad \text{for } t \in [\delta, T],\ \varepsilon \in (0, \varepsilon_1).$$

$\tilde{Y}(\tau)$ exists for all $\tau \geqslant 0$ in a bounded subdomain of $D_2$ (compare the proof of Theorem 5.4). Hence, from Theorem 5.2 we conclude that there are positive constants $M_2, k$ and $\varepsilon_2(\delta)$ such that

$$
\begin{aligned}
&|x(t, \varepsilon) - X^0| < M_2\varepsilon \\
&|y(t, \varepsilon) - \tilde{Y}(t/\varepsilon)| < M_2\varepsilon
\end{aligned}
\qquad \text{for } t \in \left[0, \varepsilon \frac{\delta}{k} \log \varepsilon^{-1}\right],\ \varepsilon \in (0, \varepsilon_2).
$$

Finally, using Theorem 5.4 we obtain that there are positive constants $M_3, \tau^*$ and $\varepsilon_3$ such that for every $\rho \in (0, 1]$

$$|y(t, \varepsilon) - Y(t)| < M_3 \varepsilon^\rho \qquad \text{for } t \in \left[\varepsilon\left(\tau^* + \rho \, \frac{2}{b} \log \, \varepsilon^{-1}\right), T\right], \varepsilon \in (0, \varepsilon_3).$$

If we take $\rho \leqslant \delta b/4k$ and $\hat{\varepsilon} \leqslant \min(\varepsilon_2, \varepsilon_3)$ such that $\tau^*/(\log \hat{\varepsilon}^{-1}) < \delta/2k$ then

$$\varepsilon \, \frac{\delta}{k} \log \, \varepsilon^{-1} > \varepsilon\left(\tau^* + \rho \, \frac{2}{b} \log \, \varepsilon^{-1}\right) \qquad \text{for all } \varepsilon \in (0, \hat{\varepsilon}).$$

The proof of Corollary 5.5 is completed by choosing $\bar{\varepsilon} := \min(\varepsilon_1, \hat{\varepsilon})$ and $M := \max(M_1, M_2, M_3)$. $\square$

So far the rigorous results of this section refer to systems of the form (5.3). On the other hand, our algorithmic approach developed in Sections 2 and 3 was designed for more general singularly perturbed systems of the form (2.1), and we have seen that even if the original system is of the form (5.3) the principal systems obtained from the original system by shift scaling transformations might be of a more general type (compare Section 4). We therefore want to state and prove a Tikhonov type result for the system

(5.24)

$$\frac{\mathrm{d}x}{\mathrm{d}t} = f(x, y, \varepsilon)$$

$$\Phi(\varepsilon) \, \frac{\mathrm{d}y}{\mathrm{d}t} = g(x, y, \varepsilon)$$

where the right-hand sides are as in Eq. (5.3) and the elements of the diagonal matrix $\Phi(\varepsilon)$ are order functions satisfying

$$\varphi_1(\varepsilon) = \varepsilon$$

and either

(5.25) $$\varphi_{j+1}(\varepsilon) = \varphi_j(\varepsilon),$$

or $$j = 1, \ldots, n - 1.$$

$$\frac{\varphi_{j+1}(\varepsilon)}{\varphi_j(\varepsilon)} \to 0 \qquad \text{as } \varepsilon \to 0,$$

*Remark*   A general system (2.1) without loss of generality can be assumed to be of this form. Since if not, one may introduce the new small parameter $\mu$ be means of $\mu := \psi_k(\varepsilon)$, where $\psi_k$ is the order function of lowest order. What is needed is that one can solve this equation for $\mu$, i.e. $\varepsilon$ can be written as $\varepsilon = \xi(\mu)$ with an order function $\xi$ satisfying $\xi \to 0$ as $\mu \to 0$; a requirement that is satisfied if the order functions are powers of $\varepsilon$.

Of course, the right-hand sides of the system (2.1) expressed in $\mu$ might then not be regular in $\mu$. But they are of the form $h^0(x, y) + \xi(\mu)\hat{h}(x, y, \mu)$ defined for $\mu \in [0, \mu_0]$; analogously the initial conditions (2.2). As can easily be seen from the proof, the result below holds for this case, but with error estimates $\xi(\mu)$ and $\max(\mu^\rho, \xi(\mu))$ instead of $\mu$ and $\mu^\rho$ if $\xi \gg \mu$ as $\mu \to 0$. The same is then of course true for Theorems 5.4 and 5.4a. The proof of the invariant manifold result used in the elegant proof given there, makes use of that regularity assumption, however. $\square$

For simplicity only and for convenience we formulate the result for the special case where $\varphi_{j+1}(\varepsilon) = \varphi(\varepsilon)$, $j = 1, \ldots, n-1$, and $\varphi(\varepsilon) \ll \varepsilon$ as $\varepsilon \to 0$. Hence, we consider the system

$$\frac{dx}{dt} = f(x, y, \varepsilon)$$

(5.26)
$$\varepsilon \frac{dy_1}{dt} = g_1(x, y, \varepsilon)$$

$$\varphi(\varepsilon) \frac{dy_2}{dt} = g_2(x, y, \varepsilon)$$

where

$$f: D \to \mathbb{R}^m$$
$$g_1: D \to \mathbb{R}^{n_1}$$
$$g_2: D \to \mathbb{R}^{n_2}$$

$D = D_1 \times D_{21} \times D_{22} \times (-\varepsilon_0, \varepsilon_0)$, $\varepsilon_0 \in (0, 1)$, being some domain in $\mathbb{R}^{m + n_1 + n_2 + 1}$, together with the initial conditions

(5.27)
$$x(0, \varepsilon) = x^0(\varepsilon), \quad y_1(0, \varepsilon) = y_1^0(\varepsilon), \quad y_2(0, \varepsilon) = y_2^0(\varepsilon)$$

$(x^0(\varepsilon), y_1^0(\varepsilon), y_2^0(\varepsilon)) \in D_1 \times D_{21} \times D_{22}$ for $\varepsilon \in (-\varepsilon_0, \varepsilon_0)$.
The order function $\varphi(\varepsilon)$ is supposed to be $o(\varepsilon)$ as $\varepsilon \to 0$ and such that if $\mu := \varphi(\varepsilon)/\varepsilon$ then $\varepsilon$ can be expressed as $\varepsilon = \xi(\mu)$ with $\xi \to 0$ as $\mu \to 0$.
The *assumptions* are similar to those in the simple case (5.3):

(A1) There is a smooth function $p_2: D_1 \times D_{21} \to D_{22}$ satisfying

$$g_2(x, y_1, p_2(x, y_1), 0) = 0 \qquad \text{for } (x, y_1) \in D_1 \times D_{21}.$$

(A2) There is a positive constant $b_2$ such that all eigenvalues of

$$B_2(x, y_1) := g_{2, y_2}(x, y_1, p_2(x, y_1), 0)$$

have real parts smaller than $-b_2$ for all $(x, y_1) \in D_1 \times D_{21}$.

This means that for every $(x^*, y_1^*) \in D_1 \times D_{21}$ $p_2(x^*, y_1^*)$ is an asymptotically stable equilibrium solution of the system

(5.28)
$$\frac{du_2}{d\tau_2} = g_2(x^*, y_1^*, u_2, 0).$$

(A3) $y_2^0(0)$ lies in the domain of attraction of the asymptotically stable equilibrium solution $p_2(x^0(0), y_1^0(0))$ of Eq. (5.28).

(B1) There is a smooth function $p_1(x):D_1 \to D_{21}$ which solves the equation

$$h_1(x, y_1) := g_1(x, y_1, p_2(x, y_1), 0) = 0 \qquad \text{for } x \in D_1.$$

(B2) There is $b_1 > 0$ such that the real parts of the eigenvalues of $B_1(x) := h_{1,y_1}(x, p_1(x))$ are smaller than $-b_1$ for all $x \in D_1$.

(B3) $y_1^0(0)$ lies in the domain of attraction of the asymptotically stable equilibrium solution $p_1(x^0(0))$ of the system

$$(5.29) \qquad \frac{\mathrm{d}u_1}{\mathrm{d}\tau_1} = h_1(x^0(0), u_1).$$

(AB4) The solution $X(t)$ of the reduced problem

$$(5.30) \qquad \frac{\mathrm{d}x}{\mathrm{d}t} = f(x, p(x), 0), \qquad x(0) = x^0(0),$$

where $p(x) := (p_1(x), p_2(x, p_1(x)))$, exists for $t \in [0, T]$, $T > 0$, with respect to $D_1$.  $\square$

From this it should be obvious what the assumptions are in the general case (5.24), (5.25) if there are more than two different order functions multiplying the derivatives of $y$.

For the IVP (5.26), (5.27) a statement similar to Theorem 5.4 (and also 5.4a) holds. Again the result (not in the sharp form given here) was first stated by A. N. Tikhonov (cf. [29] and [33]). An analogous result is given in [14]. We here again present a transparent proof which is essentially analogous to the one of Theorem 5.4, but does not make use of an invariant manifold result. The same 'direct' way of course can be used to prove Theorem 5.4.

THEOREM 5.6 Let the assumptions (A1)–(A3), (B1)–(B3), (AB4) be satisfied. Then there are positive constants $C$, $\tau^*$ and $\varepsilon^*$ such that the following assertions hold for $\varepsilon \in (0, \varepsilon^*)$:
The solution $(x(t, \varepsilon), y_1(t, \varepsilon), y_2(t, \varepsilon))$ of the IVP (5.26), (5.27) exists at least for $t \in [0, T]$ and satisfies

$$|x(t, \varepsilon) - X(t)| < C\varepsilon \qquad \text{for } t \in [0, T],$$

and for every $\rho$ in $(0, 1]$

$$|y_1(t, \varepsilon) - Y_1(t)| < C\varepsilon^\rho$$

$$\text{for } t \in \left[\varepsilon\left(\tau^* + \rho \frac{4}{b_1} \log \varepsilon^{-1}\right), T\right],$$

$$|y_2(t, \varepsilon) - Y_2(t)| < C\varepsilon^\rho$$

where $Y_1(t) := p_1(X(t))$ and $Y_2(t) := p_2(X(t), p_1(X(t)))$.  $\square$

*Proof* The proof of Theorem 5.4 includes all details. Since the philosophy of the present proof is similar we take the liberty to be less precise and mostly give just the results of the computations. In particular for the definition of the restricted domains the reader is referred to the proof of Theorem 5.4.

(i) We consider the IVP ($y := (y_1, y_2)$)

$$\frac{dx}{d\tau} = \varepsilon f(x, y, \varepsilon), \qquad x(0, \varepsilon) = x^0(\varepsilon)$$

$$(5.31) \qquad \frac{dy_1}{d\tau} = g_1(x, y, \varepsilon), \qquad y_1(0, \varepsilon) = y_1^0(\varepsilon)$$

$$\psi(\varepsilon) \frac{dy_2}{d\tau} = g_2(x, y, \varepsilon), \qquad y_2(0, \varepsilon) = y_2^0(\varepsilon)$$

where $\psi(\varepsilon) := \varphi(\varepsilon)/\varepsilon$.

By means of Assumption (B3) the solution $\tilde{Y}_1(\tau)$ of the reduced problem

$$(5.32) \qquad \frac{dy_1}{d\tau} = g_1(X^0, y_1, p_2(X^0, y_1), 0), \qquad y_1(0) = Y_1^0,$$

where $X^0 := x^0(0)$, $Y_1^0 := y_1^0(0)$, exists in $D_{21}$ for all $\tau \geq 0$. Moreover, for $r > 0$ sufficiently small (it will be specified more precisely later) there is $\tau^*(r) > 0$ such that

$$(5.33) \qquad | \tilde{Y}_1(\tau) - p_1(X^0)| < \frac{r}{2} \qquad \text{for } \tau \geq \tau^*.$$

Due to the assumptions (A1)–(A3) and due to the 'assertion (5.32)' the problem (5.31) is of Tikhonov type (for the small parameter $\mu := \psi(\varepsilon)$) on every compact $\tau$-interval $[0, k]$, $k > 0$. Hence, we may apply Theorem 5.4 and thus obtain that there are positive constants $\tilde{C}(\tau^*)$, $\tilde{\sigma}(\tau^*)$ and $\tilde{\varepsilon}(\tau^*)$ such that the solution $(\tilde{x}(\tau), \tilde{y}(\tau))$ of (5.31) exists for $\tau \in [0, \tau^*]$ and satisfies

$$
\begin{aligned}
&|\tilde{x}(\tau) - X^0| < \tilde{C}\tilde{\psi}(\varepsilon) \\
(5.34) \quad &|\tilde{y}_1(\tau) - \tilde{Y}_1(\tau)| < \tilde{C}\tilde{\psi}(\varepsilon)
\end{aligned}
\qquad \text{for } \tau \in [0, \tau^*]
$$

$$|\tilde{y}_2(\tau) - p_2(X^0, \tilde{Y}_1(\tau))| < \tilde{C}\tilde{\psi}(\varepsilon)$$

$$\text{for } \tau \in \left[\psi(\varepsilon)\left(\tilde{\sigma} + \frac{2}{b_2} \log \psi(\varepsilon)^{-1}\right), \tau^*\right],$$

$$\varepsilon \in (0, \tilde{\varepsilon}), \quad \text{where } \tilde{\psi}(\varepsilon) := \begin{cases} \varepsilon, & \text{if } \psi \lessapprox \varepsilon \\ \psi(\varepsilon), & \text{if } \psi \gg \varepsilon \end{cases}.$$

Of course, this implies that the solution $(x(t), y(t))$ of the IVP (5.26), (5.27) exists for $t \in [0, \varepsilon\tau^*]$, $\varepsilon \in (0, \tilde{\varepsilon})$, and $(x(t), y(t)) = (\tilde{x}(t/\varepsilon), (\tilde{y}(t/\varepsilon))$.

From

$$| \tilde{y}_1(\tau^*) - p_1(\tilde{x}(\tau^*)) | \leqslant | \tilde{y}_1(\tau^*) - p_1(X^0) | + | p_1(X^0) - p_1(\tilde{x}(\tau^*)) |$$

$$\leqslant | \tilde{y}_1(\tau^*) - p_1(X^0) | + L_1 | X^0 - \tilde{x}(\tau^*) |$$

where $L_1$ denotes the Lipschitz constant of $p_1(x)$ on $G_1$, we obtain, using the estimates (5.33) and (5.34), for $\varepsilon$ small enough

(5.35)          $$| \tilde{y}_1(\tau^*) - p_1(\tilde{x}(\tau^*)) | < r.$$

In an analogous way again by means of (5.33), (5.34) we derive for $\varepsilon$ small enough

(5.36)          $$| \tilde{y}_2(\tau^*) - p_2(\tilde{x}(\tau^*), \tilde{y}_1(\tau^*)) | < r.$$

(ii)  We now consider the IVP

$$\frac{du}{dt} = f(X(t) + u, p_1(X(t) + u) + v_1, p_2(X(t) + u, p_1(X(t) + u) + v_1) + v_2, \varepsilon)$$

$$- \frac{dX(t)}{dt}$$

(5.37)          $$\varepsilon \frac{dv_1}{dt} = g_1(x, p_1 + v_1, p_2(x, p_1 + v_1) + v_2, \varepsilon) - \varepsilon p_{1,x}(x) f(\dots)$$

$$\varphi(\varepsilon) \frac{dv_2}{dt} = g_2(x, p_1 + v_1, p_2(x, p_1 + v_1) + v_2, \varepsilon)$$

$$- \varphi(\varepsilon) \left[ p_{2,x}(x, p_1 + v_1) f(\dots) + p_{2,y_1}(x, p_1 + v_1) \frac{1}{\varepsilon} g_1(\dots) \right]$$

$$u(\varepsilon\tau^*) = \tilde{x}(\tau^*) - X(\varepsilon\tau^*), \quad v_1(\varepsilon\tau^*) = \tilde{y}_1(\tau^*) - p_1(\tilde{x}(\tau^*)),$$

$$v_2(\varepsilon\tau^*) = \tilde{y}_2(\tau^*) - p_2(\tilde{x}(\tau^*), \tilde{y}_1(\tau^*)).$$

To save writing we have still used the original variable $x$ instead of $X(t) + u$ in the second and third equation. We already know from above that the solution $(u(t), v_1(t), v_2(t))$ of (5.37) exists for $t \in [0, \varepsilon\tau^*]$.

Moreover, let $\delta \in (0, \delta_0]$, where $\delta_0 > 0$ is sufficiently small, and consider the following domain

$$\Omega_\delta := \{| u | < \delta\} \times \{| v_1 | < \delta\} \times \{| v_2 | < \delta\} \times \{t \in [0, T]\} \subset \mathbb{R}^{m + n_1 + n_2 + 1}$$

We suppose that for $\varepsilon$ small enough the solution $(u(t), v_1(t), v_2(t))$ with respect to $\Omega_\delta$ exists for $t \in [\varepsilon\tau^*, m_+)$. If we expand the third equation of (5.37) with respect to $v_2$ and $\varepsilon$, the second one with respect to $v_1, v_2$ and $\varepsilon$ and the first one with respect to $u, v_1, v_2$ and $\varepsilon$, we may write the system in the form (again

$$x = X(t) + u)$$

$$\frac{\mathrm{d}u}{\mathrm{d}t} = R_0(t, u, v_1, v_2, \varepsilon)$$

(5.38)
$$\varepsilon \frac{\mathrm{d}v_1}{\mathrm{d}t} = B_1(x)v_1 + R_1(t, u, v_1, v_2, \varepsilon)$$

$$\varphi(\varepsilon) \frac{\mathrm{d}v_2}{\mathrm{d}t} = \bar{B}_2(x, v_1)v_2 + R_2(t, u, v_1, v_2, \varepsilon)$$

where the matrix $\bar{B}_2(x, v_1) := B_2(x, p_1(x) + v_1)$ by means of Assumption (A2) has eigenvalues with real parts $< -b_2$ in $\Omega_{\delta_0}$, the matrix $B_1(x)$ by means of Assumption (B2) eigenvalues with real parts $< -b_1$, and the remainder terms in the Taylor formulae satisfy

(5.39)
$$\begin{aligned}
&|R_2| \leqslant K_2(|v_2|^2 + \varepsilon) \\
&|R_1| \leqslant K_1(|v_1|^2 + |v_2| + \varepsilon), \qquad \text{in } \Omega_{\delta_0} \text{ and } \varepsilon \in (0, \tilde{\varepsilon}). \\
&|R_0| \leqslant K_0(|u| + |v_1| + |v_2| + \varepsilon)
\end{aligned}$$

Let $V_2(t, s, \varepsilon)$, $V_1(t, s, \varepsilon)$ be the fundamental matrix solutions of the linear systems

$$\varphi(\varepsilon) \frac{\mathrm{d}V_2}{\mathrm{d}t} = \tilde{B}_2(t, \varepsilon)V_2, \qquad \varepsilon \frac{\mathrm{d}V_1}{\mathrm{d}t} = \tilde{B}_1(t, \varepsilon)V_1$$

on $[0, m_+)$ reducing to the identity matrices for $t = s$, where $\tilde{B}_2(t, \varepsilon) := \bar{B}_2(X(t) + u(t), v_1(t))$, $\tilde{B}_1(t, \varepsilon) := B_1(X(t) + u(t))$.

Applying a stability result of W. A. Coppel (as done for Lemma 1 in [24]) yields the estimates

(5.40)
$$\begin{aligned}
&|V_2(t, s, \varepsilon)| \leqslant C_2 e^{-\beta(t-s)/\varphi}, \\
&|V_1(t, s, \varepsilon)| \leqslant C_1 e^{-\alpha(t-s)/\varepsilon}, \qquad 0 \leqslant s \leqslant t < m_+,
\end{aligned}$$

with $\beta := b_2/2, \alpha := b_1/2$. (This type of estimate can also be obtained from a lemma by Flatto and Levinson (cf. [9]).)

We now suppose that $r$ is also so small that $C_2 r < \delta/4$ and $C_1 r < \delta/4$.

We first treat the third equation of (5.38). Using the variation of constants formula for $v_2(t)$ and applying the estimates (5.36), (5.39), (5.40) we obtain

$$|v_2(t)| \leqslant e^{-\beta(t-t^*)/\varphi} \frac{\delta}{4} + \varphi^{-1}C_2 K_2 \int_{t^*}^{t} e^{-\beta(t-s)/\varphi} (|v_2(s)|^2 + \varepsilon) \, \mathrm{d}s$$

$t \in [t^*, m_+)$, where we have put $t^* := \varepsilon\tau^*$. From this we derive

(5.41) $\quad |v_2(t)| \leqslant e^{-\beta(t-t^*)/\varphi} \dfrac{\delta}{4} + \varepsilon K + \varphi^{-1}K\delta \displaystyle\int_{t^*}^{t} e^{-\beta(t-s)/\varphi} |v_2(s)| \, \mathrm{d}s.$

We now use the following *generalized Gronwall lemma*:

If the non-negative functions $u(t)$, $a_1(t)$, $a_2(t)$ satisfy

$$u(t) \leqslant a_1(t) + \int_c^t a_2(s)u(s)\,ds \qquad \text{for } c \leqslant t \leqslant d$$

then

$$u(t) \leqslant a_1(t) + \int_c^t a_2(s)a_1(s)e^{\int_s^t a_2(\sigma)\,d\sigma}\,ds.$$

If we multiply (5.41) by $e^{\beta t/\varphi}$ we may apply this lemma and get

(5.42)
$$e^{\beta t/\varphi}\,|\,v_2(t)\,| \leqslant e^{\beta t^*/\varphi}\frac{\delta}{4} + e^{\beta t/\varphi}\,\varepsilon K + \varphi^{-1}K\delta \cdot$$

$$\cdot \int_{t^*}^t \left[ e^{\beta t^*/\varphi}\frac{\delta}{4} + e^{\beta s/\varphi}\,\varepsilon K \right] e^{K\delta(t-s)/\varphi}\,ds.$$

Let $\delta$ be so small that $K\delta - \beta =: -b < 0$ and $b \geqslant \beta/2$. Then integrating in (5.42) yields

(5.43)
$$|\,v_2(t)\,| \leqslant e^{-b(t-t^*)/\varphi}\frac{\delta}{4} + \varepsilon K + \frac{\varepsilon K^2\delta}{b}, \qquad t \in [t^*, m_+),$$

and this implies for $\varepsilon$ small enough that

$$|\,v_2(t)\,| < \frac{\delta}{2} \qquad \text{for } t \in [t^*, m_+).$$

For $v_1(t)$ we obtain analogously to (5.41)

$$|\,v_1(t)\,| \leqslant e^{-\alpha(t-t^*)/\varepsilon}\frac{\delta}{4} + \varepsilon M + \varepsilon^{-1}M\delta \int_{t^*}^t e^{-\alpha(t-s)/\varepsilon)}|\,v_1(s)\,|\,ds$$

$$+ \varepsilon^{-1}M \int_{t^*}^t e^{-\alpha(t-s)/\varepsilon}|\,v_2(s)\,|\,ds$$

and integrating the last term after taking into account (5.43) then yields for $\varepsilon$ small enough

(5.44)
$$|\,v_1(t)\,| \leqslant e^{-\alpha(t-t^*)/\varepsilon}\frac{\delta}{3} + \varepsilon\bar{M} + \varepsilon^{-1}M\delta \int_{t^*}^t e^{-\alpha(t-s)/\varepsilon}|\,v_1(s)\,|\,ds$$

which is precisely of the same form as (5.41)

In an analogous way as before we therefore get, for $\delta$ again also satisfying $M\delta - \alpha =: -a < 0$ and $a \geqslant \alpha/2$,

(5.45)
$$|\,v_1(t)\,| \leqslant e^{-a(t-t^*)/\varepsilon}\frac{\delta}{3} + \varepsilon\hat{M}, \qquad t \in [t^*, m_+),$$

and hence for $\varepsilon$ small enough

$$|v_1(t)| < \frac{\delta}{2} \qquad \text{for } t \in [t^*, m_+).$$

It remains to find an estimate for $u(t)$. The following estimate is trivial

$$(5.46) \qquad |u(t)| \leqslant C_0\varepsilon \qquad \text{for } t \in [0, t^*].$$

Hence, writing the first equation of (5.38) as an integral equation and using the estimate (5.39) we obtain

$$|u(t)| \leqslant C_0\varepsilon + K_0 \int_{t^*}^{t} (|u(s)| + |v_1(s)| + |v_2(s)| + \varepsilon)\, ds \qquad \text{for } t \in [t^*, m_+).$$

Applying the estimates (5.43) and (5.45) and integrating yields

$$|u(t)| \leqslant \bar{C}\varepsilon + K_0 \int_{t^*}^{t} |u(s)|\, ds, \qquad t \in [t^*, m_+),$$

and finally the Gronwall lemma

$$(5.47) \qquad |u(t)| \leqslant \bar{C}e^{K_0 T} \varepsilon \qquad \text{for } t \in [t^*, m_+).$$

For $\varepsilon$ small enough this of course implies that

$$|u(t)| < \frac{\delta}{2} \qquad \text{for } t \in [t^*, m_+),$$

and hence by means of the 'global existence theorem' we may conclude that the solution $(u(t), v_1(t), v_2(t))$ of the IVP (5.37) exists for $t \in [\varepsilon\tau^*, T]$ and there satisfies the estimates (5.43), (5.45) and (5.47). This of course together with (5.34) gives the existence of the solution $(x(t), y_1(t), y_2(t))$ of (5.26), (5.27) on $[0, T]$.

(iii) In a last step we have to derive estimates for $|y_1(t) - Y_1(t)|$ and $|y_2(t) - Y_2(t)|$ where $Y_1(t) := p_1(X(t))$ and $Y_2(t) := p_2(X(t), p_1(X(t)))$. They are now obtained in a trivial way just by combining the results of (ii). From

$$|y_1(t) - Y_1(t)| \leqslant |y_1(t) - p_1(x(t))| + |p_1(x(t)) - p_1(X(t))|$$
$$\leqslant |v_1(t)| + L_1 |x(t) - X(t)|$$

using (5.45) and (5.47) we get

$$(5.48) \qquad |y_1(t) - Y_1(t)| \leqslant e^{-a(t - t^*)/\varepsilon} \frac{\delta}{3} + \varepsilon M_1 \qquad \text{for } t \in [t^*, T].$$

Similarly, from

$$|y_2(t) - Y_2(t)| \leqslant |y_2(t) - p_2(x(t), y_1(t))|$$
$$+ |p_2(x(t), y_1(t)) - p_2(X(t), p_1(X(t)))|$$
$$\leqslant |v_2(t)| + L_2(|x(t) - X(t)| + |y_1(t) - Y_1(t)|)$$

using (5.43), (5.47) and (5.48) we finally have

$$(5.49) \quad |y_2(t) - Y_2(t)| \leq (L_2 e^{-a(t-t^*)/\varepsilon} + e^{-b(t-t^*)/\varphi}) \frac{\delta}{3} + \varepsilon M_2 \text{ for } t \in [t^*, T].$$

This completes the proof of Theorem 5.6.    $\square$

This proof carries over to the general case with several different order functions $\varphi_{k+1}(\varepsilon)$, $k = 1, ..., N$, and there will be the analogous steps (i), (ii), (iii). Step (i), however, is recursive in the sense that to show the assertions of (i), one needs the 'theorem of $N-1$ order functions'. In the general case, again $\varphi_2(\varepsilon)$ has to be such that $\varphi_2(\varepsilon)/\varepsilon$ admits an inverse function $\xi(\mu)$ with $\xi \to 0$ as $\mu \to 0$.

*Remarks*   —As can easily be checked in the proof, Theorems 5.4, 5.4a and 5.6 of course also hold if the initial conditions are not smooth in $\varepsilon$, i.e. if they satisfy (compare Theorem 5.1 and the remark to (5.25))

$$|x^0(\varepsilon) - x^0(0)| < M_0\varphi(\varepsilon)$$
$$|y^0(\varepsilon) - y^0(0)| < M_0\varphi(\varepsilon)$$

for $M_0 > 0$ and some order function $\varphi(\varepsilon) = o(1)$ as $\varepsilon \to 0$. Then, the $\varepsilon$ in the error estimates has to be replaced by the order function

$$\psi(\varepsilon) := \begin{cases} \varepsilon, & \text{if } \varphi \leqslant \approx \varepsilon \\ \varphi(\varepsilon), & \text{if } \varphi \gg \varepsilon \end{cases}, \text{ the } \varepsilon^\rho \text{ by } \psi_\rho(\varepsilon) := \begin{cases} \varepsilon^\rho & \text{if } \varphi \leqslant \approx \varepsilon^\rho \\ \varphi(\varepsilon), & \text{if } \varphi \gg \varepsilon^\rho \end{cases}.$$

—In the literature, instead of Assumption (2) for Eq. (5.3) one often finds the same eigenvalue condition for the matrix $B(t) := g_y(X(t), Y(t), 0)$, $Y(t) := p(X(t))$, which is of course implied by (2). Theorem 5.4 also holds for this weaker assumption which is rather technical, however.
—The above theorems also hold for a singularly perturbed system without $x$-equations (if it is non-autonomous, of course)

$$\varepsilon \frac{dy}{dt} = g(t, y, \varepsilon), \qquad y(0) = y^0(\varepsilon)$$

with the properties that there is a smooth function $p(t)$, $t \in [0, T]$, such that $g(t, p(t), 0) = 0$, $B(t) := g_y(t, p(t), 0)$ has eigenvalues with real parts smaller than $-b$ (similarly for the general case of Theorem 5.6) and $y^0(0)$ lies in the domain of attraction of the asymptotically stable equilibrium solution of the system $u' = g(t^*, u, 0)$ for every $t^* \in [0, T]$. This is a trivial consequence of the fact that we can write this non-autonomous problem in the autonomous form

$$\frac{dx}{dt} = 1, \qquad\qquad x(0) = 0$$

$$\varepsilon \frac{dy}{dt} = g(x, y, \varepsilon), \qquad y(0) = y^0(\varepsilon).$$

—Exactly the same Theorem 5.4 also holds for the case $T = \infty$ (compare also [13], [14]), if the reduced solution $(X(t), p(X(t))$ exists and is bounded for all $t \in [0, \infty)$ and if the fundamental matrix solution $U(t, s)$ of the linear system

$$\frac{dU}{dt} = A(t)U$$

where $\quad A(t) := f_x(X(t), p(X(t)), 0) - f_y(\ldots)[B(X(t))]^{-1} g_x(\ldots)$

satisfies an estimate

$$|U(t, s)| \leqslant C_0 e^{-\alpha_0(t - s)} \qquad \text{for } 0 \leqslant s \leqslant t < \infty,$$

with positive constants $C_0, \alpha_0$.

This result is easily obtained from the direct proof (i.e. analogous to the one of Theorem 5.6) if the first equation of (5.37) is expanded with respect to $u$. Due to the above assumption one then obtains an estimate analogous to (5.45) also for $|u(t)|$ instead of the estimate (5.47).

Of course the same is true for the general Theorem 5.6. But there the analogous matrix expression $A(t)$ which has to be stable gets more and more involved.

—In [14] the conditionally stable case is treated, i.e. the eigenvalues of $B(x)$ in Assumption (2) of Theorem 5.4 have eigenvalues with real parts $< -b$ and $> b$. Under these hypotheses a Tikhonov result holds for a manifold of initial values $y^0(\varepsilon)$. $\square$

An extension of the Tikhonov result for a case where the stability of the reduced manifold $\{(x, y) \mid y \in p(x), x \in D_1\}$ as required in Assumption (2) breaks down is given in the next section.

## 5.3 Breakdown of stability

As already seen before a local approximation to the solution of a singularly perturbed IVP usually is not valid for all times. In the Tikhonov case, in general, this is due to the fact that the solution of the reduced system loses its stability, i.e. it comes close to a point in $x$-space where Assumption (2) of Section 5.2 is not satisfied any more. The Tikhonov result provides estimates only for a finite distance away of such a point. However, for verifying an overlapping of the domain of validity with a subsequent local approximation one needs a sharp estimate, an interval of validity with a moving endpoint (with $\varepsilon$). Therefore, we want to generalize the Tikhonov result to such a situation where the reduced solution loses its stability in a specific way.

One of the implications of Assumption (2) is that the reduced manifold $y = p(x)$ is locally unique. In [18], [19] two kinds of bifurcation situations are considered where the branches of the reduced manifold are locally linear or quadratic, respectively, in the essential variables. Another bifurcation case is the so-called jump point situation sketched in Fig. 2, arising frequently in the context of relaxation oscillations. Typically, the trajectory follows the stable branch of the reduced manifold until it reaches a vicinity of the bifurcation point where it drops off.

In [25], [26] we have derived quite general results for the case where exactly one single eigenvalue of $B(x)$ vanishes at the point of bifurcation. In [25] we have proved the results for the orbits. The transfer of the estimates to the solutions as well as an application is given in [26]. In this paper we state the main result without proof.

We still consider Eq. (5.3). The point where the stability Assumption (2) of Section 5.2 ceases to hold we take to be the origin in $\mathbb{R}^{m+n}$; and we suppose that $f(0, 0, 0,) \neq 0$.

The precise *assumptions* are:

(A1) $\qquad f_1(0, 0, 0) > 0, \qquad f_k(0, 0, 0) = 0 \qquad (k = 2, 3, \ldots, m)$

$\qquad\qquad g(0, 0, 0) = 0$

Hence, there exist a domain $U \times V \subset \mathbb{R}^{m+n}$, containing the origin, and positive constants $\rho$ and $\varepsilon_1$ such that $f_1(x, y, \varepsilon) > \rho$ for $(x, y) \in U \times V, \varepsilon \in (-\varepsilon_1, \varepsilon_1)$.

(A2)

$$g_y(0, 0, 0) = \begin{bmatrix} 0 & 0 & . & . & . & 0 \\ 0 & & & & & \\ . & & & & & \\ . & & & \bar{A} & & \\ . & & & & & \\ 0 & & & & & \end{bmatrix}$$

where all the eigenvalues of the $(n - 1) \times (n - 1)$ matrix $\bar{A}$ have negative real parts.

Moreover, let $\bar{y} := (y_2, \ldots, y_n)$. As seen in [19], we may assume without loss of generality that $\bar{g}(x, y_1, \bar{y})$ has a Taylor formula beginning in $y_1$ with a cubic term. (A2) implies that there exists a unique solution $\bar{w}(x, y_1)$ of $\bar{g}(x, y_1, \bar{y}, 0) = 0$ in a neighbourhood $\hat{U} \times \hat{V} \subset U \times V$ containing the origin, having continuous partial derivatives there and satisfying $\bar{w}(0, 0) = 0$. For the remaining equation

$$g_1(x, y_1, \bar{w}(x, y_1), 0) = 0$$

we require

(A3) There exists a function $p_1(x)$, which is defined and continuous for $(x_1, \bar{x}) \in J \times \bar{\Omega} \subset \hat{U}$, where $J := (s_1, 0]$ for some $s_1 \in (-1, 0)$, $\bar{\Omega}$ some

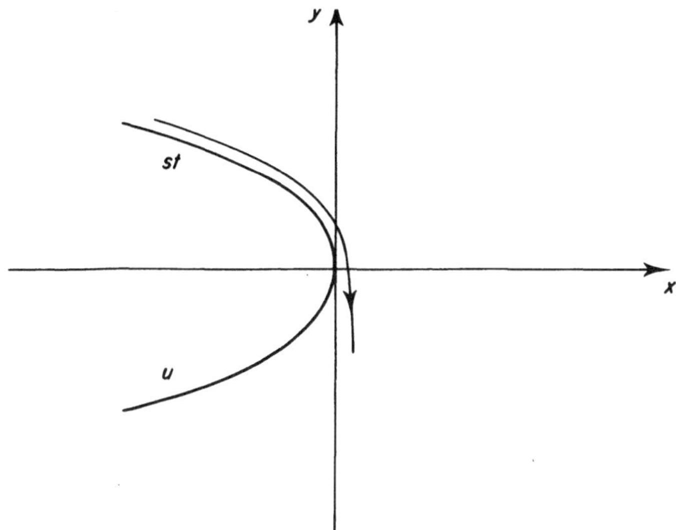



**Fig. 2.** Jump point

neighbourhood of $\bar{x} = 0$, has continuous partial derivatives with respect to $\bar{x}$ there, and is $C^1$ in $J' \times \bar{\Omega}$, $J' := (s_1, 0)$, satisfying

$$g_1(x, p_1(x), \bar{w}(x, p_1(x)), 0) = 0, \qquad x \in J \times \bar{\Omega}$$
$$p_1(0) = 0$$

and

$$p_1(x) = a_1(-x_1)^\alpha + h(x), \qquad x \in J \times \bar{\Omega}$$

$$\frac{\partial p_1}{\partial x_1}(x) = a_{11}(-x_1)^{\alpha-1} + h_1(x), \qquad x \in J' \times \bar{\Omega}$$

where $\alpha > 0$ and $a_1, a_{11} \neq 0, h(0) = 0, h(x_1, 0) = o((-x_1)^\alpha)$ and $h_1(x_1, 0) = o((-x_1)^{\alpha-1})$ as $x_1 \to 0^-$.
If $\alpha \geq 1$ we suppose that $p_1(x) \in C^1(J \times \bar{\Omega})$.

Let now $\bar{p}(x) := \bar{w}(x, p_1(x))$ and $p(x) := (p_1, \bar{p})$ and consider the initial value problem

$$(5.50) \qquad \frac{\mathrm{d}\bar{x}}{\mathrm{d}x_1} = \frac{\bar{f}(x_1, \bar{x}, p(x_1, \bar{x}), 0)}{f_1(x_1, \bar{x}, p(x_1, \bar{x}), 0)}, \qquad \bar{x}(0) = 0$$

for $(x_1, \bar{x}) \in J \times \bar{\Omega}$.

Moreover, we introduce the following initial conditions to the system (5.3):

$$(5.51) \quad x(0, \varepsilon) = x^0(\varepsilon) = X^0 + O(\varepsilon), \qquad y(0, \varepsilon) = y^0(\varepsilon) = Y^0 + O(\varepsilon)$$

where $X^0$, $Y^0$ independent of $\varepsilon$ and $(x^0(\varepsilon), y^0(\varepsilon)) \in \hat{U} \times \hat{V}$, $x^0(\varepsilon) \in J' \times \bar{\Omega}$ for all $\varepsilon \in [0, \varepsilon_1)$.

And we suppose

(A4) The solution $\bar{U}(x_1)$ of (5.50) exists for $x_1 \in J$, and

$$|\bar{x}^0(\varepsilon) - \bar{U}(x_1^0(\varepsilon))| < c_0\varepsilon$$
$$|y^0(\varepsilon) - V(x_1^0(\varepsilon))| < c_0\varepsilon$$

where $V(x_1) := p(x_1, \bar{U}(x_1))$.

There is one more condition, the essential stability condition that, together with (A2), replaces Assumption (2) for the Tikhonov case:

(A5) There are positive constants $k$ and $q$ such that

$$g_{1,y_1}(x_1, \bar{U}(x_1), V(x_1), 0) \leqslant -k(-x_1)^q, \qquad x_1 \in J. \qquad \Box$$

Let $X_1(t)$ be the solution of

$$\frac{dx_1}{dt} = f_1(x_1, \bar{U}(x_1), V(x_1), 0), \qquad x_1(0) = X_1^0$$

It exists and is unique on an interval $[0, T]$, and increases there from $X_1^0$ to 0. Then $(X(t), Y(t)) := (X_1(t), \bar{U}(X_1(t)), V(X_1(t)))$ is a solution of the reduced system of (5.3) for $t \in [0, T]$ satisfying

$$
\begin{aligned}
(5.52) \qquad & |X(0) - x^0(\varepsilon)| = O(\varepsilon) \\
& |Y(0) - y^0(\varepsilon)| = O(\varepsilon) \\
& X(T) = 0, \ Y(T) = 0.
\end{aligned}
$$

*Remarks*   —Assumption (A4), which is equivalent to the condition (5.52) for a solution of the reduced system of (5.3), is naturally satisfied if we consider a global problem whose reduced trajectory approaches the point $(x, y) = (0, 0)$ and is stable as long as it is in a finite distance from this point.
—In order to verify (A5) the asymptotic relations for $(\bar{U}(x_1), V(x_1))$ given in [25] may be used. $\Box$


To be able to formulate our result we need define the following non-negative quantities:

$$\hat{\alpha} := \min(\alpha, 1)$$

$$\bar{\alpha} := \min(2\alpha, 1)$$

$$\beta := \begin{cases} \alpha, & \text{if } \alpha < 1 \text{ and } g_1(x, y, 0) \text{ has no linear term in } x_1 \\ 0, & \text{otherwise} \end{cases}$$

moreover, $\gamma$ and $\nu$ which are defined by the estimates

$$| g_{1,\bar{x}}(x_1, \bar{U}(x_1), V(x_1), 0) | \leqslant c(-x_1)^\gamma$$
$$| g_{1,\bar{y}}(x_1, \bar{U}(x_1), V(x_1), 0) | \leqslant c(-x_1)^\nu, \qquad x_1 \in J.$$

In [25] we have shown that

$$\gamma \begin{cases} \geqslant \hat{\alpha}, & \text{if } \alpha \geqslant 1 \text{ or } g_1(x, y, 0) \text{ has no linear term in } \bar{x} \\ = 0, & \text{otherwise} \end{cases}$$

$$\nu \geqslant \hat{\alpha}.$$

Under the Assumptions (A1)–(A5) the following result holds:

THEOREM 5.7   If $q$ satisfies

$$q < \min(1 + \gamma, \hat{\alpha} + \nu)$$

then for $|x^0(\varepsilon)|, |y^0(\varepsilon)|$ sufficiently small there exist positive constants $C$ and $\varepsilon_1 \leqslant \varepsilon_0$ such that the solution $(x(t, \varepsilon), y(t, \varepsilon))$ of the initial value problem (5.3), (5.51) exists at least for $t \in [0, T - C\varepsilon^{q^*}]$, where $q^* := 1/(q_1 + q - \beta)$, $q_1 := 1 - \hat{\alpha} + q$, and

$$| x(t, \varepsilon) - X(t) | < \begin{cases} C\varepsilon \log(T - t)^{-1}, & q_1 = 1 \\ C\varepsilon(T - t)^{1 - q_1}, & q_1 > 1 \end{cases}$$

$$| y_1(t, \varepsilon) - Y_1(t) | < C\varepsilon(T - t)^{-q_1}$$

$$| \bar{y}(t, \varepsilon) - \bar{Y}(t) | < \begin{cases} C\varepsilon \log(T - t)^{-1}, & q_1 = 1 \\ C\varepsilon(T - t)^{\hat{\alpha} - q_1}, & q_1 > 1 \end{cases}$$

for $t \in [0, T - C\varepsilon^{q^*}], \varepsilon \in (0, \varepsilon_1)$.   $\square$

*Remarks*   —The above theorem treats the case where the flow along the reduced trajectory approaches the origin ('$x_1 < 0$'). The situation where the flow along a stable branch of the reduced manifold leads away from the origin ('$x_1 > 0$') is analogous (although even more technical) and is not stated here. For the precise results concerning this case the reader is referred to [25] and [26].

—Theorem 5.7 also holds for a general system of the form (5.24) if, in the case of more than two different order functions $\varphi_j(\varepsilon)$, Assumption (A2) for the matrix $\bar{A}$ is altered in the sense of Assumptions (A1), (A2), (B1), (B2) of Theorem 5.6.   $\square$

# REFERENCES

[1]  M. Abramowitz and I. A. Stegun, *Handbook of Mathematical Functions*, Dover Publications, New York, 1965.

[2] W. Altherr, An algorithm for enumerating all vertices of a convex polyhedron. *Computing*, **15** (1975).

[3] J. D. Cole, *Pertubation Methods in Applied Mathematics*, Ginn–Blaisdell, Boston, 1968.

[4] G. B. Dantzig, *Linear Programming and Extensions*, Princeton University Press, 1963.

[5] W. Eckhaus, *Matched Asymptotic Expansions and Singular Perturbations*, North Holland Publishing Co., Amsterdam, 1973.

[6] W. Eckhaus, *Asymptotic Analysis of Singular Perturbations*, North Holland Publishing Co., Amsterdam, 1979.

[7] R. J. Field, E. Körös and R. M. Noyes, Oscillations in chemical systems. II. Thorough analysis of temporal oscillation in the bromate–cerium–malonic acid system. *J. Amer. Chem. Soc.*, **94** (1972).

[8] R. J. Field and R. M. Noyes, Oscillations in chemical systems. IV. Limit cycle behaviour in a model of a real chemical reaction. *J. Chem. Phys.*, **60** (1974).

[9] L. Flatto and N. Levinson, Periodic solutions of singularly perturbed systems. *J. Rat. Mech. Anal.*, **4** (1955), 943–50.

[10] L. E. Fraenkel, On the method of matched asymptotic expansions. *Proc. Camb. Phil. Soc.*, **65** (1969).

[11] A. M. Galperin, The general solution of a finite system of linear inequalities. *Math. OR*, **1** (1976).

[12] S. P. Hastings and J. D. Murray, The existence of oscillatory solutions in the Field–Noyes model for the Belousov–Zhabotinskii reaction. *SIAM J. Appl. Math.*, **28** (1975).

[13] F. C. Hoppensteadt, Singular perturbations on the infinite interval. *Trans. Amer. Math. Soc.*, **123** (1966).

[14] F. C. Hoppensteadt, Properties of solutions of ordinary differential equations with small parameters. *Commun. Pure Appl. Math.*, **24** (1971).

[15] S. Kaplun, *Fluid Mechanics and Singular Perturbations*, P. A. Lagerstrom, L. N. Howard and C. S. Liu (eds), Academic Press, New York, 1967.

[16] J. Kevorkian, Matched asymptotic expansions or singular perturbation. Lecture Notes for Summer Institute in Dynamical Astronomy (1970).

[17] P. A. Lagerstrom and R. G. Casten, Basic concepts underlying singular perturbation techniques. *SIAM Rev.*, **14** (1972).

[18] N. R. Lebovitz and R. J. Schaar, Exchange of stabilities in autonomous systems. *Stud. Appl. Math.*, **54**, 3 (1975), 229–60.

[19] N. R. Lebovitz and R. J. Schaar, Exchange of stabilities in autonomous systems—II. Vertical bifurcation. *Stud. Appl. Math.*, **56** (1977), 1–50.

[20] J. D. Murray, *Lectures on Nonlinear Differential Equation Models in Biology*, Oxford University Press, 1977.

[21] K. Nipp, A formal method for matched asymptotic expansions applied to the Field–Noyes model of the Belousov–Zhabotinskii reaction. *Mech. Res. Commun.*, **5**, 4 (1978).

[22] K. Nipp, An algorithmic approach to singular perturbation problems in ordinary differential equations with an application to the Belousov–Zhabotinskii reaction. PhD thesis, ETH Zurich, No. 6643, 1980.

[23] K. Nipp, An extension of Tikhonov's theorem in singular perturbations for the planar case. *ZAMP*, **34** (1983), 277–90.

[24] K. Nipp, Invariant manifolds of singularly perturbed ordinary differential equations. *ZAMP*, **36** (1985), 309–20.

[25] K. Nipp, Breakdown of stability in singularly perturbed autonomous systems—I. Orbit equations. *SIAM J. Math. Anal.*, **17**, 3 (1986), 512–32.

[26] K. Nipp, Breakdown of stability in singularly perturbed autonomous systems—II. Estimates for the solutions and application. *SIAM J. Math. Anal.*, **17**, 5 (1986), 1068–85.

[27] R. E. O'Malley, Jr., *Introduction to Singular Perturbations*, Academic Press, New York and London, 1974.

[28] J. A. Stanshine and L. N. Howard, Asymptotic solutions of the Field–Noyes model for the Belousov reaction. *Stud. Appl. Math.*, **55** (1976).

[29] A. N. Tikhonov, Systems of differential equations containing small parameters multiplying the highest derivatives. *Mat. Sb.*, **31** (1952), 575–96. (In Russian.)

[30] A. N. Tikhonov, A. B. Vasil'eva and A. G. Sveshnikov, *Differential Equations*, Springer, 1985.

[31] J. J. Tyson, The Belousov–Zhabotinskii Reaction. *Lecture Notes in Biomathematics*, **10**, 1976.

[32] M. Van Dyke, *Perturbation Methods in Fluid Dynamics*, Academic Press, New York, 1964.

[33] A. B. Vasil'eva and V. M. Volosov, The work of Tikhonov and his pupils in ordinary differential equations containing a small parameter. *Russian Math. Surveys*, **31**, 4–6 (1976), 124–42.

[34] A. B. Vasil'eva, The development of the theory of ordinary differential equations with a small parameter multiplying the highest derivative during the period 1966–1976. *Russian Math. Surveys*, **31**:6 (1976), 109–31.

[35] W. Wasow, *Asymptotic Expansions of Ordinary Differential Equations*, R. E. Krieger Publishing Co., Huntington New York, 1976.

*Dynamics Reported, Volume 1*
Edited by U. Kirchgraber and H. O. Walther
© 1988 John Wiley & Sons and B. G. Teubner

# 5

# Exponential Dichotomies, the Shadowing Lemma and Transversal Homoclinic Points

**Kenneth J. Palmer**
*University of Miami, Florida*[*]
and
*University of Melbourne, Victoria*

## CONTENTS



## 1  INTRODUCTION

Smale [12, 13] studied diffeomorphisms with *transversal homoclinic points* and showed that the dynamics are *chaotic* in the neighbourhood of the orbit of such a point, in the sense that there is a compact invariant set on which the action of some iterate of the diffeomorphism is topologically conjugate to the action of the *Bernoulli shift*. One immediate consequence of this is Birkhoff's

[*]Current address.

result that the diffeomorphism has infinitely many periodic points. It also turns out that nearby diffeomorphisms must also have transversal homoclinic points and hence also infinitely many periodic points. Thus the property of having infinitely many periodic points cannot, in general, be perturbed away.

Smale proved his theorem using the famous *horseshoe* construction (see also Moser [8]). Our main aim here is to use the so-called *shadowing lemma* (Anosov [1], Bowen [2], Franke and Selgrade [5], Guckenheimer, Moser and Newhouse [6], Robinson [10]) to prove the theorem. Such a proof was sketched in Palmer [9] and carried through in detail in the case where the diffeomorphism is the period map corresponding to a periodic system of differential equations. Here we consider diffeomorphisms *per se*. Not only do we prove Smale's theorem, we also give a proof of the shadowing lemma. Our basic technique is the theory of *exponential dichotomies.*

It seems that Conley [3] got very close to a similar proof of Smale's theorem. Anyway the work done here was quite independent of his. In fact, it was inspired by a preprint of Kirchgraber [7] which was, in turn, inspired by a lecture of McGehee. Both these authors showed that one could use the shadowing lemma to prove the existence of orbits with certain 'itineraries'.

Now we summarize the contents of the paper. In Section 2 the theory of exponential dichotomies for linear difference equations is developed. The proofs run analogously to those for differential equations (Coppel [4]), except the proof of roughness (Proposition 2.10) where we have used an idea of Slyusarchuk [11]. We also give a result on the existence of bounded solutions of quasilinear systems where the linear part has an exponential dichotomy.

In Section 3 we begin the study of diffeomorphisms. Every orbit of the diffeomorphism has associated with it a linear difference equation called its *variational equation.* Roughly speaking, an invariant set is *hyperbolic* if the variational equation of each orbit in the set has an exponential dichotomy. The shadowing lemma, which is a property of hyperbolic sets, turns out to be a consequence of the result on quasilinear systems proved in Section 2.

In Section 4 transversal homoclinic points are defined. The definition of transversality is purely in terms of the variational equation of the orbit of the transversal homoclinic point. We then use the shadowing lemma to prove Smale's theorem.

A transversal homoclinic point is usually defined as a point where the *stable* and *unstable manifolds* of a *hyperbolic fixed point* meet transversally. In Section 5 we use the tools of Section 2 to prove the stable manifold theorem and show that the definition of transversal homoclinic point given in Section 4 is equivalent to the usual one. At the end of this section we show that near any transversal homoclinic point there are infinitely many others and also that the property of having a transversal homoclinic point persists under perturbation.

A preliminary version of these notes was used as the basis of informal lectures given while the author was a guest at the Mathematics Research Institute of the ETH Zürich. The author wishes to express his gratitude to J. Moser, U. Kirchgraber, K. Nipp, D. Stoffer and their co-workers for their hospitality.

## 2 EXPONENTIAL DICHOTOMIES FOR LINEAR DIFFERENCE EQUATIONS

We consider *linear difference equations*

$$(1) \qquad u_{n+1} = C_n u_n,$$

where $u_n \in \mathbb{R}^p$, $n \in J$ ($J$ is usually $\mathbb{Z}$, the non-negative integers $\mathbb{Z}^+$ or the non-positive integers $\mathbb{Z}^-$) and for each $n$ $C_n$ is an invertible $p \times p$ matrix. For $n, m \in J$ we define the *transition matrix*

$$\Phi(n, m) = \begin{cases} C_{n-1} \dots C_m & \text{if } n > m \\ I & \text{if } n = m \\ C_n^{-1} \dots C_{m-1}^{-1} & \text{if } n < m. \end{cases}$$

This has the *cocycle property*,

$$\Phi(n, k)\Phi(k, m) = \Phi(n, m) \text{ if } n, k, m \in J,$$

and the property that $u_n = \Phi(n, m)\xi$ is the unique solution of equation (1) satisfying $u_m = \xi$. Also note that as a consequence of the cocycle property

$$(2) \qquad \Phi^{-1}(n, m) = \Phi(m, n) \text{ if } n, m \in J.$$

DEFINITION 2.1  The linear difference equation (1) has an *exponential dichotomy* on $J$ if there are positive constants $K$, $\alpha$ and a family of projections $P_n$, $n \in J$, such that

$$(i) \qquad P_{n+1}C_n = C_n P_n \text{ for all } n(n < 0 \text{ if } J = \mathbb{Z}^-)$$

$$(3)\ (ii) \qquad |\Phi(n, m)P_m| \leqslant Ke^{-\alpha(n-m)} \qquad \text{for } n \geqslant m$$

$$(4) \qquad |\Phi(n, m)(I - P_m)| \leqslant Ke^{-\alpha(m-n)} \qquad \text{for } m \geqslant n.$$

By repeated application of (i) we obtain the identity

$$(5) \qquad P_n\Phi(n, m) = \Phi(n, m)P_m.$$

This means that the projections $P_n$ are invariant with respect to equation (1). That is, if $u_n$ is a solution of (1) such that $u_m$ is in the range (resp. nullspace) of $P_m$ for some $m$ then $u_n$ is in the range (resp. nullspace) of $P_n$ for all $n$.

Property (ii) says, firstly, that the $P_n$ are bounded and, secondly, that solutions $u_n$ of equation (1) which lie in the range of $P_n$ decay exponentially while those in the nullspace of $P_n$ grow exponentially.

REMARK 2.2    We note here that if, as in Definition 2.1, equation (1) has an exponential dichotomy on $J = [q, \infty)$, where $q > 0$, then it has one on $\mathbb{Z}^+$. (A similar statement holds for $J = \mathbb{Z}^-$.)

To verify this, define $P_n = \Phi(n, q)P_q\Phi(q, n)$ for $0 \leqslant n < q$. Then property (i) in Definition 2.1 is satisfied for $n \geqslant 0$. Now we know that inequalities (3) and (4) hold for $n \geqslant m \geqslant q$, $m \geqslant n \geqslant q$ respectively. Put

$$M_1 = \sup\{\,|\Phi(n, m)P_m| : 0 \leqslant n, m \leqslant q\}.$$

Then if $n \geqslant q > m \geqslant 0$

$$\begin{aligned}
|\Phi(n, m)P_m| &= |\Phi(n, q)\Phi(q, m)P_m P_m| \\
&= |\Phi(n, q)P_q\Phi(q, m)P_m| \text{ by (5)} \\
&\leqslant M_1 K e^{-\alpha(n - q)} \\
&\leqslant M_1 K e^{\alpha q} e^{-\alpha(n - m)}
\end{aligned}$$

and if $q \geqslant n \geqslant m \geqslant 0$

$$|\Phi(n, m)P_m| \leqslant M_1 \leqslant M_1 e^{\alpha q} e^{-\alpha(n - m)}.$$

Similarly if

$$M_2 = \sup\{\,|\Phi(n, m)(I - P_m)| : 0 \leqslant n, m \leqslant q\}$$

we show that if $m \geqslant q > n \geqslant 0$ or $q \geqslant m \geqslant n \geqslant 0$ then

$$|\Phi(n, m)(I - P_m)| \leqslant M_2 K e^{\alpha q} e^{-\alpha(m - n)}.$$

So noting that $M_1, M_2 \geqslant 1$ we conclude that equation (1) has an exponential dichotomy on $\mathbb{Z}^+$ with constants $\max\{M_1 K e^{\alpha q}, M_2 K e^{\alpha q}\}$ and $\alpha$.

Now it follows from (5) that when $J = \mathbb{Z}$, $\mathbb{Z}^+$ or $\mathbb{Z}^-$ then

$$\tag{6} P_n = \Phi(n, 0)P_0\Phi(0, n).$$

This shows that the rank of $P_n$ does not depend on $n$ and that $P_n$ is determined once $P_0$ is. In our first proposition we determine all possible choices of $P_0$. Note that we use the notations $\mathcal{R}, \mathcal{N}$ to denote the range and nullspace of a linear mapping.

PROPOSITION 2.3    *Let the linear difference equation* (1) *have an exponential dichotomy on* $J$ *with constants* $K$, $\alpha$ *and projections* $P_n$. *Denote by* $\Phi(n, m)$

*the transition matrix for* (1). *Then*

(i) *if* $J = \mathbb{Z}^+$,

$$(7) \qquad \mathscr{R}(P_0) = \{ \xi \in \mathbb{R}^p : \sup_{n \geqslant 0} |\Phi(n, 0)\xi| < \infty \}$$

*and if* $Q_0$ *is another projection with* $\mathscr{R}(Q_0) = \mathscr{R}(P_0)$, (1) *has an exponential dichotomy on* $\mathbb{Z}^+$ *with projections* $Q_n = \Phi(n, 0)Q_0\Phi(0, n)$;

(ii) *if* $J = \mathbb{Z}^-$,

$$(8) \qquad \mathscr{N}(P_0) = \{ \xi \in \mathbb{R}^p : \sup_{n \leqslant 0} |\Phi(n, 0)\xi| < \infty \}$$

*and if* $Q_0$ *is another projection with* $\mathscr{N}(Q_0) = \mathscr{N}(P_0)$, (1) *has an exponential dichotomy on* $\mathbb{Z}^-$ *with projections* $Q_n = \Phi(n, 0)Q_0\Phi(0, n)$;

(iii) *if* $J = \mathbb{Z}$, $\mathscr{R}(P_0)$ *is given by* (7) *and* $\mathscr{N}(P_0)$ *by* (8) *so that* $P_0$ *and hence* $P_n$ *are uniquely determined.*

*Proof*  (i) If $\xi \in \mathbb{R}^p$ it follows from (3) with $m = 0$ that for $n \geqslant 0$,

$$|\Phi(n, 0)P_0\xi| \leqslant Ke^{-\alpha n}|P_0\xi|.$$

Also using (2), (4) and (5)

$$\begin{aligned}
|(I - P_0)\xi| &= |\Phi(0, n)(I - P_n)\Phi(n, 0)(I - P_0)\xi| \\
&\leqslant |\Phi(0, n)(I - P_n)\| \Phi(n, 0)(I - P_0)\xi| \\
&\leqslant Ke^{-\alpha n}|\Phi(n, 0)(I - P_0)\xi|
\end{aligned}$$

so that for $n \geqslant 0$

$$|\Phi(n, 0)(I - P_0)\xi| \geqslant K^{-1}e^{\alpha n}|(I - P_0)\xi|.$$

Thus $\Phi(n, 0)\xi = \Phi(n, 0)P_0\xi + \Phi(n, 0)(I - P_0)\xi$ is bounded if and only if $(I - P_0)\xi = 0$, that is, if and only if $\xi \in \mathscr{R}(P_0)$. This establishes (7).

Now let $Q_0$ be any projection with $\mathscr{R}(Q_0) = \mathscr{R}(P_0)$ and define $Q_n = \Phi(n, 0)Q_0\Phi(0, n)$. Then for $n \geqslant m \geqslant 0$

$$\begin{aligned}
|\Phi(n, m)&(Q_m - P_m)| \\
&= |\Phi(n, m)\Phi(m, 0)(Q_0 - P_0)\Phi(0, m)| \text{ using (6)} \\
&= |\Phi(n, 0)(Q_0 - P_0)\Phi(0, m)| \text{ by the cocycle property} \\
&= |\Phi(n, 0)P_0(Q_0 - P_0)(I - P_0)\Phi(0, m)| \text{ since } \mathscr{R}(P_0) = \mathscr{R}(Q_0) \\
&= |\Phi(n, 0)P_0(Q_0 - P_0)\Phi(0, m)(I - P_m)| \text{ by (6) and (2)} \\
&\leqslant |\Phi(n, 0)P_0\| Q_0 - P_0 \| \Phi(0, m)(I - P_m)| \\
&\leqslant K^2 e^{-\alpha(n + m)}|Q_0 - P_0| \text{ by (3) and (4).}
\end{aligned}$$

So for $n \geqslant m \geqslant 0$,

$$|\Phi(n, m)Q_m| \leqslant |\Phi(n, m)P_m| + |\Phi(n, m)(Q_m - P_m)|$$
$$\leqslant K[1 + K|Q_0 - P_0|]e^{-\alpha(n-m)}.$$

Similarly for $m \geqslant n \geqslant 0$,

$$|\Phi(n, m)(I - Q_m)| \leqslant K[1 + K|Q_0 - P_0|]e^{-\alpha(m-n)}.$$

Finally we observe that the invariance of $Q_n$ with respect to equation (1) follows from its definition and thus complete the proof of (i).

(ii) is proved in a similar way and (iii) is an immediate consequence of (i) and (ii).  □

REMARK 2.4   It follows from Proposition 2.3 that when $J = \mathbb{Z}^+$ (resp $J = \mathbb{Z}^-$) equation (1) has no nontrivial solution $u_n$ bounded on $J$ with $P_0 u_0 = 0$ (resp. $(I - P_0)u_0 = 0$) and when $J = \mathbb{Z}$ (1) has simply no nontrivial bounded solution.

REMARK 2.5   When $C_n$ is a constant matrix, equation (1) has an exponential dichotomy on $J$ if and only if the eigenvalues of $C$ do not lie on the unit circle.

For if $C$ satisfies the latter condition, we denote by $P$ the projection with range the sum of the generalized eigenspaces of $C$ corresponding to the eigenvalues inside the unit circle and with nullspace the sum of the generalized eigenspaces corresponding to the eigenvalues outside the unit circle. Then $C$ commutes with $P$ and there exist positive constants $K$, $\alpha$ such that

$$|C^n P| \leqslant Ke^{-\alpha n}(n \geqslant 0), \quad |C^n(I - P)| \leqslant Ke^{\alpha n}(n \leqslant 0).$$

This means that the equation

$$u_{n+1} = Cu_n,$$

which has transition matrix $C^{n-m}$, has an exponential dichotomy on $\mathbb{Z}$ with constant projection $P$.

Conversely, suppose this last equation has an exponential dichotomy on $\mathbb{Z}$ with constants $K$, $\alpha$ and projections $P_n$. So

$$|C^{n-m}P_m| \leqslant Ke^{-\alpha(n-m)}(n \geqslant m), \quad |C^{n-m}(I - P_m)| \leqslant Ke^{-\alpha(m-n)}(m \geqslant n).$$

If we define $Q_n = P_{n+1}$ then we see that

$$|C^{n-m}Q_m| = |C^{n+1-(m+1)}P_{m+1}| \leqslant Ke^{-\alpha(n-m)}(n \geqslant m),$$

$$|C^{n-m}(I - Q_m)| = |C^{n+1-(m+1)}(I - P_{m+1})| \leqslant Ke^{-\alpha(m-n)}(m \geqslant n).$$

Also for all $n$, since $P_n$ is invariant,

$$Q_{n+1}C = P_{n+2}C = CP_{n+1} = CQ_n.$$

So we also have an exponential dichotomy with projections $Q_n$. It follows by Proposition 2.3(iii) that $Q_n = P_n$ for all $n$. So $P_n$ is, in fact, a constant $P$ (say).

Then it follows from the invariance of $P_n$ that $P$ commutes with $C$ and, furthermore, from the inequalities above that

$$|C^n P| \leqslant Ke^{-\alpha n}(n \geqslant 0), |C^n(I - P)| \leqslant Ke^{\alpha n}(n \leqslant 0).$$

This means that $\mathbb{R}^p = \mathscr{R}(P) \oplus \mathscr{N}(P)$ where $\mathscr{R}(P)$, $\mathscr{N}(P)$ are both invariant under $C$ and the eigenvalues of $C$ restricted to $\mathscr{R}(P)$ lie inside the unit circle and those of $C$ restricted to $\mathscr{N}(P)$ lie outside the unit circle. Hence all the eigenvalues of $C$ lie off the unit circle.

It is clear that when (1) has an exponential dichotomy on $\mathbb{Z}$ then it has one on $\mathbb{Z}^+$ and $\mathbb{Z}^-$. In our next proposition we examine the extent to which the converse of this statement is true.

PROPOSITION 2.6   *Let $C_n$ be an invertible $p \times p$ matrix function defined for all $n$. Then the linear difference equation (1) has an exponential dichotomy on $\mathbb{Z}$ if and only if it has one on both $\mathbb{Z}^+$ and $\mathbb{Z}^-$ with projection of the same rank and it has no nontrivial solution bounded on $\mathbb{Z}$.*

*Proof*   The necessity follows from the definition of exponential dichotomy and Remark 2.4.

Now suppose conversely that (1) has an exponential dichotomy on both $\mathbb{Z}^+$ and $\mathbb{Z}^-$ with projections $P_n$ and $Q_n$ respectively, both of the same rank, and that it has no nontrivial solution bounded on $\mathbb{Z}$. By Proposition 2.3 we may choose $P_0$ as any projection with range

$$V^+ = \{\xi \in \mathbb{R}^p : \sup_{n \geqslant 0} |\Phi(n, 0)\xi| < \infty\}$$

and $Q_0$ as any projection with nullspace

$$V^- = \{\xi \in \mathbb{R}^p : \sup_{n \leqslant 0} |\Phi(n, 0)\xi| < \infty\}.$$

[Here $\Phi(n, m)$ is the transition matrix for (1).] Since (1) has no nontrivial solution bounded on $\mathbb{Z}$, $V^+ \cap V^- = \{0\}$ and since $P_0$ and $Q_0$ have the same rank, $\dim V^+ + \dim V^- = p$. So $\mathbb{R}^p = V^+ \oplus V^-$ and therefore we may choose $P_0$ and $Q_0$ as the same projection, the one with range $V^+$ and nullspace $V^-$. Then an inequality like (3) holds for $n \geqslant m \geqslant 0$, $0 \geqslant n \geqslant m$ and one like (4) for $m \geqslant n \geqslant 0$, $0 \geqslant m \geqslant n$. If $n \geqslant 0 \geqslant m$, then

$$\begin{aligned}
|\Phi(n, m)P_m| &= |\Phi(n, m)\Phi(m, 0)P_0\Phi(0, m)| \text{ by (6)} \\
&= |\Phi(n, 0)P_0 P_0\Phi(0, m)| \\
&= |\Phi(n, 0)P_0\Phi(0, m)P_m| \text{ by (5)} \\
&\leqslant |\Phi(n, 0)P_0\| \Phi(0, m)P_m| \\
&\leqslant Ke^{-\alpha n}Ke^{\alpha m} \\
&= K^2 e^{-\alpha(n - m)}
\end{aligned}$$

and if $n \leqslant 0 \leqslant m$, then similarly

$$|\Phi(n, m)(I - P_m)| \leqslant |\Phi(n, 0)(I - P_0)\| \Phi(0, m)(I - P_m)| \leqslant K^2 e^{-\alpha(m-n)}.$$

Hence (1) does have an exponential dichotomy on $\mathbb{Z}$.  $\square$

Our next aim is to prove the existence of a bounded solution of a quasilinear difference equation. First we consider an inhomogeneous equation.

LEMMA 2.7 *Let the linear difference equation* (1) *have an exponential dichotomy on J with constants K, $\alpha$ and projections $P_n$. Let $h_n$, $n \in J$, be a bounded sequence of vectors.*

(i) *If $J = \mathbb{Z}$ then the inhomogeneous difference equation*

$$(9) \qquad u_{n+1} = C_n u_n + h_n$$

*has a unique solution $u_n$ bounded on $\mathbb{Z}$. Moreover for all $n$*

$$(10) \qquad |u_n| \leqslant K(1 + e^{-\alpha})(1 - e^{-\alpha})^{-1} \sup_{n \in J} |h_n|.$$

(ii) *If $J = \mathbb{Z}^+$ or $\mathbb{Z}^-$ let $\xi \in \mathbb{R}^p$. Then (9) has a unique solution $u_n$ bounded on J and satisfying*

$$P_0 u_0 = P_0 \xi (J = \mathbb{Z}^+), \quad (I - P_0)u_0 = (I - P_0)\xi (J = \mathbb{Z}^-).$$

*Moreover for $n$ in J,*

$$(11) \qquad |u_n| \leqslant K|\xi| + K(1 + e^{-\alpha})(1 - e^{-\alpha})^{-1} \sup_{n \in J} |h_n|.$$

*Proof* We prove (i) first. Denote by $\Phi(n, m)$ the transition matrix for equation (1). Then we define

$$(12) \qquad u_n = \sum_{m \in J} G(n, m + 1)h_m \qquad \text{for } n \in J$$

where $G(n, m)$ is the *Green's function* defined by

$$G(n, m) = \begin{cases} \Phi(n, m)P_m & \text{if } m \leqslant n \qquad (n, m \in J) \\ -\Phi(n, m)(I - P_m) & \text{if } n < m \end{cases}$$

By inequalities (3) and (4),

$$|G(n, m)| \leqslant K e^{-\alpha|n - m|} \text{ for } n, m \in J$$

so that

$$\sum_{m \in J} |G(n, m+1)h_m| \leqslant K \sum_{m \in J} e^{-\alpha|n-m-1|} \sup_{n \in J} |h_n|$$

$$\leqslant K(1 + e^{-\alpha})(1 - e^{-\alpha})^{-1} \sup_{n \in J} |h_n|.$$

So $u_n$ is well-defined and satisfies (10).

Now notice that for each fixed $m$ in $J$, $X_n = G(n, m)$ solves the difference equation

$$X_{n+1} = C_n X_n + H_n^{(m)},$$

where

$$H_n^{(m)} = \begin{cases} I & \text{if } n = m - 1 \\ 0 & \text{if } n \neq m - 1. \end{cases}$$

So

$$u_{n+1} = \sum_{m \in J} G(n+1, m+1)h_m$$

$$= \sum_{m \in J, m \leqslant n-1} C_n G(n, m+1)h_m + [C_n G(n, n+1) + I]h_n$$

$$+ \sum_{m \in J, m \geqslant n+1} C_n G(n, m+1)h_m$$

$$= C_n \sum_{m \in J} G(n, m+1)h_m + h_n$$

$$= C_n u_n + h_n.$$

Thus $u_n$ is certainly a solution of (9). Finally note that the uniqueness follows from Remark 2.4.

The proof of (ii) is similar. Instead of formula (12) we have

$$(13) \qquad u_n = \Phi(n, 0)P_0\xi + \sum_{m \in J} G(n, m+1)h_m$$

when $J = \mathbb{Z}^+$ and

$$u_n = \Phi(n, 0)(I - P_0)\xi + \sum_{m = -\infty}^{-1} G(n, m+1)h_m$$

when $J = \mathbb{Z}^-$. The proof that $u_n$ is a solution of (9) and satisfies the estimate (11) is essentially the same as in part (i). Now when $J = \mathbb{Z}^+$,

$$u_0 = P_0\xi + \sum_{m=0}^{\infty} G(0, m+1)h_m = P_0\xi - \sum_{m=0}^{\infty} \Phi(0, m+1)(I - P_{m+1})h_m$$

$$= P_0\xi - \sum_{m=0}^{\infty} (I - P_0)\Phi(0, m+1)h_m \text{ using (6) and (2).}$$

So $P_0 u_0 = P_0 \xi$. Similarly when $J = \mathbb{Z}^-$ we can show that $(I - P_0)u_0 = (I - P_0)\xi$. Finally the uniqueness follows from Remark 2.4. $\qquad\square$

PROPOSITION 2.8   *Let the linear difference equation (1) have an exponential dichotomy on J with constants K, $\alpha$ and projections $P_n$ and for each n in J let $g_n(x)$ be an $\mathbb{R}^p$-valued function defined for x in $\mathbb{R}^p$ near 0 such that*

$$g_n(0) = 0, \; | g_n(x_1) - g_n(x_2) | \leqslant \gamma | x_1 - x_2 |$$

*when* $| x_1 | \leqslant \Delta, | x_2 | \leqslant \Delta$. *Suppose also that*

$$2K(1 + e^{-\alpha})(1 - e^{-\alpha})^{-1}\gamma \leqslant 1.$$

(i) *When $J = \mathbb{Z}$ let $h_n, n \in \mathbb{Z}$, be a bounded sequence of vectors satisfying*

$$2K(1 + e^{\alpha})(1 - e^{-\alpha})^{-1} | h_n | \leqslant \Delta$$

*for all n. Then the nonlinear difference equation*

(14) $$u_{n+1} = C_n u_n + h_n + g_n(u_n)$$

*has a unique solution $u_n$ such that $| u_n | \leqslant \Delta$ for all n and, moreover, for all n,*

$$| u_n | \leqslant 2K(1 + e^{-\alpha})(1 - e^{-\alpha})^{-1} \sup_{n \in J} | h_n |.$$

(ii) *When $J = \mathbb{Z}^+$ or $\mathbb{Z}^-$ let $\xi \in \mathbb{R}^p$ satisfy $| \xi | \leqslant \Delta/4K$ and let $h_n, n \in J$, be a bounded sequence of vectors satisfying*

$$4K(1 + e^{-\alpha})(1 - e^{-\alpha})^{-1} | h_n | \leqslant \Delta$$

*for all n. Then equation (14) has a unique solution $u_n$ such that $| u_n | \leqslant \Delta$ for all n and*

$$P_0 u_0 = P_0 \xi \, (J = \mathbb{Z}^+), \; (I - P_0)u_0 = (I - P)\xi \, (J = \mathbb{Z}^-).$$

*Moreover, for all n,*

$$| u_n | \leqslant 2K | \xi | + 2K(1 + e^{-\alpha})(1 - e^{-\alpha})^{-1} \sup_{n \in J} | h_n |.$$

*Proof*   We prove (i) first. Denote by $l^\infty(J)$ the Banach space of $\mathbb{R}^p$-valued bounded sequences $u = \{u_n\}_{n \in J}$ with the norm

$$\| u \| = \sup_{n \in J} | u_n |.$$

Then

$$\mathscr{D} = \{u \in l^\infty(J) : \| u \| \leqslant \Delta \}$$

is a closed subset of $l^\infty(J)$ and hence a complete metric space.

We define a mapping of $\mathscr{D}$ into itself. Let $u \in \mathscr{D}$. Then by Lemma 2.7 the inhomogeneous equation

$$(15) \qquad \hat{u}_{n+1} = C_n \hat{u}_n + [h_n + g_n(u_n)]$$

has a unique bounded solution $\hat{u}_n$. Moreover by (10)

$$\|\hat{u}\| \leqslant K(1 + e^{-\alpha})(1 - e^{-\alpha})^{-1} \sup_{n \in J} |h_n + g_n(u_n)|$$

$$(16) \qquad \leqslant K(1 + e^{-\alpha})(1 - e^{-\alpha})^{-1}[\|h\| + \gamma\|u\|]$$

$$\leqslant \Delta/2 + \|u\|/2 \leqslant \Delta.$$

So the mapping $u \to \hat{u}$ certainly maps $\mathscr{D}$ into itself.

Moreover the mapping is a contraction. For if $u_1, u_2 \in \mathscr{D}$ and $\hat{u}_1, \hat{u}_2$ are their images, then $w_n = \hat{u}_{1n} - \hat{u}_{2n}$ is a bounded solution of the equation

$$w_{n+1} = C_n w_n + [g_n(u_{1n}) - g_n(u_{2n})].$$

So by (10) again,

$$\|\hat{u}_1 - \hat{u}_2\| = \|w\| \leqslant K(1 + e^{-\alpha})(1 - e^{-\alpha})^{-1} \sup_{n \in J} |g_n(u_{1n}) - g_n(u_{2n})|$$

$$\leqslant K(1 + e^{-\alpha})(1 - e^{-\alpha})^{-1}\gamma\|u_1 - u_2\|$$

$$\leqslant 1/2 \|u_1 - u_2\|.$$

Denote by $u$ the unique fixed point of the mapping $u \to \hat{u}$ in $\mathscr{D}$. Then clearly $u_n$ is a solution of (14) satisfying $|u_n| \leqslant \Delta$. Moreover it is unique since any such solution must be in $\mathscr{D}$ and a fixed point of the mapping.

Finally putting $u = \hat{u}$ in (16), we obtain

$$\|u\| \leqslant K(1 + e^{-\alpha})(1 - e^{-\alpha})^{-1}\|h\| + \tfrac{1}{2}\|u\|$$

so that

$$\|u\| \leqslant 2K(1 + e^{-\alpha})(1 - e^{-\alpha})^{-1}\|h\|.$$

Thus we complete the proof of (i).

The proof of (ii) is very similar except that we use Lemma 2.7 (ii) and (11) instead of Lemma 2.7 (i) and (10). $\hat{u}_n$ is defined as the unique bounded solution of (15) such that

$$P_0 u_0 = P_0 \xi (J = \mathbb{Z}^+), \ (I - P_0)u_0 = (I - P_0)\xi (J = \mathbb{Z}^-). \qquad \square$$

REMARK 2.9   If $g_n(x) = D_n x$, where $D_n$ is a matrix with $|D_n| \leqslant \gamma$, we get a similar result. $\gamma$ satisfies the same restriction but we need only assume that $h_n$ is bounded. When $J = \mathbb{Z}$ we find a unique bounded solution $u_n$ and when $J = \mathbb{Z}^+$ or $\mathbb{Z}^-$ a unique bounded solution $u_n$ such that $P_0 u_0 = P_0 \xi (J = \mathbb{Z}^+)$, etc. where $\xi$ is any vector in $\mathbb{R}^p$.

We now use the preceding proposition to prove the roughness theorem for exponential dichotomies.

PROPOSITION 2.10   *Let $C_n$ be a $p \times p$ invertible matrix function defined for $n \in J$ such that for all $n$*

$$|C_n^{-1}| \leqslant M$$

*and such that the linear difference equation (1) has an exponential dichotomy on $J$ with constants $K$, $\alpha$ and projections $P_n$. Suppose $0 < \delta < \alpha$ and that $D_n$ is a $p \times p$ matrix function defined for $n \in J$ and satisfying*

$$|D_n| < M^{-1},$$
$$2K(1 + e^{-\alpha})(1 - e^{-\alpha})^{-1}|D_n| \leqslant 1,$$
$$2Ke^{\alpha}(e^{-\delta} + 1)(e^{\delta} - 1)^{-1}|D_n| \leqslant 1.$$

*Then $C_n + D_n$ is invertible for all $n$ and the perturbed difference equation*

$$(17) \qquad u_{n+1} = (C_n + D_n)u_n$$

*has an exponential dichotomy on $J$ with constants $2K(1 + e^{\delta})(1 - e^{-\delta})^{-1}$, $\alpha - \delta$ and projections of the same rank as for (1).*

*Proof*   For all $n$ we have

$$|(C_n + D_n) - C_n| = |D_n| < M^{-1} \leqslant |C_n^{-1}|^{-1}.$$

So $C_n + D_n$ is invertible for $n$. [Note that the conditions $|C_n^{-1}| \leqslant M$, $|D_n| < M^{-1}$ are only used to ensure the invertibility of $C_n + D_n$.]

Now denote by $\Phi(n, m)$ the transition matrix for equation (1). For each $m$ in $J$ we define the sequence

$$H_n^{(m)} = \begin{cases} I & \text{if } n = m - 1 \\ 0 & \text{if } n \neq m - 1 \end{cases} \qquad (n \in J)$$

and we let $X_n = G(n, m)$ be the unique bounded solution of the equation

$$(18) \qquad X_{n+1} = (C_n + D_n)X_n + H_n^{(m)}$$

which satisfies $P_0 X_0 = 0$ when $J = \mathbb{Z}^+$ and $m \geqslant 1$, $P_0 X_0 = P_0$ when $J = \mathbb{Z}^+$ and $m = 0$ and $(I - P)X_0 = 0$ when $J = \mathbb{Z}^-$. This exists by Proposition 2.8 and Remark 2.9 and satisfies

$$|G(n, m)| \leqslant 2K(1 + e^{-\alpha})(1 - e^{-\alpha})^{-1}$$

for $n$, $m$ in $J$. [Note that when $J = \mathbb{Z}^+$ and $m = 0$, $H_n^{(m)} = 0$ for all $n$ and so $|G(n, 0)| \leqslant 2K$.] The remainder of the proof consists in showing that $G(n, m)$ is a 'Green's function' for equation (17) (cf. the proof of Lemma 2.7).

For fixed $m$ in $J$ consider the modified equation

$$X_{n+1} = e^{\varepsilon_n^{(m)}(\alpha - \delta)}(C_n + D_n)X_n + H_n^{(m)},$$

where

$$\varepsilon_n^{(m)} = \begin{cases} 1 & \text{if } n \geqslant m \\ -1 & \text{if } n < m. \end{cases}$$

This equation can be rewritten as

$$(19) \qquad X_{n+1} = e^{\varepsilon_n^{(m)}(\alpha - \delta)} C_n X_n + e^{\varepsilon_n^{(m)}(\alpha - \delta)} D_n X_n + H_n^{(m)}.$$

Now the equation

$$u_{n+1} = e^{\varepsilon_n^{(m)}(\alpha - \delta)} C_n u_n$$

has an exponential dichotomy on $J$ with constants $K$, $\delta$ and projections $P_n$. So by Proposition 2.8 and Remark 2.9 again, equation (19) has a unique bounded solution $X_n = G_\delta(n, m)$ satisfying $P_0 X_0 = 0$ when $J = \mathbb{Z}^+$ and $m \geqslant 1$, etc. Moreover we have the estimate

$$|G_\delta(n, m)| \leqslant 2K(1 + e^{-\delta})(1 - e^{-\delta})^{-1}.$$

Then one sees easily that

$$\begin{aligned}
X_n &= e^{-\varepsilon_n^{(m)}(\alpha - \delta)(n - m)} G_\delta(n, m) \\
&= e^{-(\alpha - \delta)|n - m|} G_\delta(n, m)
\end{aligned}$$

is a bounded solution of (18) satisfying $P_0 X_0 = 0$ when $J = \mathbb{Z}^+$ and $m \geqslant 1$, etc. So by uniqueness

$$G(n, m) = e^{-(\alpha - \delta)(n - m)} G_\delta(n, m).$$

Hence we have the estimate

$$(20) \qquad |G(n, m)| \leqslant 2K(1 + e^{-\delta})(1 - e^{-\delta})^{-1} e^{-(\alpha - \delta)|n - m|}$$

for $n, m \in J$.

From the equation (18) that $G(n, m)$ satisfies we see that for $n \geqslant m$

$$\begin{aligned}
G(n, m) &= (C_{n-1} + D_{n-1}) G(n - 1, m) \\
&= (C_{n-1} + D_{n-1}) \dots (C_m + D_m) G(m, m).
\end{aligned}$$

Now if we take $n = m - 1$ in (18) with $X_n = G(n, m)$, we obtain

$$G(m, m) = (C_{m-1} + D_{m-1}) G(m - 1, m) + I.$$

So for $n < m$

$$\begin{aligned}
-(I - G(m, m)) &= (C_{m-1} + D_{m-1}) G(m - 1, m) \\
&= (C_{m-1} + D_{m-1}) \dots (C_n + D_n) G(n, m).
\end{aligned}$$

Hence if we denote by $\Psi(n, m)$ the transition matrix for equation (17) and write $Q_m = G(m, m)$, we have shown that

$$G(n, m) = \begin{cases} \Psi(n, m) Q_m & \text{for } n \geqslant m \\ -\Psi(n, m)(I - Q_m) & \text{for } n < m. \end{cases}$$

In view of (20) all that remains to be shown is that the $Q_m$ are projections invariant with respect to equation (17).

To show that $Q_m$ is a projection we fix $m$ in $J$ and define for $n \in J$

$$X_n = \begin{cases} \Psi(n, m)Q_m^2 = G(n, m)Q_m & \text{for } n \geqslant m \\ -\Psi(n, m)(I - Q_m^2) = G(n, m)(I + Q_m) & \text{for } n < m. \end{cases}$$

$X_n$ is bounded since $G(n, m)$ is. Also when $n \neq m - 1$ it is easy to see that

$$X_{n+1} = (C_n + D_n)X_n.$$

When $n = m - 1$,

$$\begin{aligned} X_{n+1} = X_m = Q_m^2 &= -(I - Q_m^2) + I \\ &= -(C_{m-1} + D_{m-1})\Psi(m - 1, m)(I - Q_m^2) + I \\ &= (C_{m-1} + D_{m-1})X_{m-1} + I. \end{aligned}$$

So $X_n$ is a solution of (18). Finally when $J = \mathbb{Z}^+$ and $m \geqslant 1$, $P_0 X_0 = 0$, etc. since $G(n, m)$ has these properties. It follows by uniqueness that $X_n = G(n, m)$ for $n \in J$. Taking $n = m$, we have $Q_m^2 = Q_m$. So $Q_m$ is indeed a projection.

The invariance of the $Q_m$ is proved by a similar kind of argument. Fix $m$ in $J(m \leqslant -1$ if $J = \mathbb{Z}^-)$ and define for $n \in J$

$$X_n = \begin{cases} \Psi(n, m)Q_m\Psi(n, m + 1) & \text{for } n > m \\ -\Psi(n, m)(I - Q_m)\Psi(m, m + 1) & \text{for } n \leqslant m. \end{cases}$$

$X_n = G(n, m)\Psi(m, m + 1)$ for $n \neq m$ and so it is bounded. When $n \neq m$ it is easy to see that

$$X_{n+1} = (C_n + D_n)X_n.$$

When $n = m$,

$$\begin{aligned} X_{n+1} = X_{m+1} = \Psi(m + 1, m)Q_m\Psi(m, m + 1) &= (C_m + D_m)[X_m + \Psi(m, m + 1)] \\ &= (C_m + D_m)X_m + I. \end{aligned}$$

So $X_n$ is a solution of (18) with $m + 1$ instead of $m$. Finally when $J = \mathbb{Z}^+$ and $m \geqslant 1$, $P_0 X_0 = 0$ and when $J = \mathbb{Z}^-$, $(I - P_0)X_0 = 0$ since $G(n, m)$ has these properties. When $J = \mathbb{Z}^+$ and $m = 0$,

$$P_0 X_0 = -P_0(I - G(0, 0))\Psi(0, 1) = 0 \text{ since } P_0 G(0, 0) = P_0.$$

So by uniqueness $X_n = G(n, m + 1)$ for $n, m \in J(m \leqslant -1$ if $J = \mathbb{Z}^-)$. In particular, taking $n = m + 1$ we have

$$\Psi(m + 1, m)Q_m\Psi(m, m + 1) = G(m + 1, m + 1) = Q_{m+1}.$$

That is,

$$(C_m + D_m)Q_m = Q_{m+1}(C_m + D_m).$$

So the projections $Q_m$ are indeed invariant with respect to equation (17) and the proof is finished.  $\square$

REMARK 2.11   One can also get an estimate for $|Q_n - P_n|$ in the following way. The Green's function for the unperturbed system (1) is defined as

$$G_0(n, m) = \begin{cases} \Phi(n, m)P_m & \text{for } n \geqslant m \\ -\Phi(n, m)(I - P_m) & \text{for } n < m \end{cases}$$

and is for each $m$ the unique bounded solution of the equation

$$X_{n+1} = C_n X_n + H_n^{(m)}$$

such that $P_0 X_0 = 0$ when $J = \mathbb{Z}^+$ and $m \geqslant 1$, etc. So $W_n = G(n, m) - G_0(n, m)$ is the unique bounded solution of

$$W_{n+1} = C_n W_n + D_n W_n + D_n G_0(n, m)$$

such that $P_0 W_0 = 0$ when $J = \mathbb{Z}^+$ and $(I - P_0)W_0 = 0$ when $J = \mathbb{Z}^-$. Then by Proposition 2.8 and Remark 2.9 we have the estimate

$$|W_n| \leqslant 2K(1 + e^{-\alpha})(1 - e^{-\alpha})^{-1} \sup_{n \in J} |D_n| K.$$

Take $n = m$ and we have for all $m$ in $J$,

$$|Q_m - P_m| \leqslant 2K^2(1 + e^{-\alpha})(1 - e^{-\alpha})^{-1} \sup_{n \in J} |D_n|.$$

## 3   HYPERBOLIC SETS AND THE SHADOWING LEMMA

Let $f: \mathbb{R}^p \to \mathbb{R}^p$ be a $C^1$ diffeomorphism. A subset $S$ of $\mathbb{R}^p$ is said to be *invariant* if $f(S) = S$.

DEFINITION 3.1   A compact invariant set $S$ is said to be *hyperbolic* if for each $x$ in $S$ the *'variational equation'* along the orbit $\{f^n(x)\}$

$$(21) \qquad u_{n+1} = Df(f^n(x))u_n$$

has an exponential dichotomy on $\mathbb{Z}$ with both the constants $K$, $\alpha$ and the rank of the projection $P_n(x)$ independent of $x$.

The following proposition gives an alternative definition of hyperbolic set.

PROPOSITION 3.2   *Let $f: \mathbb{R}^p \to \mathbb{R}^p$ be a $C^1$ diffeomorphism and $S$ a compact invariant set. Then $S$ is hyperbolic if and only if there are positive constants $K$, $\alpha$ and a projection matrix valued function $\mathscr{P}(x)$, $x \in S$, of constant rank such that for all $x$ in $S$*

$$(22) \qquad \mathscr{P}(f(x))Df(x) = Df(x)\mathscr{P}(x),$$

$$(23)\, |Df^n(x)\mathscr{P}(x)| \leqslant Ke^{-\alpha n}(n \geqslant 0),\, |Df^n(x)(I - \mathscr{P}(x))| \leqslant Ke^{\alpha n}(n \leqslant 0).$$

*Proof*  Let us note that the transition matrix for equation (21) is given by

$$\Phi(n, m)(x) = Df^{n-m}(f^m(x)).$$

First we prove the sufficiency. We define

$$P_n(x) = \mathscr{P}(f^n(x))$$

and show that (21) has an exponential dichotomy on $\mathbb{Z}$ with constants $K$, $\alpha$ and projections $P_n(x)$. $P_n(x)$ is invariant since

$$\begin{aligned}
P_{n+1}(x)Df(f^n(x)) &= \mathscr{P}(f^{n+1}(x))Df(f^n(x)) \\
&= Df(f^n(x))\mathscr{P}(f^n(x)) \text{ by (22)} \\
&= Df(f^n(x))P_n(x).
\end{aligned}$$

Also for $n \geqslant m$

$$\begin{aligned}
|\Phi(n, m)(x)P_m(x)| &= |Df^{n-m}(f^m(x))\mathscr{P}(f^m(x))| \\
&\leqslant Ke^{-\alpha(n-m)}
\end{aligned}$$

and for $n \leqslant m$

$$\begin{aligned}
|\Phi(n, m)(x)(I - P_m(x))| &= |Df^{n-m}(f^m(x))(I - \mathscr{P}(f^m(x)))| \\
&\leqslant Ke^{-\alpha(m-n)},
\end{aligned}$$

where we have used (23). So the sufficiency is proved.

To prove the necessity, let $K$, $\alpha$ and $P_n(x)$ be as in Definition 3.1. Note first that for any integer $m$ the variational equation

$$u_{n+1} = Df(f^n(f^m(x)))u_n$$

has an exponential dichotomy on $\mathbb{Z}$ with projections $P_n(f^m(x))$. But this difference equation can also be written as

$$u_{n+1} = Df(f^{n+m}(x))u_n$$

from which it is clear by comparison with (21) that it has an exponential dichotomy with projections $P_{n+m}(x)$. By uniqueness of the projections (Proposition 2.3 (iii)) it follows that $P_{n+m}(x) = P_n(f^m(x))$. In particular we have the identity

(24) $$P_n(x) = P_0(f^n(x)).$$

We define

$$\mathscr{P}(x) = P_0(x).$$

Then $\mathscr{P}(x)$ is of constant rank and it follows from identity (24) that

$$P_n(x) = \mathscr{P}(f^n(x)).$$

So for all $x$

$$\mathscr{P}(f(x))Df(x) = P_1(x)Df(f^0(x))$$
$$= Df(f^0(x))P_0(x) \text{ by the invariance of } P_n(x)$$
$$= Df(x)\mathscr{P}(x).$$

Thus we have the invariance property (22). Also for all $x \in S$ and $n \geqslant 0$

$$|Df^n(x)\mathscr{P}(x)| = |\Phi(n,0)(x)P_0(x)|$$
$$\leqslant Ke^{-\alpha n}$$

and for $n \leqslant 0$,

$$|Df^n(x)(I - \mathscr{P}(x))| = |\Phi(n,0)(x)(I - P_0(x))|$$
$$\leqslant Ke^{\alpha n}.$$

This establishes (23) and completes the proof of the proposition. $\square$

PROPOSITION 3.3  *Let $\mathscr{P}(x)$ be as in Proposition 3.2. Then $\mathscr{P}(x)$ is continuous.*

*Proof*  It follows from (23) with $n = 0$ that

$$|\mathscr{P}(x)| \leqslant K$$

for all $x$ in $S$. Since $S$ is compact, to show that $\mathscr{P}$ is continuous it suffices to show that the mapping $x \to \mathscr{P}(x)$ has closed graph. So let $x_k \to x$ and $\mathscr{P}(x_k) \to Q$. We have to show that $Q = \mathscr{P}(x)$. Now, from (23), we have the inequalities

$$(25) \qquad |Df^{n-m}(f^m(x_k))\mathscr{P}(f^m(x_k))| \leqslant Ke^{-\alpha(n-m)} \qquad (n \geqslant m)$$
$$|Df^{n-m}(f^m(x_k))(I - \mathscr{P}(f^m(x_k)))| \leqslant Ke^{-\alpha(m-n)} \qquad (m \geqslant n).$$

Also, by repeated application of (22),

$$(26) \qquad \mathscr{P}(f^n(x_k)) = Df^n(x_k)\mathscr{P}(x_k)Df^{-n}(f^n(x_k)).$$

Letting $k \to \infty$ in (25), (26) we obtain

$$|Df^{n-m}(f^m(x))Q_m| \leqslant Ke^{-\alpha(n-m)}(n \geqslant m), |Df^{n-m}(f^m(x))(I - Q_m)|$$
$$\leqslant Ke^{-\alpha(m-n)}(m \geqslant n),$$

where $Q_n = Df^n(x)QDf^{-n}(f^n(x))$. This means that the linear difference equation (21), which has transition matrix $Df^{n-m}(f^m(x))$, has an exponential dichotomy on $\mathbb{Z}$ with projections $Q_n$. But we know already that it has an exponential dichotomy with projections $P_n(x) = \mathscr{P}(f^n(x))$. So by uniqueness (Proposition 2.3 (iii)) $Q_n = \mathscr{P}(f^n(x))$ for all $n$. In particular, taking $n = 0$, $Q = \mathscr{P}(x)$. $\square$

DEFINITION 3.4    A doubly infinite sequence $\{y_n\}$ in $\mathbb{R}^p$ is said to be a $\delta$ *pseudo-orbit* of a $C^1$ diffeomorphism $f: \mathbb{R}^p \to \mathbb{R}^p$ if for all integers $n$

$$|y_{n+1} - f(y_n)| \leqslant \delta.$$

An orbit $\{f^n(x)\}$ is said to $\varepsilon$-*shadow* the $\delta$ pseudo-orbit $\{y_n\}$ if for all integers $n$

$$|f^n(x) - y_n| \leqslant \varepsilon.$$

We now state and prove the so-called '*shadowing lemma*'.

THEOREM 3.5    *Let S be a compact hyperbolic set for the $C^1$ diffeomorphism $f: \mathbb{R}^p \to \mathbb{R}^p$. Then given $\varepsilon > 0$ sufficiently small there exists $\delta > 0$ such that every $\delta$ pseudo-orbit in S has a unique $\varepsilon$-shadowing orbit.*

*Proof*    Let $K$, $\alpha$ and $\mathscr{P}(x)$ be as in Definition 3.1 and Proposition 3.2. Denote by $B$ a bounded closed ball containing $S$ and put

$$M = \sup_{x \in B} |Df(x)|.$$

Denote by $\omega_1(\cdot)$ the modulus of continuity for $Df(x)$ with $x$ in $B$. That is, if $\Delta > 0$,

$$\omega_1(\Delta) = \sup\{|Df(x) - Df(y)| : x, y \in B, |x - y| \leqslant \Delta\}.$$

Also by Proposition 3.3 $\mathscr{P}(x)$ is continuous on the set $S$. Denote by $\omega_2(\cdot)$ its modulus of continuity.

We have to solve the nonlinear difference equation

$$x_{n+1} = f(x_n)(n \in \mathbb{Z})$$

for a sequence $\{x_n\}$ such that $|x_n - y_n| \leqslant \varepsilon$ for all $n$. Write

$$x_n = y_n + z_n.$$

Then $\{z_n\}$ is a solution of

(27)
$$z_{n+1} = C_n z_n + h_n + g_n(z_n),$$

where

$$C_n = Df(y_n),$$
$$g_n(z) = f(y_n + z) - f(y_n) - Df(y_n)z$$

and
$$h_n = f(y_n) - y_{n+1}.$$

Define

$$P_n = \mathscr{P}(y_n), \quad Q_{n+1} = \mathscr{P}(f(y_n))$$

for $n$ in $\mathbb{Z}$. Then it follows from (22) that for all $n$

$$(28) \qquad Q_{n+1}C_n = C_n P_n.$$

Now define

$$D_n = P_{n+1}C_n P_n + (I - P_{n+1})C_n(I - P_n).$$

It is easily verified that

$$P_{n+1}D_n = D_n P_n.$$

Also

$$
\begin{aligned}
|C_n - D_n| &= |(I - P_{n+1})C_n P_n + P_{n+1}C_n(I - P_n)| \\
&= |(I - P_{n+1})Q_{n+1}C_n + P_{n+1}(I - Q_{n+1})C_n| \text{ by (28)} \\
&= |[(Q_{n+1} - P_{n+1})Q_{n+1} + P_{n+1}(P_{n+1} - Q_{n+1})]C_n| \\
&\qquad \text{since } P_{n+1}, \ Q_{n+1} \text{ are projections} \\
&\leqslant \{|Q_{n+1}| + |P_{n+1}|\}|C_n\|Q_{n+1} - P_{n+1}| \\
&\leqslant 2KM\omega_2(\delta).
\end{aligned}
$$

Thus for all $n$,

$$(29) \qquad |C_n - D_n| \leqslant 2KM\omega_2(\delta).$$

We now rewrite (27) as

$$(30) \qquad z_{n+1} = D_n z_n + h_n + [(C_n - D_n)z_n + g_n(z_n)]$$

and aim to apply Proposition 2.8. In order to do this, we must first show that the linear equation

$$(31) \qquad z_{n+1} = D_n z_n$$

has an exponential dichotomy on $\mathbb{Z}$. We shall, in fact, show that the equation has an exponential dichotomy with projections $P_n$. The invariance of the $P_n$ follows from the definition of $D_n$ and from the equality $P_{n+1}D_n = D_n P_n$.

Using the inequality

$$
\begin{aligned}
|y_{n+k+1} - f^{k+1}(y_n)| &\leqslant |y_{n+k+1} - f(y_{n+k})| + |f(y_{n+k}) - f(f^k(y_n))| \\
&\leqslant \delta + M|y_{n+k} - f^k(y_n)|,
\end{aligned}
$$

it is easy to prove by induction on $k$ that

$$(32) \qquad |y_{n+k} - f^k(y_n)| \leqslant (1 + M + \cdots + M^{k-1})\,\delta$$

for all $n$ and $k \geqslant 1$.

Denote by $\Psi(n, m)$ the transition matrix for (31). Now fix $n$ and put

$$d_k = |\Psi(n + k, n)P_n - Df^k(y_n)P_n|,$$

where $k \geqslant 1$. Choose the smallest positive integer $N$ such that

$$Ke^{-\alpha N} \leqslant 1/2.$$

Our aim is to estimate $d_N$. We do this by getting an estimate for $d_{k+1}$ in terms of $d_k$ for $1 \leqslant k \leqslant N-1$. First we must assume

$$(33) \qquad (1 + M + \cdots + M^{N-1})\, \delta \leqslant \operatorname{dist}(S, B^c).$$

This ensures that $f^k(y_n) \in B$ for all $n$ and $1 \leqslant k \leqslant N$, using (32). Then

$$
\begin{aligned}
d_{k+1} &= |\Psi(n+k+1, n)P_n - Df^{k+1}(y_n)P_n| \\
&\leqslant |\Psi(n+k+1, n)P_n - Df(f^k(y_n))\Psi(n+k, n)P_n| \\
&\quad + |Df(f^k(y_n))\Psi(n+k, n)P_n - Df(f^k(y_n))Df^k(y_n)P_n| \\
&\hspace{6cm}\text{using the chain rule}
\end{aligned}
$$

$$
\begin{aligned}
&\leqslant |D_{n+k} - Df(f^k(y_n))|\,\|\Psi(n+k, n)P_n| \\
&\quad + |Df(f^k(y_n))|\,\|\Psi(n+k, n)P_n - Df^k(y_n)P_n| \\
&\leqslant \{|D_{n+k} - C_{n+k}| + |Df(y_{n+k}) - Df(f^k(y_n))|\}\,|\Psi(n+k, n)P_n| \\
&\quad + |Df(f^k(y_n))|\,\|\Psi(n+k, n)P_n - Df^k(y_n)P_n| \\
&\leqslant \{2KM\omega_2(\delta) + \omega_1((1 + M + \cdots + M^{k-1})\,\delta)\}2^k M^k K^{2k} K + Md_k
\end{aligned}
$$

where we have used (29), (32) and the fact that $2MK^2$ is an upper bound for $|D_n|$. Hence for $1 \leqslant k \leqslant N-1$,

$$d_{k+1} \leqslant \omega_3(\delta) + Md_k$$

where

$$\omega_3(\delta) = \sup_{1 \leqslant k \leqslant N-1} 2^k M^k K^{2k+1}\{2KM\omega_2(\delta) + \omega_1((1 + M + \cdots + M^{k-1})\,\delta)\}.$$

It follows by induction on $k$ that

$$d_N \leqslant (1 + M + \cdots + M^{N-2})\omega_3(\delta) + M^{N-1}d_1.$$

Now

$$d_1 = |[D_n - C_n]P_n| \leqslant 2K^2 M\omega_2(\delta) \text{ by (29)}.$$

So

$$d_N \leqslant (1 + M + \cdots + M^{N-2})\omega_3(\delta) + 2K^2 M^N \omega_2(\delta)$$

$$(34) \qquad\qquad \leqslant 1/4,$$

if $\delta$ is small enough.

Then it follows that for all $n$

$$
\begin{aligned}
|\Psi(n+N, n)P_n| &\leqslant d_N + |Df^N(y_n)P_n| \\
&= d_N + |Df^N(y_n)\mathscr{P}(y_n)| \\
&\leqslant d_N + Ke^{-\alpha N} \text{ by (23)} \\
&\leqslant \tfrac{1}{4} + \tfrac{1}{2}.
\end{aligned}
$$

So for all $n$,

$$|\Psi(n+N,n)P_n| \leqslant 3/4.$$

We want to extend this last inequality to one for $|\Psi(n,m)P_n|$ where $n \geqslant m$. This is a unique non-negative integer $k$ such that

$$m + kN \leqslant n < m + (k+1)N.$$

Then

$$|\Psi(n,m)P_m|$$
$$= |\Psi(n,m+kN)\Psi(m+kN,m+(k-1)N)\ldots\Psi(m+N,m)P_m|$$
$$\text{by the cocycle property}$$
$$= |\Psi(n,m+kN)\Psi(m+kN,m+(k-1)N)P_{m+(k-1)N}\ldots\Psi(m+N,m)P_m|$$
$$\text{using the identity like (5)}$$
$$\leqslant |\Psi(n,m+kN)\|\Psi(m+kN,m+(k-1)N)P_{m+(k-1)N}|\ldots$$
$$|\Psi(m+N,m)P_m|$$
$$\leqslant (2MK^2)^{N-1}(\tfrac{3}{4})^k \leqslant (2MK^2)^{N-1}(\tfrac{3}{4})^{((n-m)/N)-1}.$$

So for $n \geqslant m$,

$$|\Psi(n,m)P_m| \leqslant K_1 e^{-\alpha_1(n-m)},$$

where

$$K_1 = \tfrac{4}{3}(2MK^2)^{N-1}, \alpha_1 = \frac{\log 4/3}{N}.$$

Similarly we can show that for $m \geqslant n$

$$|\Psi(n,m)(I-P_m)| \leqslant K_1 e^{-\alpha_1(m-n)}.$$

Hence we have shown that when $\delta > 0$ is so small that (33) and (34) are satisfied the linear equation (31) has an exponential dichotomy on $\mathbb{Z}$ with constants $K_1$, $\alpha_1$ and projections $P_n$.

Now we apply Proposition 2.8 to equation (30) to prove the existence of a unique solution $z_n$ such that $|z_n| \leqslant \varepsilon$ for all $n$. The nonlinear term in (30) has Lipschitz constant

$$|C_n - D_n| + \omega_1(\varepsilon)$$

provided we restrict $z$ to $|z| \leqslant \varepsilon$ and assume $\varepsilon \leqslant \text{dist}(S, B^c)$. So the condition on the Lipschitz constant in Proposition 2.8 will be fulfilled provided

$$(35) \qquad 4K_1(1 + e^{-\alpha_1})(1 - e^{-\alpha_1})^{-1}2KM\omega_2(\delta) \leqslant 1$$

and

$$(36) \qquad 4K_1(1 + e^{-\alpha_1})(1 - e^{-\alpha_1})^{-1}\omega_1(\varepsilon) \leqslant 1,$$

where we have used (29). Also the condition on $h_n$ will be fulfilled provided

$$(37) \qquad 2K_1(1 + e^{-\alpha_1})(1 - e^{-\alpha_1})^{-1}\delta \leqslant \varepsilon.$$

Hence if $\varepsilon > 0$ satisfies $\varepsilon \leqslant \operatorname{dist}(S, B^c)$ and (36) and then $\delta > 0$ and is chosen so small that (33), (34), (35) and (37) are satisfied, the difference equation (27) has a unique solution $z_n$ such that $|z_n| \leqslant \varepsilon$ for all $n$.  $\square$

REMARK 3.6    The variational equation of the shadowing orbit $x_n$ also has an exponential dichotomy on $\mathbb{Z}$ provided $\varepsilon$ and $\delta$ are sufficiently small. The variational equation is, in fact,

$$(38) \qquad u_{n+1} = D_n u_n + \{C_n - D_n + Dg_n(z_n)\}u_n.$$

We know that equation (31) has an exponential dichotomy on $\mathbb{Z}$ with constants $K_1, \alpha_1$. So by Proposition 2.10 equation (38) will also have an exponential dichotomy on $\mathbb{Z}$ with projections of the same rank provided

$$\sup_{n \in \mathbb{Z}} \{|C_n - D_n| + |Dg_n(z_n)|\} \leqslant 2KM\omega_2(\delta) + \omega_1(\varepsilon)$$

is sufficiently small and this can be achieved by making $\delta$ and $\varepsilon$ small.


# 4   CHAOTIC BEHAVIOUR IN THE NEIGHBOURHOOD OF A TRANSVERSAL HOMOCLINIC POINT: SMALE'S THEOREM

Let $f: \mathbb{R}^p \to \mathbb{R}^p$ be a $C^1$ diffeomorphism.


DEFINITION 4.1    $x_0 \in \mathbb{R}^p$ is said to be a *hyperbolic fixed point* of $f$ if $f(x_0) = x_0$ and the eigenvalues of $Df(x_0)$ lie off the unit circle.

Let $x_0$ be a hyperbolic fixed point of $f$ and denote by $P$ the projection with range the sum of the generalized eigenspaces of $Df(x_0)$ corresponding to the eigenvalues inside the unit circle and with nullspace the sum of the generalized eigenspaces corresponding to the eigenvalues outside the unit circle. Then $Df(x_0)$ commutes with $P$ and there exist positive constants $K$, $\alpha$ such that

$$(39) \qquad |Df^n(x_0)P| \leqslant Ke^{-\alpha n}(n \geqslant 0), |Df^n(x_0)(I - P)| \leqslant Ke^{\alpha n}(n \leqslant 0).$$

This means that the variational equation along the orbit $\{x_0\}$

$$(40) \qquad u_{n+1} = Df(x_0)u_n$$

has an exponential dichotomy on $\mathbb{Z}$ with the constant projection $P$. So if $x_0$ is a hyperbolic fixed point the set $\{x_0\}$ is a hyperbolic set in the sense of Definition 3.1.

DEFINITION 4.2   $y_0 \in \mathbb{R}^p$ is said to be a *homoclinic point* with respect to the hyperbolic fixed point $x_0$ if $y_0 \neq x_0$ and $f^n(y_0) \to x_0$ as $|n| \to \infty$.

The following proposition examines the variational equation along the orbit of a homoclinic point.


PROPOSITION 4.3   *Let $y_0$ be a homoclinic point with respect to a hyperbolic fixed point $x_0$ of a $C^1$ diffeomorphism $f: \mathbb{R}^p \to \mathbb{R}^p$. The the variational equation*

$$(41) \qquad\qquad u_{n+1} = Df(f^n(y_0))u_n$$

*has an exponential dichotomy on both $\mathbb{Z}^+$ and $\mathbb{Z}^-$ with the rank of the projection, in both cases, being equal to the number of eigenvalues of $Df(x_0)$ inside the unit circle.*


*Proof*   Equation (40) has an exponential dichotomy on $\mathbb{Z}$ with the constant projection $P$, the rank of $P$ being equal to the number of eigenvalues of $Df(x_0)$ inside the unit circle.

For fixed $q \geqslant 0$ consider the equation

$$(42) \qquad \begin{aligned} u_{n+1} &= Df(f^{n+q}(y_0))u_n \\ &= [Df(x_0) + (Df(f^{n+q}(y_0)) - Df(x_0))]\, u_n. \end{aligned}$$

We can make $|Df(f^{n+q}(y_0)) - Df(x_0)|$ uniformly small for $n \geqslant 0$ by choosing $q$ sufficiently large. So it follows from Proposition 2.10 that if $q$ is sufficiently large, equation (42) has an exponential dichotomy on $\mathbb{Z}^+$ with projections of the same rank as $P$. By comparing the transition matrices for (41) and (42) one sees that this implies that equation (41) has an exponential dichotomy on $[q, \infty)$ and hence by Remark 2.2 on $\mathbb{Z}^+$, with projections of the same rank as $P$.

Similarly we can show that (41) also has an exponential dichotomy on $\mathbb{Z}^-$ with projections of the same rank as $P$.   $\square$


DEFINITION 4.4   Let $y_0$ be a homoclinic point with respect to a hyperbolic fixed point $x_0$ of a $C^1$ diffeomorphism $f: \mathbb{R}^p \to \mathbb{R}^p$. $y_0$ is said to be a *transversal homoclinic point* if the variational equation (41) has no nontrivial solution bounded on $\mathbb{Z}$.


PROPOSITION 4.5   *Let $y_0$ be a transversal homoclinic point with respect to a hyperbolic fixed point $x_0$ of a $C^1$ diffeomorphism $f: \mathbb{R}^p \to \mathbb{R}^p$. Then the set*

$$\{x_0\} \cup \{f^n(y_0) : n \in \mathbb{Z}\}$$

*is a hyperbolic set.*

*Proof* By Proposition 4.3 the variational equation (41) has an exponential dichotomy on both $\mathbb{Z}^+$ and $\mathbb{Z}^-$ with projections of the same rank. Also since $y_0$ is transversal (41) has no nontrivial solution bounded on $\mathbb{Z}$. It follows by Proposition 2.6 that (41) has an exponential dichotomy on $\mathbb{Z}$. Also if $q$ is a fixed integer by comparing the transition matrix for the equation

$$(43) \qquad u_{n+1} = Df(f^n(f^q(y_0)))u_n = Df(f^{n+q}(y_0))u_n$$

with the transition matrix for (41), we see that (43) has an exponential dichotomy on $\mathbb{Z}$ with the same constants and with projections of the same rank as for (41). Moreover since $x_0$ is a hyperbolic fixed point its variational equation (40) has an exponential dichotomy on $\mathbb{Z}$ with projections having rank equal to the number of eigenvalues of $Df(x_0)$ inside the unit circle. Since the projections for (43) have this as rank also and since there are only two possible choices for the constants, it follows from Definition 3.1 that the set $\{x_0\} \cup \{f^n(y_0) : n \in \mathbb{Z}\}$, which is clearly compact and invariant, is a hyperbolic set. $\quad\square$

Again let $y_0$ be a transversal homoclinic point with respect to a hyperbolic fixed point $x_0$ of a $C^1$ diffeomorphism $F : \mathbb{R}^p \to \mathbb{R}^p$. Then we know that the set

$$(44) \qquad S = \{x_0\} \cup \{f^n(y_0) : n \in \mathbb{Z}\}$$

is hyperbolic. Our aim now is to use the shadowing lemma to show that the dynamics of $f$ near $S$ are chaotic.


DEFINITION 4.6    Let $S$ be as in (44). An *orbit segment of length l* is a finite sequence of $l$ points either of the form

$$\{x_0, x_0, ..., x_0\}$$

or of the form

$$\{f^{q+1}(y_0), f^{q+2}(y_0), ..., ...f^{q+l}(y_0)\}.$$

We consider doubly infinite sequences

$$..., C_{-1}, C_0, C_1, ...$$

of orbit segments. We string the points in the segments together to form a doubly infinite sequence of points.

Now by the shadowing lemma (Theorem 3.5) given $\varepsilon > 0$ sufficiently small there is a $\delta > 0$ such that any $\delta$ pseudo-orbit lying in $S$ has a unique $\varepsilon$-shadowing orbit (this latter orbit may not be in $S$). Fix such an $\varepsilon$ and the corresponding $\delta$ and consider doubly infinite sequences formed from orbit segments with the property that the left endpoint and the image under $f$ of the right endpoint lie in the ball of radius $\delta/2$ centred at $x_0$. (This condition is always satisfied when the orbit segment contains only the point $x_0$ and in the other cases can be satisfied by choosing $q$ and $l$ sufficiently large.) Then each such doubly infinite sequence has a unique $\varepsilon$-shadowing orbit. So what we get are orbits which fluctuate between staying near $x_0$ and following the orbit of $y_0$ around, in a random sort of manner.

Smale [12, 13] has given a more exact mathematical description of the chaotic dynamics near $S$.

DEFINITION 4.7   Denote by $\mathscr{S}_m(m \geqslant 2)$ the set of doubly infinite sequences

$$a = (..., a_{-1}, a_0, a_1, a_2, ...)$$

where $a_i \in \{1, 2, ..., m\}$. So $\mathscr{S}_m = \{1, 2, ..., m\}^{\mathbb{Z}}$. We give the set $\{1, 2, ..., m\}$ the discrete topology and $\mathscr{S}_m$ the product topology. The *Bernoulli shift* $\beta : \mathscr{S}_m \to \mathscr{S}_m$ is the homeomorphism defined by

$$[\beta(a)]_i = a_{i+1}.$$

THEOREM 4.8   *Let $y_0$ be a transversal homoclinic point with respect to a hyperbolic fixed point $x_0$ of a $C^1$ diffeomorphism $F: \mathbb{R}^p \to \mathbb{R}^p$. Then there is an integer $l$ and a homeomorphism $\phi$ of $\mathscr{S}_m$ onto a compact subset of $\mathbb{R}^p$ which is invariant under $f$ and such that*

$$f^l \circ \phi = \phi \circ \beta,$$

*that is, the action of $f^l$ on $\phi(\mathscr{S}_m)$ is conjugate to the action of $\beta$ on $\mathscr{S}_m$.*

*Proof*   Let $p_1, p, ..., p_r$, where $r = m - 1$ or $m$, be distinct integers. Choose and fix $\varepsilon > 0$ as small as is required when the shadowing lemma is applied to the hyperbolic set $S = \{x_0\} \cup \{f^n(y_0) : n \in \mathbb{Z}\}$ and also so that

(45)
$$\varepsilon < \tfrac{1}{2} \min_{i \neq j} | f^{p_i}(y_0) - f^{p_j}(y_0)|.$$

When $r = m - 1$ we also require that

(46)
$$\varepsilon < \tfrac{1}{2} \min_i | f^{p_i}(y_0) - x_0|.$$

Now let $\delta > 0$ correspond to $\varepsilon$ as in the shadowing lemma and choose $N$ so that for $i = 1, ..., r$

$$|f^n(y_0) - x_0| < \delta/2 \quad \text{when } n = N + p_i + 1 \text{ or } p_i - N.$$

For $i = 1, \ldots, r$ we define the orbit segment

$$C_i = \{ f^{p_i - N}(y_0), \ldots, f^{p_i}(y_0), \ldots, f^{p_i + N}(y_0) \}$$

and if $r = m - 1$, we define

$$C_m = \{ x_0, \ldots, x_0 \},$$

where there are $2N + 1$ $x_0$'s.

Now let

$$a = (\ldots, a_{-1}, a_0, a_1, \ldots)$$

be in $\mathscr{S}_m$. Then, as in the discussion before Definition 4.7, the doubly infinite sequence formed by stringing the orbit segments

$$\ldots, C_{a_{-1}}, C_{a_0}, C_{a_1}, \ldots$$

together is a $\delta$ pseudo-orbit and hence has a unique $\varepsilon$-shadowing orbit. We define $\phi(a)$ to be the point in this orbit which 'shadows' the midpoint of the orbit segment $C_{a_0}$. So if we denote by $z_{ij}(a)$ the $j$th point of $C_{a_i}$, $\phi(a)$ is the *unique* point such that

$$(47) \qquad | f^{N + (i - 1)(2N + 1) + j}(\phi(a)) - z_{ij}(a) | \leqslant \varepsilon$$

for all $i$ and $j = 1, \ldots, 2N + 1$. If we replace $i$ by $i + 1$ we get

$$| f^{N + i(2N + 1) + j}(\phi(a)) - z_{i+1, j}(a) | \leqslant \varepsilon.$$

But this can be rewritten as

$$| f^{N + (i - 1)(2N + 1) + j}(f^{2N + 1}(\phi(a))) - z_{ij}(\beta(a)) | \leqslant \varepsilon,$$

since $a_{i+1} = [\beta(a)]_i$. This last inequality holds for all $i$ and $j = 1, \ldots, 2N + 1$. By comparison with (47) we see by uniqueness that this means

$$\phi(\beta(a)) = f^{2N + 1}(\phi(a)).$$

This holds for all $a \in \mathscr{S}_m$ and so

$$\phi \circ \beta = f^{2N + 1} \circ \phi.$$

This also shows that $\phi(\mathscr{S}_m)$ is invariant under $f^{2N + 1}$.

Now we show that $\phi$ is one-to-one. If $a \neq a'$ then $a_i \neq a_i'$ for some $i$. Then

$$| f^{N + (i - 1)(2N + 1) + (N + 1)}(\phi(a)) - f^{N + (i - 1)(2N + 1) + (N + 1)}(\phi(a')) |$$

$$\geqslant | z_{i, N+1}(a) - z_{i, N+1}(a') | - | f^{N + (i - 1)(2N + 1) + (N + 1)}(\phi(a)) - z_{i, N+1}(a) |$$

$$- | f^{N + (i - 1)(2N + 1) + (N + 1)}(\phi(a')) - z_{i, N+1}(a') |$$

$$\geqslant | z_{i, N+1}(a) - z_{i, N+1}(a') | - 2\varepsilon \text{ using (47)}$$

$$> 0 \text{ using (45) and (46).}$$

So $\phi(a) \neq \phi(a')$, as required.

Next we show that $\phi$ is continuous. If not, there is a point $a$ in $\mathscr{S}_m$, a number $\Delta > 0$ and a sequence $a^{(k)}$ in $\mathscr{S}_m$ such that for all $k$,

$$a_i^{(k)} = a_i \text{ for } |i| \leqslant k$$

but

(48)
$$|\phi(a^{(k)}) - \phi(a)| \geqslant \Delta.$$

$\phi(\mathscr{S}_m)$ lies in the $\varepsilon$-neighbourhood of $S$ and so is certainly bounded. So taking a subsequence if necessary, we may assume without loss of generality that $\phi(a^{(k)}) \to x$ (say). Then from (48)

$$|x - \phi(a)| \geqslant \Delta.$$

Now using (47) we have

$$|f^{N + (i-1)(2N+1) + j}(\phi(a^{(k)})) - z_{ij}(a^{(k)})| \leqslant \varepsilon$$

for all $k$, $i$ and $j = 1, \ldots, 2N + 1$. But $C_{a_i^{(k)}} = C_{a_i}$ if $|i| \leqslant k$. So if we fix $i$ and let $k \to \infty$ we get

$$|f^{N + (i-1)(2N+1) + j}(x) - z_{ij}(a)| \leqslant \varepsilon$$

for all $i$ and $j = 1, \ldots, 2N + 1$. By uniqueness, we must have $x = \phi(a)$ which is a contradiction.

Hence $\phi$ is one to one and continuous. Since $\mathscr{S}_m$ is compact Hausdorff it follows that $\phi(\mathscr{S}_m)$ is compact and that $\phi$ is a homeomorphism.   $\square$

REMARK 4.9   The theorem was proved with $l = 2N + 1$. Of course, it is not necessary that $l$ be odd. It was only chosen so for the sake of convenience.

REMARK 4.10   We can easily arrange, as Smale did, that the transversal homoclinic point $y_0$ be in $\phi(\mathscr{S}_m)$. Take $r = m - 1$ and $p_1 = 0$ and choose $\delta$ and $N$ so that they satisfy the additional conditions

$$\delta \leqslant 2\varepsilon$$

and

$$|f^n(y_0) - x_0| < \delta/2 \quad \text{when } n > N \text{ or } n \leqslant -N.$$

Now the pseudo-orbit corresponding to

$$a = (\ldots, m, m, 1, m, m, \ldots)$$

is formed from the orbit segments

$$\ldots, C_m, C_m, C_1, C_m, C_m, \ldots,$$

where

$$C_1 = \{f^{-N}(y_0), \ldots, y_0, \ldots, f^N(y_0)\}$$

and

$$C_m = \{x_0, \ldots, x_0, \ldots, x_0\}.$$

It is clear that the orbit of $y_0$, $\{f^n(y_0)\}$, $\varepsilon$-shadows this pseudo-orbit and that $y_0$ shadows the midpoint of $C_{a_0} = C_1$. Hence $\phi(a) = y_0$ and $y_0 \in \phi(\mathscr{S}_m)$, as asserted.

Since $\phi(\mathscr{S}_m)$ is $f^{2N+1}$-invariant all the iterates of $y_0$ under $f^{2N+1}$ are in $\phi(\mathscr{S}_m)$ also and since $\phi(\mathscr{S}_m)$ is closed, it follows that $x_0$ is in $\phi(\mathscr{S}_m)$ too.


## 5    EQUIVALENCE OF DEFINITIONS OF TRANSVERSAL HOMOCLINIC POINT

The main purpose of this section is to show that the definition of transversal homoclinic point given in Section 4 above is equivalent to the usual definition in terms of stable and unstable manifolds.

First we prove the existence of a local stable manifold of a hyperbolic fixed point.


PROPOSITION 5.1    *Let $x_0$ be a hyperbolic fixed point of a $C^1$ diffeomorphism $f: \mathbb{R}^p \to \mathbb{R}^p$. So there exist positive constants $K$, $\alpha$ and a projection $P$ commuting with $Df(x_0)$ such that inequalities (39) hold.*

*Let $\Delta > 0$ have the property that*

$$(49) \qquad |Df(x) - Df(x_0)| \leqslant \gamma = \tfrac{1}{2} K^{-1}(1 + e^{-\alpha})^{-1}(1 - e^{-\alpha})$$

*when $|x - x_0| \leqslant \Delta$.*

*Then there is a $C^1$ mapping $v: \{\xi \in \mathscr{R}(P): |\xi| < \Delta/4K\} \to \mathbb{R}^p$ with the following properties:*

(i)  *$x = v(\xi)$ is the unique $x$ such that*

$$(50) \qquad P(x - x_0) = \xi, |f^n(x) - x_0| \leqslant \Delta \text{ for } n \geqslant 0;$$

(ii)  *$|v(\xi) - x_0| \leqslant \Delta, |Dv(\xi)| \leqslant 2K$ for all $\xi$;*
(iii)  *$v(0) = x_0, Dv(0) = P$;*
(iv)  *$v$ is one to one.*


*Proof*    Given $\xi \in \mathscr{R}(P)$ with $|\xi| < \Delta/4K$, we seek $x$ such that conditions (50) hold. Put

$$f^n(x) = x_0 + z_n.$$

Then $z_n$ has to be a solution of the nonlinear difference equation

$$(51) \qquad z_{n+1} = Df(x_0)z_n + g(z_n),$$

where

$$(52) \qquad g(z) = f(x_0 + z) - f(x_0) - Df(x_0)z.$$

Also we want

$$(53) \qquad Pz_0 = \xi(= P\xi) \text{ and } |z_n| \leqslant \Delta \text{ for } n \geqslant 0.$$

Note that since

$$|Dg(z)| = |Df(x_0 + z) - Df(x_0)|$$

it follows from (49) that

$$(54) \qquad |g(z_1) - g(z_2)| \leqslant \gamma |z_1 - z_2|$$

for $|z_1|$, $|z_2| \leqslant \Delta$. Moreover,

$$g(0) = 0.$$

So all the conditions of Proposition 2.8 (ii) are satisfied and we conclude that there is a unique solution of equation (51) satisfying conditions (53). Denote this solution as $z_n(\xi)$.

Note that when $\xi = 0$, $z_n = 0$ for all $n$ is a solution satisfying conditions (53). So by uniqueness

$$z_n(0) = 0 \text{ for } n \geqslant 0.$$

Now we define

$$v(\xi) = x_0 + z_0(\xi)$$

so that for $n \geqslant 0$

$$(55) \qquad f^n(v(\xi)) = x_0 + z_n(\xi).$$

Then (i) and the first items in (ii), (iii) follow immediately. The first condition in (53) gives the identity

$$P(v(\xi) - x_0) = \xi$$

from which the one-oneness of $v$ follows at once.

All that remains is to prove the smoothness of $v$. First we show that $z_n(\xi)$ is Lipschitzian in $\xi$. Let $\xi_1$, $\xi_2$ be in the domain of $v$. Then $w_n = z_n(\xi_1) - z_n(\xi_2)$ is a solution of the inhomogeneous equation

$$w_{n+1} = Df(x_0)w_n + [g(z_n(\xi_1)) - g(z_n(\xi_2))]$$

which is bounded on $\mathbb{Z}^+$ and satisfies $Pw_0 = P(\xi_1 - \xi_2)$. It follows from Lemma 2.7 (ii) and (54) that

$$\| w \| \leqslant K | \xi_1 - \xi_2 | + K(1 + e^{-\alpha})(1 - e^{-\alpha})^{-1}\gamma \| w \|$$

where

$$\| w \| = \sup_{n \geqslant 0} | w_n | .$$

Thus

$$\| w \| \leqslant K | \xi_1 - \xi_2 | + 1/2 \| w \|$$

and so

$$\| w \| \leqslant 2K | \xi_1 - \xi_2 | .$$

Hence for $n \geqslant 0$,

$$(56) \qquad | z_n(\xi_1) - z_n(\xi_2) | \leqslant 2K | \xi_1 - \xi_2 | .$$

Now consider the matrix equation

$$(57) \qquad Z_{n+1} = [Df(x_0) + Dg(z_n(\xi))] Z_n,$$

where $| Dg(z_n(\xi)) | \leqslant \gamma$. By Proposition 2.8 (ii) and Remark 2.9 this equation has a unique solution $Z_n(\xi)$ which is bounded on $\mathbb{Z}^+$ and satisfies

$$PZ_0(\xi) = P.$$

We show that $Z_n(\xi) = Dz_n(\xi)$. To this end consider

$$w_n = z_n(\xi + h) - z_n(\xi) - Z_n(\xi)h$$

where $\xi, h \in \mathscr{R}(P)$ with $| \xi | < \Delta/4K, | h | < \Delta/4K - | \xi |$. We see that $w_n$ is a solution to the equation

$$w_{n+1} = Df(x_0)w_n + [ g(z_n(\xi + h)) - g(z_n(\xi)) - Dg(z_n(\xi))Z_n(\xi)h ]$$

which is bounded on $\mathbb{Z}^-$ and satisfies $Pw_0 = 0$. The norm of the inhomogeneous term is

$$\left| \int_0^1 Dg(z_n(\xi) + \theta(z_n(\xi + h) - z_n(\xi))) \, d\theta \, [z_n(\xi + h) - z_n(\xi)] - Dg(z_n(\xi))Z_n(\xi)h \right|$$

$$\leqslant \int_0^1 \left| Dg(z_n(\xi) + \theta(z_n(\xi + h) - z_n(\xi))) - Dg(z_n(\xi)) \right| d\theta \, | z_n(\xi + h) - z_n(\xi) |$$

$$+ | Dg(z_n(\xi) \| w_n |$$

$$\leqslant \omega(2K | h |)2K | h | + \gamma \| w \|,$$

where $\|w\| = \sup_{n \geqslant 0} |w_n|$, $\omega(\cdot)$ is the modulus of continuity of $Df(x)$ in $|x - x_0| \leqslant \Delta$ and we have used (56). From Lemma 2.7 (ii) we have the estimate

$$\|w\| \leqslant K(1 + e^{-\alpha})(1 - e^{-\alpha})^{-1}[\omega(2K|h|)2K|h| + \gamma\|w\|]$$
$$\leqslant 2K^2(1 + e^{-\alpha})(1 - e^{-\alpha})^{-1}\omega(2K|h|)|h| + \tfrac{1}{2}\|w\|.$$

Hence

$$\|w\| \leqslant 4K^2(1 + e^{-\alpha})(1 - e^{-\alpha})^{-1}\omega(2K|h|)|h|.$$

Thus

$$|z_n(\xi + h) - z_n(\xi) - Z_n(\xi)h| = 0(|h|) \text{ as } |h| \to 0,$$

*uniformly* with respect to $\xi$. From this it follows that for $n \geqslant 0$ $Dz_n(\xi) = Z_n(\xi)$ and, moreover, that $Dz_n(\xi)$ is continuous in $\xi$. Also the Lipschitz condition (56) implies that

$$|Dz_n(\xi)| \leqslant 2K.$$

Finally when $\xi = 0$, $z_n(\xi) = 0$ and so equation (57) takes the form

$$Z_{n+1} = Df(x_0)Z_n.$$

Now $Z_n = Df^n(x_0)P$ is a solution of this equation which is bounded on $\mathbb{Z}^+$ and satisfies $PZ_0 = P$. So by uniqueness,

$$Z_n(0) = Df^n(x_0)P \ (n \geqslant 0).$$

Taking $n = 0$ and recalling the definition of $v(\xi)$, we see that we have shown that $v(\xi)$ is $C^1$ and that the second items in (ii) and (iii) hold. $\quad\square$

DEFINITION 5.2   Let the hypotheses of Proposition 5.1 hold. Then we define the *local stable manifold* of $x_0$ to be the set

$$W^s_{\mathrm{loc}}(x_0) = \{v(\xi) : \xi \in \mathscr{R}(P), |\xi| < \Delta/4K\}.$$

We define the *global stable manifold* of $x_0$ to be the set

$$W^s(x_0) = \{x \in \mathbb{R}^p : f^n(x) \to x_0 \text{ as } n \to \infty\}.$$

Now we want to study the relationship between the local and global stable manifolds. For this we need the following lemma.

LEMMA 5.3 *Let the hypotheses of Proposition 5.1 hold. Suppose $x \in \mathbb{R}^p$ satisfies*

(58) $$|f^n(x) - x_0| \leqslant \Delta \text{ for } n \geqslant 0.$$

*Then we have the estimate*

$$|f^n(x) - x_0| \leqslant (2K + e^\alpha)e^{-[\alpha - \log(1 + 2K\gamma e^\alpha)]n}|x - x_0| \quad for \ n \geqslant 0,$$

*where*

$$\alpha - \log(1 + 2K\gamma e^\alpha) > 0.$$

*Proof* Note first using (49) that

$$1 + 2K\gamma e^\alpha = 1 + e^\alpha(1 - e^{-\alpha})(1 + e^{-\alpha})^{-1} < e^\alpha,$$

which implies $\alpha - \log(1 + 2K\gamma e^\alpha) > 0$.

Now let $x$ satisfy (58). Then $z_n = f^n(x) - x_0$ is a solution of the difference equation (51) satisfying $|z_n| \leqslant \Delta$ for $n \geqslant 0$. Now the Green's function (see Lemma 2.7) for the equation

$$z_{n+1} = Df(x_0)z_n$$

is given by

$$G(n, m) = \begin{cases} Df^{n-m}(x_0)P & \text{for } m \leqslant n \\ -Df^{n-m}(x_0)(I - P) & \text{for } n < m. \end{cases}$$

$z_n$ is a solution of the inhomogeneous equation

$$z_{n+1} = Df(x_0)z_n + g(z_n)$$

which is bounded on $n \geqslant 0$ and so it follows from (13) in the proof of Lemma 2.7 that

$$z_n = Df^n(x_0)Pz_0 + \sum_{m=0}^{n-1} Df^{n-m-1}(x_0)Pg(z_m)$$

$$- \sum_{m=n}^{\infty} Df^{n-m-1}(x_0)(I - P)g(z_m)$$

for $n \geqslant 0$. So for $n \geqslant 0$ we have the inequality

$$(59) \qquad |z_n| \leqslant Ke^{-\alpha n}|z_0|$$

$$+ K\gamma\left\{\sum_{m=0}^{n-1} e^{-\alpha(n-m-1)}|z_m| + \sum_{m=n}^{\infty} e^{-\alpha(m+1-n)}|z_m|\right\}.$$

For $n \geqslant 0$ put

$$\mu_n = \sup_{m \geqslant n} |z_m|.$$

Given $n \geqslant 0$ and $\varepsilon > 0$ we can find $n' \geqslant n$ such that

$$(1 - \varepsilon)\mu_n \leqslant |z_{n'}|.$$

Then from (59) with $n'$ instead of $n$,

$$(1 - \varepsilon)\mu_n \leqslant Ke^{-\alpha n'}|z_0| + K\gamma\left\{\sum_{m=0}^{n'-1} e^{-\alpha(n'-m-1)}|z_m|\right.$$

$$\left. + \sum_{m=n'}^{\infty} e^{-\alpha(m+1-n')}|z_m|\right\}$$

$$\leqslant Ke^{-\alpha n}|z_0| + K\gamma\left\{\sum_{m=0}^{n-1} e^{-\alpha(n-m-1)}|z_m|\right.$$

$$+ \sum_{m=n}^{n'-1} e^{-\alpha(n'-m-1)}|z_m|$$

$$\left. + \sum_{m=n'}^{\infty} e^{-\alpha(m+1-n')}|z_m|\right\}$$

$$\leqslant Ke^{-\alpha n}|z_0| + K\gamma\sum_{m=0}^{n-1} e^{-\alpha(n-m-1)}|z_m|$$

$$+ K\gamma(1 + e^{-\alpha})(1 - e^{-\alpha})^{-1}\mu_n$$

$$\leqslant Ke^{-\alpha n}|z_0| + K\gamma\sum_{m=0}^{n-1} e^{-\alpha(n-m-1)}|z_m| + \tfrac{1}{2}\mu_n.$$

So

$$(\tfrac{1}{2} - \varepsilon)\mu_n \leqslant Ke^{-\alpha n}|z_0| + K\gamma\sum_{m=0}^{n-1} e^{-\alpha(n-m-1)}|z_m|.$$

This holds for all $\varepsilon > 0$. We let $\varepsilon \to 0$ to get

$$|z_n| \leqslant \mu_n \leqslant 2Ke^{-\alpha n}|z_0| + 2K\gamma\sum_{m=0}^{n-1} e^{-\alpha(n-m-1)}|z_m|$$

for $n \geqslant 0$. Put $w_n = e^{\alpha n}|z_n|$ and we have

$$w_n \leqslant 2K|z_0| + 2Ke^{\alpha}\gamma\sum_{m=0}^{n-1} w_m$$

for $n \geqslant 0$. By induction this implies for $n \geqslant 1$ that

$$w_n \leqslant (2K|z_0| + 2Ke^{\alpha}\gamma w_0)(1 + 2Ke^{\alpha}\gamma)^{n-1}.$$

Hence for $n \geqslant 1$

$$|z_n| \leqslant (2K + 2Ke^{\alpha}\gamma))e^{-\alpha n}(1 + 2Ke^{\alpha}\gamma)^{n-1}|z_0|$$

$$= \frac{2K + 2Ke^{\alpha}\gamma}{1 + 2Ke^{\alpha}\gamma} e^{-[\alpha - \log(1 + 2Ke^{\alpha}\gamma)]n}|z_0|$$

$$\leqslant (2K + e^{\alpha})e^{-[\alpha - \log(1 + 2Ke^{\alpha}\gamma)]n}|z_0|. \qquad \square$$

PROPOSITION 5.4     *Let the hypotheses of Proposition 5.1 hold. Then*

(i) *$W^s_{loc}(x_0)$ is a submanifold of $\mathbb{R}^p$;*
(ii) *$x \in W^s(x_0)$ if and only if there is an integer $N$ such that*

$$f^N(x) \in W^s_{loc}(x_0);$$

*also $f^{-N}(W^s_{loc}(x_0))$ is a submanifold of $\mathbb{R}^p$ containing $x$ and contained in $W^s(x_0)$. Its tangent space at $x$ is denoted by $T_x W^s(x_0)$ and is given by*

$$T_x W^s(x_0) = \{\xi \in \mathbb{R}^p : \sup_{n \geq 0} |Df^n(x)\xi| < \infty\}.$$

*Proof*   $W^s_{loc}(x_0)$ is the image of the $C^1$ mapping $v(\xi)$ defined for $\xi$ in $R^p$, $|\xi| < \Delta/4K$. By Proposition 5.1 $v$ is one to one. Also from the proof of the proposition, $PDv(\xi) = P$ and so $Dv(\xi)$ is of full rank for all $\xi$. Thus $W^s_{loc}(x_0)$ is indeed a submanifold of dimension equal to the rank of $P$.

Now it follows from Proposition 5.1 that if $x \in W^s_{loc}(x_0)$ then for $n \geq 0$

$$|f^n(x) - x_0| \leq \Delta.$$

But then it follows from Lemma 5.3 that $|f^n(x) - x_0| \to 0$ as $n \to \infty$ so that $x \in W^s(x_0)$. Hence

$$W^s_{loc}(x_0) \subset W^s(x_0).$$

More generally, if $f^N(x) \in W^s_{loc}(x_0)$ for some $N$ then by what we have just proved $f^N(x) \in W^s(x_0)$ and so $x \in W^s(x_0)$, since from its definition $W^s(x_0)$ is clearly an invariant set.

Conversely, suppose $x \in W^s(x_0)$. Then there exists $N$ such that

$$|f^n(x) - x_0| < \Delta/4K^2 \text{ for } n \geq N.$$

This implies that

$$|f^n(f^N(x)) - x_0| < \Delta/4K^2 < \Delta \text{ for } n \geq 0.$$

In particular, $\xi = P(f^N(x) - x_0)$ satisfies $|\xi| < \Delta/4K$. It follows from Proposition 5.1 (i) that $f^N(x) = v(\xi)$ with $\xi = P(f^N(x) - z_0)$. So $f^N(x) \in W^s_{loc}(x_0)$. This completes the proof that $x \in W^s(x_0)$ if and only if $f^N(x) \in W^s_{loc}(x_0)$.

Now let $x \in W^s(x_0)$. Then from what we have just proved there is an integer $N$ such that $f^N(x) \in W^s_{loc}(x_0)$. Then $f^{-N}(W^s_{loc}(x_0))$ contains $x$, it is a submanifold of $\mathbb{R}^p$ since $f^{-N}$ is a diffeomorphism and it is contained in $W^s(x_0)$ by the first part. We denote its tangent space at $x$ by $T_x W^s(x_0)$ and then we have

(60)     $$T_x W^s(x_0) = Df^{-N}(f^N(x))\{T_{f^N(x)} W^s_{loc}(x_0)\}.$$

Now for any $y$ in $W^s_{loc}(x_0)$, $y = v(\xi)$ for some $\xi \in \mathcal{R}(P)$, $|\xi| < \Delta/4K$ and

$$T_y W^s_{loc}(x_0) = \mathcal{R}(Dv(\xi)) = \mathcal{R}(Z_0(\xi)),$$

where $Z_n(\xi)$ is the unique solution of (57), which is bounded on $\mathbb{Z}^+$ and satisfies $PZ_0(\xi) = P$. Using (52) and (55), we see that (57) can be rewritten as

$$Z_{n+1} = Df(f^n(y))Z_n.$$

Since $f^n(y) \to x_0$ as $n \to \infty$ it follows as in the proof of Proposition 4.3 that the variational equation

$$(61) \qquad u_{n+1} = Df(f^n(y))u_n$$

has an exponential dichotomy on $\mathbb{Z}^+$ with projections $P_n$ having the same rank as $P$. For any fixed vector $\eta$, $Z_n(\xi)\eta$ is a solution of (61) which is bounded on $\mathbb{Z}^+$ and so by Proposition 2.3

$$Z_0(\xi)\eta \in \mathcal{R}(P_0).$$

This holds for all $\eta$ and so

$$\mathcal{R}(Z_0(\xi)) \subset \mathcal{R}(P_0).$$

However, since $PZ_0(\xi) = P$, $Z_0(\xi)$ has the same rank as $P$ and hence the same rank as $P_0$. Thus

$$\mathcal{R}(Z_0(\xi)) = \mathcal{R}(P_0).$$

So

$$T_y W^s_{\text{loc}}(x_0) = \mathcal{R}(P_0)$$

$$= \left\{ \xi \in \mathbb{R}^p : \sup_{n \geq 0} |Df^n(y)\xi| < \infty \right\}$$

using Proposition 2.3 and the fact that the transition matrix for (61) is $Df^{n-m}(f^m(y))$.

Then, using (60), we have

$$T_x W^s(x_0) = \left\{ Df^{-N}(f^N(x))\xi : \xi \in \mathbb{R}^p, \sup_{n \geq 0} |Df^n(f^N(x))\xi| < \infty \right\}.$$

Now

$$Df^n(f^N(x))\xi = Df^{n+N}(x)Df^{-N}(f^N(x))\xi$$

and so

$$T_x W^s(x_0) = \left\{ \xi \in \mathbb{R}^p : \sup_{n \geq 0} |Df^{n+N}(x)\xi| < \infty \right\}$$

which clearly equals

$$\left\{ \xi \in \mathbb{R}^p : \sup_{n \geq 0} |Df^n(x)\xi| < \infty \right\}. \qquad \square$$

REMARK 5.5   Let $x_0$ be a hyperbolic fixed point of a $C^1$ diffeomorphism $f: \mathbb{R}^p \to \mathbb{R}^p$. We define the *global unstable manifold* of $x_0$ to be the set

$$W^{\mathrm{u}}(x_0) = \{ x \in \mathbb{R}^p : f^n(x) \to x_0 \text{ as } n \to -\infty \}.$$

Similarly to Proposition 5.1 we can prove the existence of a $C^1$ mapping $w: \{ \xi \in \mathcal{N}(P) : |\xi| < \Delta/4K \} \to \mathbb{R}^p$ with the property that $x = w(\xi)$ is the unique $x$ such that

$$(I - P)(x - x_0) = \xi, \, |f^n(x) - x_0| \leqslant \Delta \text{ for } n \leqslant 0$$

and also with properties analogous to (ii), (iii) and (iv). Then as in Proposition 5.4 we can show that $x \in W^{\mathrm{u}}(x_0)$ if and only if there is an integer $N$ such that $f^N(x) \in W^{\mathrm{u}}_{\mathrm{loc}}(x_0)$, where the latter is the *local unstable manifold*, that is, the range of the mapping $w$. Moreover, if we define $T_x W^{\mathrm{u}}(x_0)$ as the tangent space to $f^{-N}(W^{\mathrm{u}}_{\mathrm{loc}}(x_0))$ at $x$, then

$$T_x W^{\mathrm{u}}(x_0) = \left\{ \xi \in \mathbb{R}^p : \sup_{n \leqslant 0} |Df^n(x)\xi| < \infty \right\}.$$

PROPOSITION 5.6   *Let $x_0$ be a hyperbolic fixed point of a $C^1$ diffeomorphism $f: \mathbb{R}^p \to \mathbb{R}^p$ and let $W^{\mathrm{s}}(x_0)$, $W^{\mathrm{u}}(x_0)$ be the global stable and unstable manifolds of $x_0$. Then $y_0$ is a homoclinic point with respect to $x_0$ if and only if $y_0 \neq x_0$ and*

$$y_0 \in W^{\mathrm{s}}(x_0) \cap W^{\mathrm{u}}(x_0).$$

*$y_0$ is transversal (according to Definition 4.4) if and only if*

$$T_{y_0} W^{\mathrm{s}}(x_0) \cap T_{y_0} W^{\mathrm{u}}(x_0) = \{0\}.$$

*Proof*   The fact that $y_0$ is a homoclinic point (according to Definition 4.2) if and only if $y_0 \neq x_0$ and $y_0 \in W^{\mathrm{s}}(x_0) \cap W^{\mathrm{u}}(x_0)$ follows at once from the definitions. Now the solution $u_n$ of the variational equation (41) satisfying $u_0 = \xi$ is given by $u_n = Df^n(y_0)\xi$. Clearly then $u_n$ is bounded on $\mathbb{Z}$ if and only if

$$\xi \in T_{y_0} W^{\mathrm{s}}(x_0) \cap T_{y_0} W^{\mathrm{u}}(x_0).$$

So (41) has a nontrivial solution bounded on $\mathbb{Z}$, that is, $y_0$ is a transversal homoclinic point according to Definition 4.4, if and only if

$$T_{y_0} W^{\mathrm{s}}(x_0) \cap T_{y_0} W^{\mathrm{u}}(x_0) = \{0\}. \quad \square$$

REMARK 5.7   This is a remark that should be appended to Theorem 4.8 but is put here because a result from this section is needed.

We want to show that near any transversal homoclinic point there are infinitely many other such points. So let $y_0$ be a transversal homoclinic point with respect to a hyperbolic fixed point $x_0$.

Let $\Delta > 0$ be as in Proposition 5.1. In the proof of Theorem 4.8 take $r = m - 1$ and $p_1 = 0$. Choose $N_1$ so that

$$|f^n(y_0) - x_0| \leqslant \frac{|x_0 - y_0|}{2} \text{ if } |n| > N_1.$$

Also choose $\varepsilon > 0$ so that it satisfies the additional conditions

$$\varepsilon \leqslant \Delta, \, \varepsilon \leqslant \frac{|x_0 - y_0|}{4}, \, \varepsilon \leqslant \tfrac{1}{2} \min\{|f^i(y_0) - y_0| : 0 < |i| \leqslant N_1\}.$$

Let $q$ be a positive integer and consider the sequence in $\mathscr{S}_m$

$$a = (\dots, m, m, 1, \dots, 1, m, m, \dots)$$

where there are $q$ 1's and $a_0$ is the first one. By the shadowing lemma there is a unique orbit which $\varepsilon$-shadows the pseudo-orbit formed by the orbit segments

$$\dots, C_m, C_{a_0} = C_1, \dots, C_1, C_m, \dots .$$

Let $x^{(q)}$ be the point on the orbit which shadows the midpoint $y_0$ of

$$C_{a_0} = C_1 = \{f^{-N}(y_0), \dots, y_0, \dots, f^N(y_0)\}.$$

Then

$$|x^{(q)} - y_0| < \varepsilon.$$

Since

$$C_m = \{x_0, \dots, x_0, \dots, x_0\},$$

for $|n|$ sufficiently large

$$|f^n(x^{(q)}) - x_0| < \varepsilon \leqslant \Delta.$$

So by Lemma 5.3 $f^n(x^{(q)}) \in W^s(x_0)$ for $n > 0$ sufficiently large. Since $W^s(x_0)$ is invariant this means that $x^{(q)} \in W^s(x_0)$. Similarly we prove that $x^{(q)} \in W^u(x_0)$. Hence

$$x^{(q)} \in W^s(x_0) \cap W^u(x_0).$$

Also since $|x^{(q)} - y_0| < \varepsilon$ and $|x_0 - y_0| \geqslant 4\varepsilon$, $x^{(q)} \neq x_0$. So $x^{(q)}$ is a homoclinic point.

That the variational equation

$$u_{n+1} = Df^n(x^{(q)})u_n$$

has an exponential dichotomy on $\mathbb{Z}$ follows from Remark 3.6 ($\varepsilon$ and $\delta$ may have to be made smaller to ensure this). So $x^{(q)}$ is a transversal homoclinic point.

Now each point on the orbit of $x^{(q)}$ $\varepsilon$-shadows either $x_0$ or one of the points $f^i(y_0), |i| \leq N$. Our assumptions on $\varepsilon$ and $N_1$ imply that if a point $\varepsilon$-shadows $x_0$ or $f^i(y_0)$ where $0 < |i| \leq N$, then it cannot $\varepsilon$-shadow $y_0$. On the other hand $y_0$ appears $q$ times in the pseudo orbit and so there are $q$ points on the orbit which do $\varepsilon$-shadow $y_0$. So in fact there are *exactly* $q$ points in the orbit of $x^{(q)}$ which $\varepsilon$-shadow $y_0$. Hence if $q_1 \neq q_2$ the orbits of $x^{(q_1)}$ and $x^{(q_2)}$ do not intersect.

Thus we have shown that the ball of radius $\varepsilon$ centered at $y_0$ contains countably many transversal homoclinic points, none of which lies on the orbit of any other.


REMARK 5.8    Finally we show that if the $C^1$ diffeomorphism $f: \mathbb{R}^p \to \mathbb{R}^p$ has a transversal homoclinic point and $g: \mathbb{R}^p \to \mathbb{R}^p$ is another diffeomorphism which is $C^1$ near $f$, then $g$ also has a transversal homoclinic point.

So let $x_0$ be a hyperbolic fixed point for $f$. Then there are constants $K, \alpha$ and a projection $P$ such that inequalities (39) hold. Suppose also that $y_0$ is a transversal homoclinic point with respect to $x_0$ so that $f^n(y_0) \to x_0$ as $|n| \to \infty$ and the variational equation (41) has an exponential dichotomy on $\mathbb{Z}$ with constants which we may assume to be $K, \alpha$ also.

Now let $g: \mathbb{R}^p \to \mathbb{R}^p$ be a $C^1$ diffeomorphism such that

(62) $$|g(x) - f(x)| \leq \varepsilon \text{ for } x \in S$$

(63) $$|Dg(x) - Df(x)| \leq \varepsilon \text{ for } |x - y| \leq \Delta_0, y \in S,$$

where $S = \{x_0\} \cup \{f^n(y_0): n \in \mathbb{Z}\}$, $\Delta_0$ is a positive constant and the conditions on $\varepsilon$ will be determined below.

We first show that when $\varepsilon$ is sufficiently small, $g$ has a hyperbolic fixed point near $x_0$. We do this by showing that the difference equation

(64) $$x_{n+1} = g(x_n)$$

has a unique bounded solution $x_n$ near $x_0$. If we put

$$x_n = x_0 + z_n,$$

then $z_n$ has to be a solution of the equation

(65) $$z_{n+1} = Df(x_0)z_n + G(z_n),$$

where

$$G(z) = f(x_0 + z) - f(x_0) - Df(x_0)z + g(x_0 + z) - f(x_0 + z).$$

We see that

$$|G(0)| = |g(x_0)| \leq \varepsilon$$

and that when $|z| \leqslant \Delta_0$,

$$|DG(z)| \leqslant |Df(x_0 + z) - Df(x_0)| + |Dg(x_0 + z) - Df(x_0 + z)|$$
$$\leqslant \omega(|z|) + \varepsilon,$$

where $\omega(\cdot)$ is the modulus of continuity of $Df(x)$ in the set $\{x \in \mathbb{R}^p : |x - y| \leqslant \Delta_0 \text{ for some } y \in S\}$.

We assume that

$$(66) \qquad \Delta(\varepsilon) = 2K(1 + e^{-\alpha})(1 - e^{-\alpha})^{-1}\varepsilon \leqslant \Delta_0.$$

Then if

$$(67) \qquad 2K(1 + e^{-\alpha})(1 - e^{-\alpha})^{-1}(\omega(\Delta(\varepsilon)) + \varepsilon) \leqslant 1$$

it follows from Proposition 2.8 that equation (65) has a unique solution $z_n$ such that

$$|z_n| \leqslant \Delta(\varepsilon)$$

for all $n$. This means that $x_n = x_0 + z_n$ is the unique solution of equation (64) such that

$$|x_n - x_0| \leqslant \Delta(\varepsilon)$$

for all $n$. But $x_{n+1}$ is clearly also such a solution. By uniqueness, we must have $x_{n+1} = x_n$ for all $n$. So $x_n$ is, in fact, a constant $x_0'$. $x_0'$ is then a fixed point of $g$ satisfying

$$|x_0' - x_0| \leqslant \Delta(\varepsilon).$$

Moreover

$$|Dg(x_0') - Df(x_0)| \leqslant |Dg(x_0') - Df(x_0')| + |Df(x_0') - Df(x_0)|$$
$$\leqslant \varepsilon + \omega(\Delta(\varepsilon)).$$

Then it follows from Proposition 2.10 with $\delta = \alpha/2$ that if (67) holds and also

$$(68) \qquad 2Ke^{\alpha}(e^{-\alpha/2} + 1)(e^{\alpha/2} - 1)^{-1}[\varepsilon + \omega(\Delta(\varepsilon))] \leqslant 1,$$

the linear equation

$$u_{n+1} = Dg(x_0')u_n$$

has an exponential dichotomy on $\mathbb{Z}$ with constants $K_1 = 2K(1 + e^{-\alpha/2})(1 - e^{-\alpha/2})^{-1}$, $\alpha/2$. Then Remark 2.5 tells us that the eigenvalues of $Dg(x_0')$ lie off the unit circle and that if $P'$ is the projection with range the sum of the generalized eigenspaces corresponding to the eigenvalues inside the unit circle and with nullspace the sum of the generalized eigenspaces corresponding to the eigenvalues outside the unit circle, the inequalities

$$|Dg^n(x_0')P'| \leqslant K_1 e^{-\alpha n/2}(n \geqslant 0), \; |Dg^n(x_0')(I - P')| \leqslant K_1 e^{\alpha n/2}(n \leqslant 0)$$

are satisfied. So certainly $x_0'$ is a hyperbolic fixed point of $g$.

We now show that for $\varepsilon$ sufficiently small, $g$ has a transversal homoclinic point $y_0'$ near $y_0$. We do this by finding a solution $x_n$ of equation (64) near $f^n(y_0)$. If we write

$$x_n = f^n(y_0) + z_n$$

then $z_n$ must be a solution of the equation

$$(69) \qquad z_{n+1} = Df(f^n(y_0))z_n + G_n(z_n),$$

where

$$G_n(z) = f(f^n(y_0) + z) - f(f^n(y_0)) - Df(f^n(y_0))z + g(f^n(y_0) + z)$$
$$- f(f^n(y_0) + z).$$

Note that

$$|G_n(0)| = |g(f^n(y_0)) - f(f^n(y_0))| \leqslant \varepsilon$$

and if $|z| \leqslant \Delta_0$,

$$|DG_n(z)| \leqslant |Df(f^n(y_0) + z) - Df(f^n(y_0))|$$
$$+ |Dg(f^n(y_0) + z) - Df(f^n(y_0 + z)|$$
$$\leqslant \omega(|z|) + \varepsilon.$$

Then it follows from Proposition 2.8 that if conditions (66), (67) are satisfied, equation (69) has a unique solution $z_n$ such that

$$|z_n| \leqslant \Delta(\varepsilon)$$

for all $n$. Then $y_0' = y_0 + z_0$ is the unique point such that

$$|g^n(y_0') - f^n(y_0)| \leqslant \Delta(\varepsilon)$$

for all $n$.

Now choose $\Delta_1 > 0$ so that

$$4K_1(1 + e^{-\alpha/2})(1 - e^{-\alpha/2})^{-1}\omega(\Delta_1) \leqslant 1$$

and assume that

$$(70) \qquad 8K_1(1 + e^{-\alpha/2})(1 - e^{-\alpha/2})^{-1}\varepsilon \leqslant 1,$$

$$(71) \qquad 3\Delta(\varepsilon) \leqslant \Delta_1.$$

Then if $|x - x_0'| \leqslant \Delta_1$,

$$|Dg(x) - Dg(x_0')| \leqslant |Dg(x) - Df(x)| + |Df(x) - Df(x_0')|$$
$$+ |Df(x_0') - Dg(x_0')|$$
$$\leqslant \varepsilon + \omega(\Delta_1) + \varepsilon$$
$$\leqslant \gamma_1 = \tfrac{1}{2}K_1^{-1}(1 + e^{-\alpha/2})^{-1}(1 - e^{-\alpha/2}).$$

This means we can apply Proposition 5.1 and Lemma 5.3 to $g$ and $x_0'$ with the quantities $K_1$, $\alpha/2$, $\Delta_1$, $\gamma_1$ instead of $K$, $\alpha$, $\Delta$, $\gamma$.

There exists a positive integer $N$ such that

$$|f^n(y_0) - x_0| \leqslant \Delta_1/3$$

for $|n| \geqslant N$. Then if $|n| \geqslant N$,

$$\begin{aligned} |g^n(y_0') - x_0'| &\leqslant |g^n(y_0') - f^n(y_0)| + |f^n(y_0) - x_0| + |x_0' - x_0| \\ &\leqslant \Delta(\varepsilon) + \Delta_1/3 + \Delta(\varepsilon) \\ &\leqslant \Delta_1, \text{ using (71)}. \end{aligned}$$

It follows then from Lemma 5.3 that $g^N(y_0') \in W^s(x_0')$ and hence that $y_0' \in W^s(x_0')$. Similarly, $y_0' \in W^u(x_0')$. Also if we assume

$$(72) \qquad\qquad 2\Delta(\varepsilon) < |y_0 - x_0|,$$

then

$$|y_0' - x_0'| \geqslant |y_0 - x_0| - |y_0' - y_0| - |x_0' - x_0| \geqslant |y_0 - x_0| - 2\Delta(\varepsilon) > 0.$$

So $y_0' \neq x_0'$ and $y_0'$ is therefore a homoclinic point with respect to $x_0'$.

Moreover for all $n$

$$\begin{aligned} |Dg(g^n(y_0')) - Df(f^n(y_0))| &\leqslant |Dg(g^n(y_0')) - Df(g^n(y_0'))| \\ &\quad + |Df(g^n(y_0')) - Df(f^n(y_0))| \\ &\leqslant \varepsilon + \omega(\Delta(\varepsilon)). \end{aligned}$$

Since (41) has an exponential dichotomy on $\mathbb{Z}$ with constants $K$, $\alpha$ and (67), (68) are satisfied, it follows from Proposition 2.10 that the variational equation

$$u_{n+1} = Dg(g^n(y_0'))u_n$$

has an exponential dichotomy on $\mathbb{Z}$. Hence $y_0'$ is a transversal homoclinic point.

In conclusion, we have shown that if $g : \mathbb{R}^p \to \mathbb{R}^p$ is a $C^1$ diffeomorphism satisfying (62), (63), where $\varepsilon$ satisfies (66), (67), (68), (70), (71) and (72) then $g$ has a transversal homoclinic point.


## REFERENCES

[1] D. V. Anosov, Geodesic flows on compact Riemannian manifolds of negative curvature. *Proc. Steklov Inst. Math.*, **90** (1967); English transl., *Amer. Math. Soc. Transl.* (1969).

[2] R. Bowen, *On Axiom A Diffeomorphisms*, NSF-CBMS Regional Conf. Series, No. 35, Amer. Math. Soc., Providence, RI, 1978.

[3] C. C. Conley, Hyperbolic invariant sets and shift automorphisms, in *Dynamical Systems Theory and Applications*, J. Moser (ed.), Lecture Notes in Physics No. 38, pp. 539–49. Springer-Verlag, Berlin, 1975.

[4] W. A. Coppel, *Dichotomies in Stability Theory*, Lecture Notes in Mathematics No. 629. Springer-Verlag, Berlin, 1978.

[5] J. Franke and J. Selgrade, Hyperbolicity and chain recurrence, *J. Differential Equations*, **26** (1977), 27–36.

[6] J. Guckenheimer, J. Moser and S. Newhouse, *Dynamical Systems*, Birkhäuser, Boston, 1980.

[7] U. Kirchgraber, Erratische Lösungen der periodisch gestörten Pendelgleichung, preprint, University of Würzburg, 1982.

[8] J. Moser, *Stable and Random Motions in Dynamical Systems*. Princeton Univ. Press, Princeton, NJ, 1973.

[9] K. J. Palmer, Exponential dichotomies and transversal homoclinic points. *J. Differential Equations*, **55** (1984), 225–56.

[10] C. Robinson, Stability theorems and hyperbolicity in dynamical systems. *Rocky Mountain J. Math.*, **7** (1977), 425–37.

[11] V. E. Slyusarchuk, Exponential dichotomy for solutions of discrete systems. *Ukranian Math. J.*, **35** (1983), 98–103.

[12] S. Smale, Diffeomorphisms with many periodic points, in *Differential and Combinatorial Topology*, S. Cairns (ed.), Princeton Univ. Press, Princeton, NJ, 1965, pp. 63–80.

[13] S. Smale, Differentiable dynamical systems. *Bull. Amer. Math. Soc.*, **73**, (1967), 747–817.